MATERIALS SCIENCE AND TECHNOLOGIES

THERMOPLASTIC AND THERMOSETTING POLYMERS AND COMPOSITES

MATERIALS SCIENCE AND TECHNOLOGIES

Additional books in this series can be found on Nova's website
under the Series tab.

Additional E-books in this series can be found on Nova's website
under the E-books tab.

MATERIALS SCIENCE AND TECHNOLOGIES

THERMOPLASTIC AND THERMOSETTING POLYMERS AND COMPOSITES

LINDA D. TSAI
AND
MATTHEW R. HWANG
EDITORS

Nova Science Publishers, Inc.
New York

For permission to use material from this book please contact us:
Telephone 631-231-7269; Fax 631-231-8175
Web Site: http://www.novapublishers.com

Additional color graphics may be available in the e-book version of this book.

LIBRARY OF CONGRESS CATALOGING-IN-PUBLICATION DATA

Thermoplastic and thermosetting polymers and composites / editors, Linda D.
Tsai and Matthew R. Hwang.
p. cm.
Includes index.
ISBN 978-1-61209-264-5 (hardcover)
1. Thermoplastic composites. 2. Thermoplastics. I. Tsai, Linda D. II.
Hwang, Matthew R.
TA418.9.C6T4625 2010
668.4'23--dc22
2010048386

Published by Nova Science Publishers, Inc. † New York

CONTENTS

PREFACE

This new book presents current research in the study of thermoplastic and thermosetting polymers, including the fabrication of aromatic polyester composites containing carbon nanotubes; thermoplastic and thermosetting composites with natural fibers; phase separation of PMMA-modififed vinyl-ester thermosets; thermoplastic and thermosetting polymers from vegetable oils and the mechanisms of impregnation during compression molding of thermoplastic matrix composites.

Chapter 1 - This chapter describes the fabrication of aromatic polyester composites containing carbon nanotube (CNT), and the effect of modified CNT on the overall properties of aromatic polyester nanocomposites. Modification of CNT to introduce functional groups on the surface was performed to enhance intermolecular interactions between CNT and polymer matrix through hydrogen bonding formation. Morphological observations revealed that the modified CNT was uniformly dispersed in the polymer matrix and increased interfacial adhesion between the nanotubes and the polymer, as compared to the unmodified CNT. The crystallization behavior of aromatic polyester nanocomposites highly depends on the modified CNT and cooling rate. The variations of the nucleation activity and activation energy for crystallization reflected the enhancement of the crystallization of aromatic polyester nanocomposites induced by the modified CNT. Combined Avrami and Ozawa analysis was found to be effective in describing the non-isothermal crystallization of aromatic polyester nanocomposites in the presence of the modified CNT. Furthermore, a very small quantity of the modified CNT substantially improved thermal stability and mechanical properties of aromatic polyester nanocomposites. For thermotropic liquid crystal polymer (TLCP) nanocomposites, there is also a significant dependence of the mechanical, rheological, and thermal properties of TLCP nanocomposites on the uniform dispersion and interfacial adhesion, and their synergistic effect was more effective at low content of the modified CNT. The key to improve the overall properties of TLCP nanocomposites depends on the optimization of the unique geometry and dispersion state of CNT and the interfacial interactions in TLCP nanocomposites during the melt processing. This chapter demonstrates that the overall properties of aromatic polyester nanocomposites are strongly dependent on the uniform dispersion of CNT and the interaction between CNT and polyester, which can be enhanced by surface modification of CNT, providing a design guide of CNT-reinforced polyester nanocomposites with a great potential for industrial uses.

Chapter 2 - During the last decades, there has been a continuous increase in the production of commodity plastic products because of their low production cost, excellent mechanical properties, as well as chemical, weather and biodegradation resistance. The duration of use of plastic materials is relatively short compared to their life time; it ranges from several months to few years, depending on the conditions of use and environment. Now a day the scientific world is focusing is the development of starch based composites. These materials are available largely from natural resources and are 100% degradable. This chapter presents a brief description about the different biodegradable matrices as well as the different components, structures, transformations and properties of starch-based matrices and biofibers. The reinforcing effect of macro-fibers s well as like nano fibers like whiskers or nanocrystals on starch as matrix will be discussed.

Chapter 3 - Nowadays, great attention has been dedicated to the development of natural fiber reinforced composites. Natural fibers provide with interesting properties the final composite, especially those related to environment the protection such as their capacity to be recyclable, renewable raw material, and less abrasive and harmful behavior. Some advantages associated to the use of natural fibers as reinforcement in plastics are their non-abrasive nature, biodegradability, low energy consumption, low cost, low density and high specific properties. The specific mechanical properties of natural fibers are comparable to those of traditional reinforcements.

However, certain drawbacks such as incompatibility with a hydrophobic polymer matrix, the tendency to form aggregates during processing and poor resistance to moisture greatly reduce the potential of natural fibers to be used as reinforcement in polymers. On the other hand, various treatments are being used to improve fiber-matrix compatibility. This process is considered critical as development phase of these materials due to strong interfiber hydrogen bonding, which holds the fibers together. Methods for surface modification can be physical or chemical according to superficial modification approach of the fiber. Others frequently used treatments are bleaching, acetylation and alkali treatment.

In this chapter, the main results presented in the literature are summarized, focusing attention on the properties in terms of physical and chemical structure of natural fibers, thermal and mechanical properties, processing behavior and final properties of natural fibers with thermoplastics and thermosetting matrixes paying attention to the use of physical and chemical treatments for the improvement of fiber-matrix interaction.

Chapter 4 - In this Chapter, the initial miscibility, the developed morphologies, and the final properties of styrene(St)/vinyl-ester(VE) thermosets modified with poly(methyl methacrylate) (PMMA) are discussed. The effect of changing the molecular weight and the polydispersity of the VE oligomer and the PMMA modifier are presented. Firstly, the miscibility of binary and ternary physical mixtures of the components involved in the different studied formulations is analyzed. The experimental liquid-liquid equilibrium curves (e.g. cloud-point curves) allow computing the binary interaction parameters, χ, in the framework of the Flory-Huggins theory for polydisperse systems. These parameters are used to model the quasiternary phase diagram that represents the initial thermodynamic state of each particular system. This miscibility behavior originates quite different morphologies in the cured materials, generated by polymerization induced phase separation (PIPS) mechanism. For instance, dispersion of thermoplastic-rich particles in a thermoset-rich matrix, cocontinuous structure, dispersion of thermoset-rich particles in a thermoplastic-rich matrix (phase-inverted structure), or typical macrophase morphology characterized by droplets-like domains with secondary phase separation inside the droplets can be observed. These morphological structures are directly related to the thermal and mechanical properties, as well as the volume shrinkage of the final systems. The evaluation of the dynamic mechanical behavior, flexural modulus, compressive yield stress, and fracture toughness shows that the addition of PMMA increases the fracture resistance without significantly compromising the thermal or mechanical properties of the vinyl-ester networks, which is inevitable when using elastomeric additives. The reason for the existence of an optimum modifier concentration is also discussed.

Chapter 5 - Vegetable oils are excellent renewable raw materials for polymers. This chapter discusses the application of environmentally friendly and high efficient processes to plant oils and several approaches to new biobased materials and is organized as a function of the thermoplastic or thermosetting nature of the final polymers. Starting from undecylenic or oleic acid, the authors obtained difunctional telechelic diols by thiol-ene "click" coupling, that can were further polymerized to linear polyurethanes. Moreover, the authors synthesized thermoplastic flame retardant phosphorus-containing polyesters by ADMET polymerization. To obtain thermosetting materials, they explored the cationic polymerization of triglyceride double bonds of the fatty acid chain. The authors carried out the copolymerization of soybean oil with styrenic monomers containing silicon, boron or phosphorus producing materials with improved mechanical and flame retardant properties. Moreover, they obtained organic-inorganic hybrid materials with promising properties for optical applications by the hydrosilylation of alkenyl-terminated fatty acid derivatives. The presence of double bonds in triglycerides makes possible to attach some functional groups through chemical modification and they described various chemical pathways for functionalising triglycerides and fatty acids. Epoxidation is one of the most interesting chemical modifications that leads to epoxidized vegetable oils. The authors reported the preparation of biobased polyhedral oligomeric silsesquioxanes-nanocomposites from epoxidized linseed oil. Moreover, they obtained new fatty-acid derived compounds that could find applications as flame retardant materials in biobased epoxy resins.

The authors described the preparation of a new family of epoxidized methyl oleate-based polyether polyols which were used in the synthesis of polyurethanes with specific applications: silicon-containing polyurethanes with enhanced flame-retardant properties and polyurethane networks with potential applications in biomedicine. An enone-containing triglyceride derivative was obtained by an environmentally friendly chemical procedure from high oleic sunflower oil, that could be crosslinked with diamines. In a similar way, triglycerides containing secondary allylic alcohols can be obtained, that can be further functionalised with acrylate or phosphorus-containing derivatives to obtain flame retardant thermosets.

Chapter 6 - Continuous fiber reinforced thermoplastic matrix composites, based on commodity polymers, are attracting a growing interest in many industrial application due to distinct advantages over thermosetting matrix composites. Commingled yarn semi-pregs, or stackings of thermoplastic films and dry reinforcements, lead to fast processing by compression molding. In the forming process the material is placed between matched die, to promote the flow of the matrix into the interstices of the fibers, due to the applied pressure and temperature.

The aim of this paper is to study the different phenomena involved in the consolidation of thermoplastic matrix composites, and to model the impregnation process in both commingled yarn semipregs and in composites produced by film stacking technique. Optical microscopy showed that for both systems there are two mechanisms of impregnation: macro-scale impregnation, associated with the formation of a homogeneous molten pool around each fiber bundle, and micro-scale impregnation, associated to the flow of the matrix inside each bundle. Since the two mechanisms occur at different temperatures, their contribution to the reduction of void fraction can be considered separately.

Macro-scale and micro-scale impregnation during film stacking were simulated by two different finite element models, taking into account the non Newtonian rheological behavior of the matrix. The micro-scale impregnation of fibers was simulated by using a randomly spaced and non-overlapping unidirectional filaments. The results obtained showed that at low molding pressures the polymer melt exhibits a Newtonian behavior during micro-scale impregnation, which makes it possible to predict the tow permeability by the Darcy law. At high molding pressures and during macro-scale impregnation, the high shear rates and non-Newtonian behavior of the melt required the introduction of a permeability coefficient that is also dependent on the rheological properties of the melt.

The combination of on-line consolidation measurement, thermomechanical analysis and numerical analysis showed that the macro-scale impregnation during molding of commingled yarn fabrics, taking place at lower temperatures, is governed by matrix fiber deformation and sintering. At a temperature higher than the onset of the flow region of the matrix, micro-scale impregnation occurs, due to matrix flow inside each bundle.

Chapter 7 - Composites consisting of epoxies and inorganic nanoparticles are of high technological importance. Despite the fact that these nanocomposites are already widely used, fundamental understanding of the physical and chemical processes before and during epoxy network formation in the presence of nanoparticles is still missing. The present work presents investigations of the thermal and chemical glass transition in epoxies filled with two different types of nanoparticles: hydrophilic alumina and hydrophobic silica. It is shown that macroscopic investigations of static and dynamic thermodynamic properties of the glass transition behaviour allow elucidating network formation in epoxies in the presence of nanoparticles.

In: Thermoplastic and Thermosetting Polymers and Composites ISBN: 978-1-61209-264-5
Editors: Linda D. Tsai and Matthew R. Hwang

Chapter 1

AROMATIC POLYESTER NANOCOMPOSITES CONTAINING MODIFIED CARBON NANOTUBE

Jun Young Kim*

Corporate Research & Development Center, Samsung SDI Co. Ltd., Republic of Korea,
Department of Materials Science and Engineering, Massachusetts Institute of Technology, USA

ABSTRACT

This chapter describes the fabrication of aromatic polyester composites containing carbon nanotube (CNT), and the effect of modified CNT on the overall properties of aromatic polyester nanocomposites. Modification of CNT to introduce functional groups on the surface was performed to enhance intermolecular interactions between CNT and polymer matrix through hydrogen bonding formation. Morphological observations revealed that the modified CNT was uniformly dispersed in the polymer matrix and increased interfacial adhesion between the nanotubes and the polymer, as compared to the unmodified CNT. The crystallization behavior of aromatic polyester nanocomposites highly depends on the modified CNT and cooling rate. The variations of the nucleation activity and activation energy for crystallization reflected the enhancement of the crystallization of aromatic polyester nanocomposites induced by the modified CNT. Combined Avrami and Ozawa analysis was found to be effective in describing the non-isothermal crystallization of aromatic polyester nanocomposites in the presence of the modified CNT. Furthermore, a very small quantity of the modified CNT substantially improved thermal stability and mechanical properties of aromatic polyester nanocomposites. For thermotropic liquid crystal polymer (TLCP) nanocomposites, there is also a significant dependence of the mechanical, rheological, and thermal properties of TLCP nanocomposites on the uniform dispersion and interfacial adhesion, and their synergistic effect was more effective at low content of the modified CNT. The key to improve the overall properties of TLCP nanocomposites depends on the optimization of the unique geometry and dispersion state of CNT and the interfacial interactions in TLCP nanocomposites during the melt processing. This chapter demonstrates that the overall

* Corresponding author. E-mail address: junykim74@hanmail.net (J. Y. Kim)

properties of aromatic polyester nanocomposites are strongly dependent on the uniform dispersion of CNT and the interaction between CNT and polyester, which can be enhanced by surface modification of CNT, providing a design guide of CNT-reinforced polyester nanocomposites with a great potential for industrial uses.

1. Introduction

1.1 Carbon Nanotube

Carbon nanotube (CNT), which was discovered by Iijima [1] in 1991, has attracted of great interest as advanced reinforcements in a wide range of potential scientific and industrial applications. Moreover, this discovery has created a high level of activity in materials research, leading to a practical realization of the extraordinary properties of CNT. The CNTs were first synthesized as a by-product in arc-discharge method in the synthesis of fullerenes and are currently being prepared by various methods, including arc-discharge [2-4], laser ablation [5-7], high-pressure CO conversion [8,9], chemical vapor deposition [10-14], electrolysis [15], solar energy [16] methods.

The CNT consisting of concentric cylinders of graphite layers is a new form of carbon and can be classified into three types [17-19]: single-walled carbon nanotubes (SWCNT), double-walled carbon nanotubes (DWCNT), and multi-walled carbon nanotubes (MWCNT). SWCNT consists of a single layer of carbon atoms through the thickness of the cylindrical wall with the diameters of 1.0-1.4 nm, two such concentric cylinders forms DWCNT, and MWCNT consists of several layers of coaxial carbon tubes, the diameters of which range from 10 to 50 nm with the length of more than 10 μm [17-19]. The graphite nature of the nanotube lattice results in a fiber with high strength, stiffness, and conductivity, and higher aspect ratio represented by very small diameter and long length makes it possible for CNTs to be ideal nanoreinforcing fillers in advanced polymer nanocomposites [20]. Both theoretical and experimental approaches suggest the exceptional mechanical properties of CNTs ~100 times higher than the strongest steel at a fraction of the weight [21-25 The Young's modulus, strength, and toughness of SWCNT shows 0.32~1.47 TPa of Young's modulus, 10-52 GPa of strength, and ~770 J/g of toughness, respectively [24]. For MWCNT, the values of strength, Young's modulus, and toughness were found to be 11-63 GPa, 0.27-0.95 TPa, and ~1240 J/g, respectively [25]. In addition, CNTs exhibit excellent electrical properties and electric current carrying capacity ~1000 times higher than copper wires [26].

In general, MWCNTs show inferior mechanical performance as compared to SWCNTs. However, MWCNTs have a cost advantage, in that they can be produced in much larger quantities at lower cost compared with the SWNT. In addition, MWCNTs are usually individual, longer, and more rigid than SWCNTs. Because of their remarkable physical properties such as high aspect ratio and excellent mechanical strength, MWCNTs are regarded as prospective reinforcing fillers in high performance polymer nanocomposites. For these reasons, extensive research and development have been directed towards the potential applications of CNTs for novel composite materials in a wide range of industrial fields. The fundamental research progressed to date on applications of CNTs also suggests that CNTs can be utilized as promising reinforcements in new kinds of polymer nanocomposites with remarkable physical/chemical characteristics [19].

1.2. Aromatic Polyester

Aromatic polyester resins are one of the most versatile of all engineering plastics providing good performance in various industrial applications. The continued technological development in the polymer composites leads to the expanded capability of aromatic polyester reins in material-consuming industries. In particular, the melt blends based on poly(ethylene terephthalate) (PET) and poly(ethylene 2,6 naphthalate) (PEN) has been attractive because of significant improvement in the overall properties of the polymer composites by the combination of excellent properties of PEN with the economical efficiency of PET from an industrial perspective [27-29].

PET, that is semi-crystalline polymer with high mechanical properties and good processability, has been widely used as engineering plastics for the automotive, electronics, and structural materials in a broad range of industries [30,31]. However, there are continuing practical demands for achieving excellent properties of PET with various processing conditions, in order to be utilized in advanced industrial fields that should be required for high performance and low manufacturing cost.

PEN is transparent aromatic polyester having a similar chemical structure to PET, a widely used polyester resin in conventional industry, with the exception that PEN has naphthalene ring in the main chain instead of benzene ring, and it is of great industrial importance due to its high performance, good physical properties, and low cost. As the introduction of the naphthalene ring into the main chains stiffens the polymer chains and improves their physical properties, PEN typically exhibits enhanced thermal, mechanical, and gas barrier properties as compared to PET. PEN thus holds a potential for industrial applications, including food packaging materials, high performance industrial fibers, magnetic recording tapes, and flexible printed circuits [32-37]. In this regard, research and development has been extensively performed to date both to displace the PET and to develop commercial applications of PEN, such a high performance polymer [38-44]. Although promising, however, insufficient mechanical properties and thermal stability of PEN have often hindered its practical application in a broad range of industry.

1.3 CNT/Polymer Nanocomposites

Polymer nanocomposites that are a new class of the materials based on the reinforcement of polymers using nanofillers have attracted a great deal of interest in fields from basic science to the industrial applications because of remarkable improvements in the physical and mechanical properties at low filler loadings. Recently, the fabrication of the polymer nanocomposites by incorporating the nano-scaled reinforcements into the polymer matrix is believed to become a key technology on advanced composite materials [45]. As described previously, CNTs have attracted of great interest as advanced materials due to the combination of their uniquely excellent properties with high aspect ratio and small size [46]. In particular, excellent mechanical, thermal, and electrical properties of CNTs, have led to their use in promising nanoreinforcing fillers in the polymer nanocomposites [47]. This feature has motivated considerable efforts to fabricate polymer nanocomposites containing CNTs, with the benefit of nanotechnology, in the development of advanced polymer nanocomposites for next generation.

For the fabrication of CNT/polymer nanocomposites, major goals to realize the potential applications of CNTs as nanoreinforcing fillers are (a) homogeneous dispersion of CNTs in the polymer matrix and (b) strong interfacial adhesion between CNTs and polymer matrix [48–50]. Typically, CNTs tend to bundle together and to form some agglomeration because of intrinsic van der Waals attraction between the individual tubes [51]. Weak interfacial bonding between the nanotubes and the polymer matrix has limited the efficient load transfer to the polymer matrix, playing a limited reinforcement role in the polymer nanocomposites [52-54]. The functionalization of CNTs, which can be considered as an effective method to achieve the uniform dispersion of CNTs and their compatibility with the polymers, can lead to the enhancement of the interfacial adhesion between CNTs and polymer matrix, thereby improving the overall properties of CNT/polymer nanocomposites [55-58].

Another challenge for achieving high performance polymer nanocomposites is to optimize the processing of CNT/polymer nanocomposites with low cost. Currently, four processing techniques are in common use to fabricate CNT/polymer nanocomposites [59-76]: direct mixing, solution method, *in situ* polymerization, and melt compounding. Of these processing techniques, the melt compounding has been accepted as the simplest and the most effective method from a commercial perspective, because this process makes it possible to fabricate high performance polymer nanocomposites at low processing cost, and also facilitates commercial scale-up [66-76]. Furthermore, the combination of very small quantity of relatively expensive CNTs with conventional cheap thermoplastic polymers provides attractive possibilities for improving the physical properties of polymer nanocomposites using a cost-effective method [66-76].

As the mechanical properties of the polymer nanocomposites are influenced by their morphology and crystallization behavior and the introduction of CNTs plays a significant role in those [69-73,76- 78], it is very instructive to characterize the nucleation and crystallization behavior of CNT/polymer nanocomposites as a function of processing condition for achieving high performance polymer nanocomposites as well as realize full potential applications of CNTs in thermoplastic polymer-based nanocomposites. Therefore, the crystallization behavior and structural development of CNT-reinforced polymer nanocomposites should be analyzed to realize the full potential of the CNT for application in thermoplastic matrix-based polymer nanocomposites. It is also very important to understand the non-isothermal crystallization behavior of polymers or polymer nanocomposites, particularly if processing techniques for the fabrication of engineering plastics under non-isothermal conditions are being considered. The processing of polymer nanocomposites involves complex deformation behaviors, which may affect the nucleation and crystallization behavior of polymer composites.

The rheological behavior of polymer nanocomposites as a function of processing conditions is of great importance in polymer processing, particularly for the analysis and design of processing operations, as well as understanding structure-property relationships of polymer nanocomposites. In this regard, the rheological properties of CNT/polymer nanocomposites should be characterized both to realize a great potential of CNTs as promising nanoreinforcing fillers for polymer nanocomposites under the optimized processing conditions. The thermal stability of polymer nanocomposites plays a crucial role in determining their processing and applications because it affects the final properties of polymer nanocomposites, such as the upper-limit use temperature and dimensional stability. For the development of polymer nanocomposites with better balance in processing and performance, it is also instructive to characterize thermal stability and decomposition behavior of polymer nanocomposites. A generalized understanding of thermal stability and thermal degradation behavior of polymer nanocomposites makes it possible to develop and extend their applications in a broad range of industry.

This chapter focuses on the fabrication and characterization of aromatic polyester nanocomposites reinforced with very small quantity of CNTs to clarify the possibilities for achieving high performance polyester nanocomposites and to attempt to overcome the limitation of conventional polyester composites. Aromatic polyester nanocomposites containing CNTs were prepared by simple melt blending in a twin-screw extruder to create advanced nanocomposite materials with low processing cost for practical applications in various industrial fields. The effects of modified CNTs on the overall properties of aromatic polyester nanocomposites are clarified in detail. In addition, the fabrication of thermotropic liquid crystal copolyester nanocomposites and the influence of modified CNTs on the physical properties of the nanocomposites were discussed. This chapter will help in preliminary evaluation and understanding of CNT-reinforced polyester nanocomposites, and provide a design guide of aromatic polyester nanocomposites with a great potential for industrial uses This chapter also suggests a simple and cost-effective method that will facilitate the industrial realization of aromatic polyester nanocomposites containing CNTs with enhanced physical properties.

2. Fabrication of Aromatic Polyester Nanocomposites

2.1. Materials

Aromatic polyester resins used was poly(ethylene 2,6-naphthalate) (PEN) with an intrinsic viscosity of 0.97 dl/g, supplied by Hyosung Corp., Korea. Thermotropic liquid crystal polymer (TLCP) used was a flexible semi-aromatic copolyester synthesized from poly(p-hydroxybenzoate) and PET with a molar ratio of 80:20, purchased from Unitika Co. Ltd., Japan. According to the supplier, TLCP has an intrinsic viscosity of 0.55 dl/g, determined at 30°C in a phenol/tetrachloroethane (50/50, v/v) mixture. The nanotubes used were multiwall CNT (degree of purity > 95%) synthesized by a thermal chemical vapor deposition (CVD) process, purchased from Iljin Nanotech Co., Korea. The diameter and length of CNT were in the rage of 10-30 nm and 10-50 μm, respectively.

2.2. Preparation of Aromatic Polyester Nanocomposites

The CNT was modified by the following steps: CNTs were treated with the mixture of concentrated sulfuric acid (H_2SO_4, 98%) and nitric acid (HNO_3, 68%) with a volumetric ratio of 3:1 at 80°C for 4 h to create the carboxylic acid groups on the nanotube surface. After this mixture was cooled to room temperature, it was diluted with deionized water and then vacuum-filtered through 0.22 μm milipore PTFE membranes, and washed with an excess of distilled water until the pH value of the filtrate reached approximately 7. The filtrated solid was dried *in vacuo* at 100°C for at least 24 h, yielding the modified CNT. The carboxylic acid groups on the nanotubes were effectively induced via this chemical modification [55] in order to increase the CNT's chemical affinity with the PEN, yet to decrease the π–π stacking effect among the aromatic rings of the nanotubes, which often leads to the formation of their agglomeration [79].

All the materials were dried at 120°C *in vacuo* for at least 24 h before use, to minimize the effects of moisture. Aromatic polyester nanocomposites were prepared by a melt blending process in a Haake rheometer (Haake Technik GmbH, Germany) equipped with a twin-screw (intermeshing co-rotating type). The temperature of the heating zone, from the hopper to the die, was set to 280, 290, 295, and 285°C, and the screw speed was fixed at 20 rpm. For the fabrication of PEN nanocomposites, PEN was melt blended with the addition of CNT content, specified as 0.1, 0.5, and 1.0 wt.% in the polymer matrix, respectively. For TLCP nanocomposites, the zone temperatures were set to 290, 300, 305, and 295°C, and the screw speed was fixed at 40 rpm. Prior to melt blending, TLCP and CNT were physically premixed before being fed into hopper of the extruder to achieve better dispersion of CNTs with TLCP matrix. The TLCP contents were 0.5, 1.0, 1.5 wt.% in the polymer matrix. Upon completion of melt blending, the extruded strands were allowed to cool in the water-bath, and then cut into pellets with constant diameter and length using a rate-controlled PP1 pelletizer (Haake Technik GmbH, Germany).

3. Poly (Ethylene 2,6-Naphthalate) Nanocomposite

3.1. Modification of CNT

FT-IR spectra of pristine and modified CNTs are shown in Figure 1A. The broad shoulder peak ranged in 3200–3400 cm^{-1} was attributed to the O–H stretching band induced by the hydroxyl groups attached at CNTs. The characteristic peak observed at approximately 1570 cm^{-1} was attributed to the IR-phonon mode of multiwall CNTs [80,81]. For the modified CNT, the peaks observed at near 1190 and 1730 cm^{-1} were assigned to the C=O and C–O stretching vibrations of the carboxylic acid groups [81,82]. This result demonstrates that carboxylic acid groups on the surface of the modified CNT were effectively induced via the chemical treatment. As shown in Figure 1B, similar patterns observed in the Raman spectra of pristine and modified CNTs indicated that chemical modification did not affect the graphite structure of CNTs. This chemical modification of CNTs in the HNO_3/H_2SO_4 mixture can be used to produce carboxylic acid groups at local defect sites on CNTs and their ends [55]. Raman spectra of pristine and modified CNTs exhibited three characteristic peaks observed at near 1350, 1570, and 2690 cm^{-1}, respectively, which were termed D-band, G-band, and D^*-band (overtone of D-band), respectively [83]. The G-band, corresponding to the Raman-allowed phonon high-frequency mode, was related to the structural intensity of the sp^2-

hybridized carbon atoms of CNTs [84,85]. The D-band reflected the disorder-induced carbon atoms, resulting from the defects in CNTs and their ends, and its intensity decreases with the degree of the graphitization of CNTs [84,85]. The relative intensity of D-band was slightly increased in modified CNT, indicating that the sp^2 hybrid decreased and the sp^3 hybrid increased after the chemical modification of the nanotubes. In addition, the relative intensity ratio of G-band to D-band (I_G/I_D) of modified CNT was slightly decreased as compared to pristine CNT, which was attributed to the increase in the degree of disorder and the presence of defects on the surface of modified CNT by chemical modification [86].

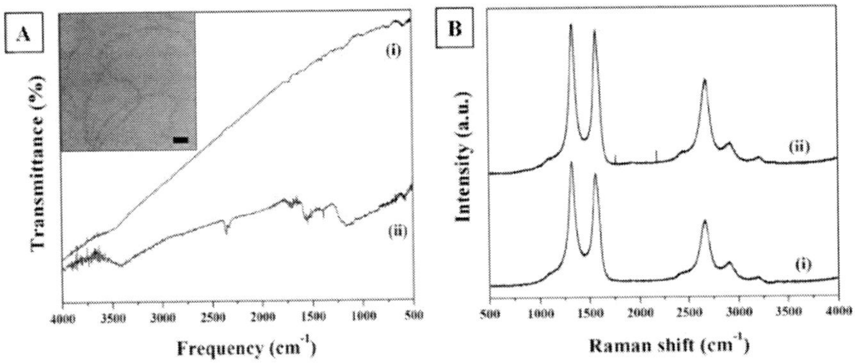

Figure 1. (A) FT-IR and (B) Raman spectra of (i) pristine CNT and (ii) modified CNT. The inset of Figure 1A shows representative TEM micrograph of modified CNT (Scale bar = 50 nm. Reproduced with permission from Ref. [71]. © 2008 Elsevier Ltd.

As shown in the inset of Figure 1A, the modified CNT exhibited loosely entangled organization instead of showing the agglomerated structures or bundles observed in pristine CNT, implying more effective enhancement of the dispersion of modified CNT than that of pristine CNT. However, it was reported that there was a possibility to occur the slight damage and the decreased length in the nanotube structure induced by the strong acid treatment [87,88].

The elemental compositions of PEN nanocomposites were characterized by X-ray photoelectron spectroscopy (XPS). In XPS survey spectra of PEN nanocomposites (Figure 2A), two characteristic peaks corresponding to C_{1s} and O_{1s} at approximately 285 and 532 eV, respectively, were observed, which were assigned to the sp^2 carbon and the ether-type group with oxygen bonded to carbon [89,90]. High resolution C_{1s} XPS spectra of PEN nanocomposites containing modified CNT are shown in Figure 2B. The C_{1s} spectra of pure PEN exhibited three characteristic peaks at 285, 286.5, and 288.7 eV, corresponding to C–C, C–O, and C=O groups, respectively [91]. As compared to pure PEN, the C_{1s} spectra of PEN nanocomposites containing modified CNT shift to higher binding energy, and also revealed the presence of three peaks, corresponding to C–C (285.2 eV), C–O (289.1 eV), and C=O (291.3 eV) groups, respectively. This slight shifting of characteristic peaks to higher binding energy for PEN nanocomposites containing modified CNT was attributed to the interaction of functional groups formed on the surface of modified CNT with PEN molecules as well as good dispersion of modified CNT in the PEN matrix. It should be noted that the functional groups formed on the surface of modified CNT contribute to the enhancement of the interaction in PEN nanocomposites through hydrogen bonding formation, as shown in Figure 2C.

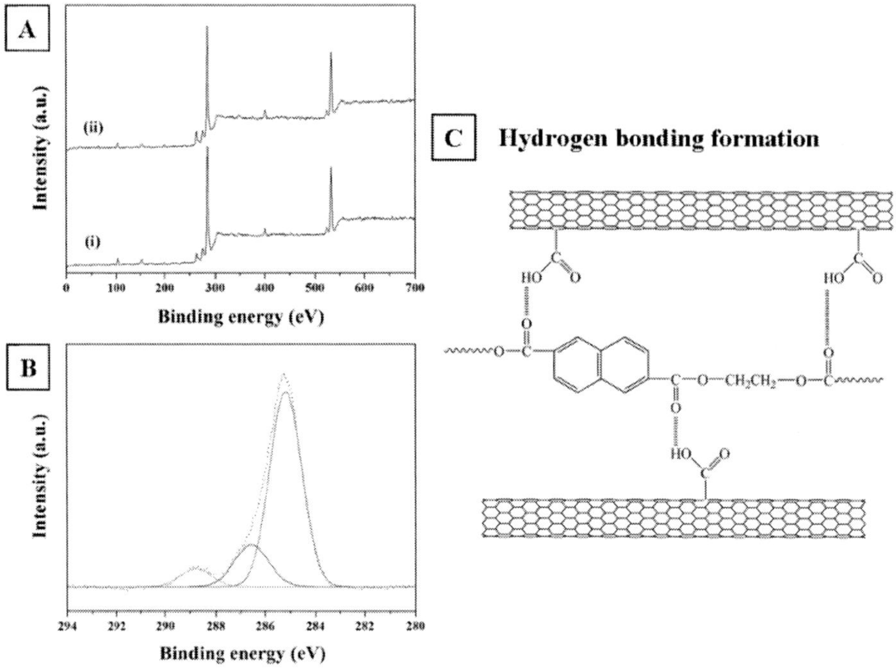

Figure 2. (A) XPS survey spectra of (i) PEN and (ii) PEN nanocomposites. (B) High resolution C_{1s} XPS spectra of PEN nanocomposites containing modified CNT. (C) Schematic showing possible interactions of hydrogen bonding between modified CNT and PEN. Reproduced with permission from Ref. [71]. © 2008 Elsevier Ltd.

3.2. Thermal Behavior of PEN Nanocomposites

TGA was conducted to identify the thermal stability of PEN nanocomposites containing CNTs, and the results are shown in Figure 3A. For modified CNT, the first weight loss observed below 100°C in the TGA traces was attributed to the loss of water molecules in the nanocomposite samples [92], and further gradual decrease in the residual weight was attributed to organic decomposition of thermally unstable functional groups formed on the surface of modified CNT [93]. The incorporation of CNTs into the PEN matrix increases the thermal decomposition temperatures and the residual yields of PEN nanocomposites and this enhancing effect was more significant in PEN nanocomposites containing modified CNT. The presence of CNTs could lead to the stabilization of PEN matrix, resulting in the enhanced thermal stability of PEN nanocomposites. The CNT can effectively act as physical barriers to prevent the transport of volatile decomposed products out of PEN nanocomposites during thermal decomposition. In addition, good interfacial adhesion between modified CNT and PEN matrix may restrict the thermal motion of PEN molecules [94], resulting in further improvement in thermal stability of PEN nanocomposites containing modified CNT.

TGA kinetic analysis was conducted on PEN nanocomposites to clarify the effects of CNTs on the thermal stability of PEN nanocomposites. The activation energy for thermal decomposition (E_a) of PEN nanocomposites can be estimated from the TGA thermograms by the Horowitz–Metzger integral kinetic method [95] as follows:

$$\ln\left[\ln(1-\alpha)^{-1}\right]=\frac{E_a\theta}{RT_{dm}^2} \tag{1}$$

where α is the weight loss; E_a is the activation energy for thermal decomposition; T_{dm} is the temperature at the maximum rate of weight loss; θ is the variable auxiliary temperature defined as $\theta = T - T_{dm}$; and R is the universal gas constant. The value of E_a can be determined from the slope of the plot of $\ln[\ln(1-\alpha)^{-1}]$ versus θ as shown in Figure 3B. The E_a values of PEN nanocomposites containing pristine CNT and modified CNT were 293.5 and 305.3 kJ/mol, respectively. As compared to PEN nanocomposites, higher E_a value of PEN nanocomposites containing modified CNT suggested that PEN nanocomposites with more uniform dispersion of modified CNT were more thermally stable than those containing pristine CNT. This feature was also attributed to the interactions between modified CNT and PEN matrix, which increase the activation of thermal decomposition of PEN matrix relative to pristine CNT. Because of excellent thermal conductivity of CNT [96,97], the enhanced interfacial interaction between modified CNT and PEN matrix resulted in the increased thermal conductivity of PEN nanocomposites, leading to the improvement in the thermal stability. In addition, it can be deduced that the E_a values of PEN nanocomposites calculated from the Horowitz–Metzger method exhibited good reliance on describing the thermal decomposition kinetics of PEN nanocomposites, which was confirmed by the values of the correlation coefficient (r^2) greater than 0.99.

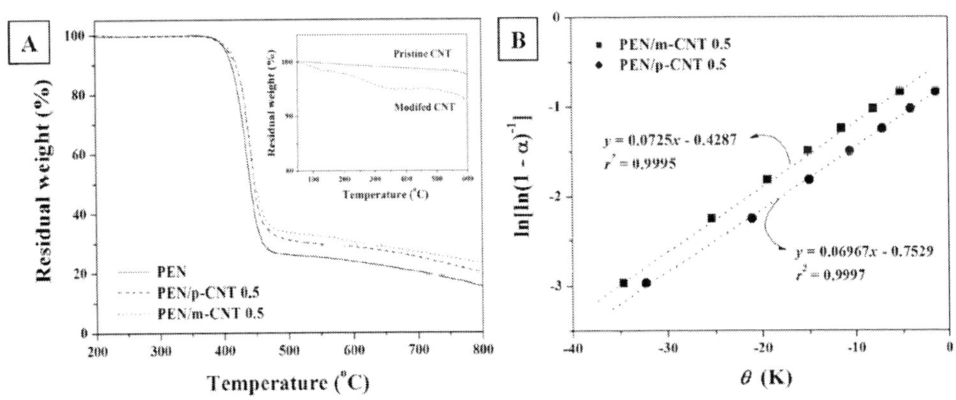

Figure 3. (A) TGA thermograms of PEN nanocomposites. (B) Plots of ln[ln(1 - α)-1] versus θ as shown for PEN nanocomposites. By the equation (1), the slope provides an estimate the activation energy for the thermal decomposition (Ea) of PEN nanocomposites. Reproduced with permission from Ref. [71]. © 2008 Elsevier Ltd.

3.3. Rheological Behavior of PEN nanocomposites

Complex viscosity ($|\eta^*|$), storage modulus (G'), and loss modulus (G'') of PEN nanocomposites as a function of frequency (ω) are shown in Figure 4. The $|\eta^*|$ of PEN nanocomposites decreased with increasing frequency, indicating a non-Newtonian behavior over the whole frequency range measured. The shear thinning behavior of PEN nanocomposites was attributed to random orientation and entangled molecular chains in the nanocomposites during applied shear force. PEN nanocomposites containing pristine and modified CNTs exhibited higher $|\eta^*|$ value than that of pure PEN at low frequency, indicating that the interconnected or network structures formed as a result of particle–particle and particle–polymer interactions [71,75]. The $|\eta^*|$ of PEN nanocomposites containing modified CNT further increased due to the increased interactions between PEN and nanotubes.

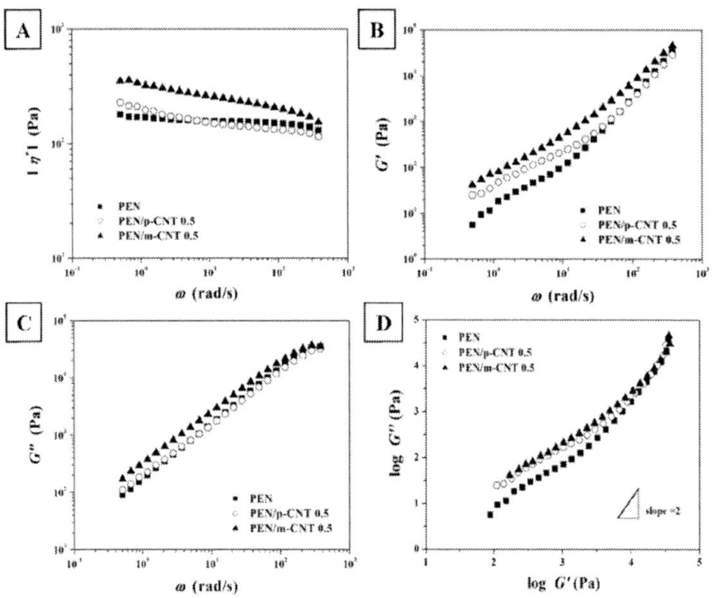

Figure 4. (A) Complex viscosity ($|\eta^*|$), (B) storage modulus (G′), and (C) loss modulus (G″) of PEN nanocomposites as a function of frequency (ω). (D) Plots of log G′ versus log G″ for PEN nanocomposites. Reproduced with permission from Ref. [71]. © 2008 Elsevier Ltd.

Furthermore, PEN nanocomposites containing modified CNT exhibited higher $|\eta^*|$ and more distinct shear thinning behavior over the whole frequency range, relative to the other systems, suggesting either better dispersion of the nanotubes or stronger nanotube–polymer interactions. The variations of the shear thinning exponent (n) estimated from the relationship of $|\eta^*| \approx \omega^n$ [98] also indicated that the shear thinning behavior of PEN nanocomposites significantly depended on the presence of pristine and modified CNTs (Table 1). The *n* values of PEN nanocomposites slightly decreased with the incorporation of pristine and modified CNTs and this effect was more pronounced in the case of modified CNT. This feature was attributed to the enhanced interfacial interactions between modified CNT and PEN matrix as well as the uniform dispersion of modified CNT. Pinnavaia and Beall [99] suggested that the better the nanofillers are dispersed, the stronger is the reinforcing effect on the polymer nanocomposites at a given content. As shown in Table 1, PEN nanocomposites exhibited slightly different *n* values even at the same CNT content, indicating the influence of the modification on the extent of CNT dispersion. In addition, PEN nanocomposites containing pristine and modified CNTs show the relationship between the shear thinning exponent and the mechanical properties for PEN nanocomposites: the more the shear thinning behavior, the better the reinforcement effect on the mechanical properties. Similar observation has been reported in that the tensile modulus of the modified clay/poly(butylene terephthalate) nanocomposites prepared via melt compounding was improved with lower values of the shear exponent [100].

Table 1. Variations of low-frequency slopes of $|\eta^*|$, G', and G'' versus ω for PEN nanocomposites.

Materials	Slope of $|\eta^*|$ versus ω	Slope of G' versus ω	Slope of G'' versus ω
PEN	-0.13	1.26	0.95
PEN/pristine CNT	-0.17	1.02	0.89
PEN/modified CNT	-0.19	0.93	0.85

As shown in Figures 4B and C, the values of G' and G'' for PEN nanocomposites increased with increasing frequency and this enhancing effect was more significant at low-frequency region. This rheological response is similar to the relaxation behavior of typical filled-polymer composite systems [101]: if the polymer chains are fully relaxed and exhibit a characteristic homopolymer-like terminal behavior, the flow curves of polymers can be expressed by a power law of $G' \propto \omega^2$ and $G'' \propto \omega$ [102]. Krisnamoorti and Giannelis [103] reported that the slopes of $G'(\omega)$ and $G''(\omega)$ for the polymer/layered silicate nanocomposites were much smaller than 2 and 1, respectively, suggesting that large deviations in the presence of a small quantity of layered silicate were caused by the formation of a network-like structures in the molten state. The non-terminal behavior with the power-law dependence for G' and G'' of PEN nanocomposites was observed (Table 1). The decrease in the slope of G' and G'' for PEN nanocomposites with the introduction of CNTs can be explained by the fact that the nanotube–nanotube or the nanotube–polymer interactions can lead to the formation of the interconnected or network-like structures, resulting in the pseudo solid-like behavior of PEN nanocomposites. The extent of the increase in G' of PEN nanocomposites was higher than that of G'' over the frequency range measured. The values of G' and G'' for PEN nanocomposites were higher than those of pure PEN, particularly at low frequency, and this enhancing effect was more pronounced in the case of modified CNT (Figures 4B and C). The higher G' and G'' values of PEN nanocomposites at low frequency demonstrated the formation of the interconnected or network structures via particle–particle and particle–polymer interactions by the presence of CNTs, resulting in more elasticity than pure PEN [71,75]. As the applied frequency increased, the interconnected or network structures were broken down due to high levels of shear force and PEN nanocomposites exhibited almost similar or slightly higher G' and G'' values than that of pure PEN at high frequency. Furthermore, the values of G' and G'' for PEN nanocomposites containing modified CNT were higher than in the case of pristine CNT, suggesting the increase in the interactions between modified CNT and PEN matrix. The plots of log G' versus log G'' for PEN nanocomposites are shown in Figure 4D. In general, the Cole–Cole plot provides a master curve with a slope of 2 for isotropic and homogeneous polymer melts, irrespective of temperatures [104]. The PEN nanocomposites did not provide a perfect single master curve, and exhibited the shifting and the change of the slope of the plot with the introduction of pristine and modified CNTs. The slopes in the terminal regime of PEN nanocomposites were less than 2, indicating that PEN nanocomposite systems were heterogeneous and they underwent some chain conformational changes due to the interconnected or network structures by the presence of CNTs. In addition, the slope of PEN nanocomposites containing modified CNT was lower than in the case of pristine CNT. This phenomenon may be attributed to the existence of the interfacial interactions between modified CNT and PEN. However, over the G'' values of approximately 10^4, the slope of PEN nanocomposites increased and approached to similar slope of pure PEN, indicating that the interconnected or network structures formed by the nanotube–nanotube and the polymer–nanotube interactions were collapsed by high levels of shear force [71].

3.4. Morphology and Mechanical Properties of PEN nanocomposites

Modified CNT exhibits less entangled structures due to the functional groups formed on the nanotube surfaces via chemical modification as compared to pristine CNT showing more aggregated structures (Figure 1A). In general, the drawbacks related to the homogeneous dispersion of the nanotubes in the polymer matrix resulted from intrinsic van der Waals attractions between the individual nanotubes in combination with high aspect ratio and large surface area, making it difficult for the nanotubes to disperse in the polymer matrix. The interfacial adhesion between the nanotubes and the polymer matrix plays an important role in improving the overall properties of polymer nanocomposites. As shown in Figure 5, modified CNT was dispersed well in PEN nanocomposites, which was explained by the fact that modified CNT stabilizes its dispersion by good interactions with PEN, resulting from the increased polarity by the functional groups formed on the nanotube surfaces as well as good interactions of the –COOH groups with the C=O groups of PEN.

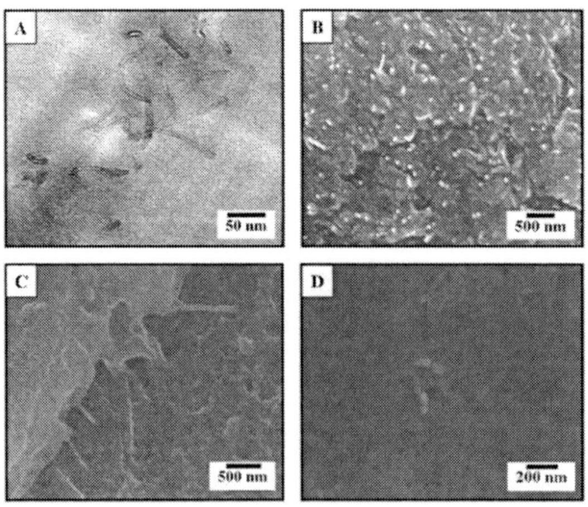

Figure 5. (A) TEM micrograph of PEN nanocomposites containing 0.1 wt.% of modified CNT. (B-D) SEM micrographs of PEN nanocomposites containing 0.5 wt.% of modified CNT. Reproduced with permission from Ref. [71]. © 2008 Elsevier Ltd.

This interfacial adhesion between modified CNT and PEN matrix is crucial for improving the mechanical properties of PEN nanocomposites. For PEN nanocomposites, two ends of modified CNT were covered by PEN (Figures 5C and D), and this feature indicated that modified CNT had a good wetting with PEN matrix, suggesting the existence of strong interactions between them. The presence of the functional groups on the surfaces of modified CNT results in the interfacial interaction between the polymer matrix and the nanotubes in PEN nanocomposites. For instance, the hydrogen atoms at the –COOH groups of modified CNT form the hydrogen bonding with the C=O groups of PEN macromolecular chains as illustrated in Figure 2B. This result can be considered as the evidence for efficient load transfer between PEN matrix and modified CNT during tensile testing.

The incorporation of a very small quantity of CNT significantly improved the mechanical properties of PEN nanocomposites due to the nanoreinforcing effect of CNT with high aspect ratio (Table 2). In comparison with pure PEN, PEN nanocomposites exhibited higher tensile strength and tensile modulus, and this enhancing effect was more pronounced in the case of PEN nanocomposites containing modified CNT. The enhancement of the mechanical properties of PEN nanocomposites containing modified CNT was attributed to the better interfacial bonding between modified CNT and PEN as well as the better dispersion of modified CNT in PEN matrix. The incorporation of modified CNT containing C–C bond defects and –COOH groups into PEN resulted in the good interfacial adhesion between modified CNT and PEN matrix, suggesting that the functional groups formed on the surfaces of modified CNT were helpful for improving the interfacial interaction with C=O groups in the PEN matrix, thus being favorable to more effective load transfer from the polymer matrix to the nanotubes. Therefore, the enhancement of homogeneous dispersion and interfacial adhesion in PEN nanocomposites, resulting from strong interactions between modified CNT and PEN chains, can lead to significant improvement in the overall mechanical properties of PEN nanocomposites containing modified CNT. As shown in Table 2, the elongation at break for PEN nanocomposites decreased with the introduction of pristine and modified CNTs, indicating that PEN nanocomposites became somewhat brittle as compared to pure PEN because of the increased stiffness of PEN nanocomposites and the micro voids formed around the nanotubes during tensile testing. However, PEN nanocomposites containing modified CNT exhibited higher value than in the case of pristine CNT, which was attributed to the enhancement of homogeneous dispersion of modified CNT and the interfacial interactions between modified CNT and PEN matrix.

Table 2. Mechanical properties of pure PEN and PEN nanocomposites containing 0.5 wt.% of CNT.

Materials	Tensile strength (MPa)	Tensile modulus (GPa)	Elongation at break (%)
PEN	66.3 ± 4.8	1.68 ± 0.032	253.2 ± 24
PEN/pristine CNT	84.9 ± 3.6	1.87 ± 0.025	126.4 ± 20
PEN/modified CNT	87.8 ± 3.4	1.98 ± 0.021	145.8 ± 21

For characterizing the effect of modified CNT on the mechanical properties of the PEN nanocomposites, it is also instructive to compare the experimental results with the values predicted from the theoretical models. It is known that the modified Halpin–Tsai equation has been used to predict the modulus of the CNT-filled polymer nanocomposites [105,106]. Assuming the random oriented discontinuous distribution of modified CNT in PEN matrix, the modified Halpin–Tsai equation can be expressed as follows [105–108].

$$E_C = \left[\left(\frac{3}{8}\right)\left(\frac{1+2(l_{CNT}/d_{CNT})\eta_L V_{CNT}}{1-\eta_L V_{CNT}}\right)+\left(\frac{5}{8}\right)\left(\frac{1+2\eta_T V_{CNT}}{1-\eta_T V_{CNT}}\right)\right]E_{PEN} \tag{2}$$

where $\eta_L = \dfrac{(E_{CNT}/E_{PEN})-1}{(E_{CNT}/E_{PEN})+2(l_{CNT}/d_{CNT})}$ and $\eta_T = \dfrac{(E_{CNT}/E_{PEN})-1}{(E_{CNT}/E_{PEN})+2}$

where E_C, E_{PEN}, and E_{CNT} are the tensile modulus of nanocomposites, PEN, and CNT, respectively; l_{CNT}/d_{CNT} is the ratio of length to diameter for CNT, and V_{CNT} is the volume fraction of CNT in the nanocomposites. To fit the equation (2) to the experimental results for PEN nanocomposites, the weight fraction was transformed to the volume fraction, taking the densities of PEN (1.407 g/cm^3) [109] and the perfectly graphitized CNT (2.16 g/cm^3) [105,106]. The theoretical values of the nanocomposite modulus can be estimated assuming the aspect ratio of ~1000 and E_{CNT} of ~450 GPa. The E_{CNT} values used represent a mid-range value in the modulus ranges of CNT previously measured [110].

The tensile strength of PEN nanocomposites containing modified CNT can be estimated from the following equation [111]:

$$\sigma_C = \sigma_{CNT} \cdot V_{CNT} + \sigma_{PEN} \cdot V_{PEN} \tag{3}$$

where σ_C, σ_{CNT}, and σ_{PEN} are the tensile strength of nanocomposites, CNT, and PEN, respectively. The theoretical values of the nanocomposite strength can be estimated based on the σ_{CNT} value that is ~11 GPa based on previous literature [25].

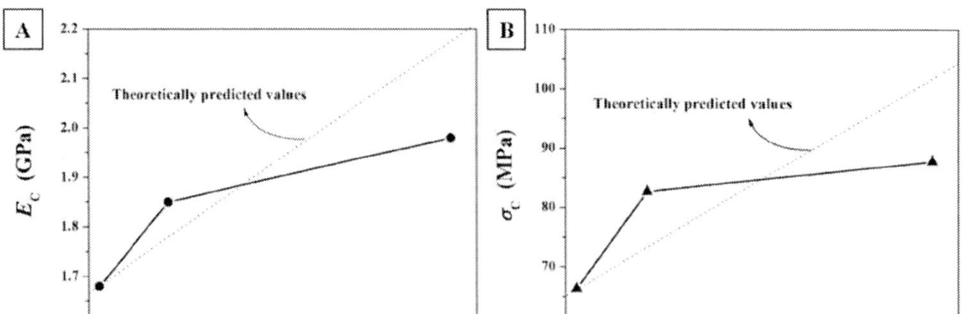

Figure 6. Theoretically predicted values and the experimental results for (A) the nanocomposite modulus and (B) the nanocomposites strength of PEN nanocomposites containing modified CNT. The dotted lines represent the theoretically predicted values based on the Equations (2) and (3). Reproduced with permission from Ref. [71]. © 2008 Elsevier Ltd.

The theoretically predicted values and the experimental data for the modulus and the strength of PEN nanocomposites containing modified CNT are compared in Figure 6. The experimental results for the mechanical properties of PEN nanocomposites were lower than those of theoretically predicted values. At lower CNT content PEN nanocomposites containing modified CNT exhibited higher values than those of theoretically predicted values, suggesting that the interfacial interaction was more effective in the enhancement of mechanical properties of PEN nanocomposites at 0.1 wt.% of modified CNT content than at 0.5 wt.% content. Similar effect has been also observed in MWCNT-grafted polyhedral oligomeric silsequioxane (MWCNT-*g*-POSS) and poly(L-lactide) (PLLA) nanocomposite systems [48] in that for MWCNT-*g*-POSS/PLLA nanocomposites, the interfacial interactions were more effective in strengthening the nanocomposites at low MWCNT-*g*-POSS loading than at high MWCNT-*g*-POSS loading. Thus, the synergistic effect of the combined strong interfacial interaction and homogenous dispersion was more effective at lower CNT content for improving the mechanical properties of PEN nanocomposites. In summary, the introduction of modified CNT into PEN matrix results in the enhancement of interfacial adhesion between modified CNT and PEN as well as good dispersion of modified CNT, thereby improving significantly the overall mechanical properties of PEN nanocomposites.

3.5. Crystallization Behavior of PEN Nanocomposite

The incorporation of pristine has little effect on the glass transition temperature (T_g) and melting temperature (T_m) of PEN nanocomposites (Table 3). However, the T_g of PEN nanocomposites slightly increased with the introduction of modified CNT. The increase in the T_g of PEN nanocomposites containing modified CNT was attributed to the hindrance of segmental motions of PEN macromolecular chains with the presence of modified CNT [112]. The crystallization temperatures were significantly increased with the introduction of pristine and modified CNTs, together with the fact that PEN nanocomposites have a lower degree of supercooling (ΔT) for crystallization with the introduction of CNT, indicating the efficiency of them as strong nucleating agents for the crystallization of PEN. This result suggests the enhancement of the crystallization of PEN nanocomposites with the presence of pristine and modified CNTs and this enhancing effect is more pronounced in the case of modified CNT.

Table 3. Thermal properties of pure PEN and PEN nanocomposites containing 0.5 wt.% of CNT.

Materials	T_g [a] (°C)	T_{cc} [a] (°C)	T_m [a] (°C)	T_{mc} [b] (°C)	ΔT [c] (°C)
PEN	119.1	199.8	263.0	190.0	73.0
PEN/pristine CNT	118.8	186.7	264.2	218.3	45.9
PEN/modified CNT	123.2	184.6	263.8	218.1	45.7

[a] Values obtained from the DSC heating traces at a scanning rate of 10 °C/min.

[b] Crystallization temperatures measured from the DSC cooling traces at a scanning rate of 10 °C/min.

[c] The degree of supercooling, $\Delta T = T_m - T_{mc}$.

Under the non-isothermal crystallization, as the cooling rate increases the crystallization peak temperature range becomes broader and shifts to lower temperatures, indicating that the lower the cooling rate, the earlier crystallization occurs. When the sample was cooled quickly, more supercooling was required to initiate crystallization, because the motion of polymer molecules can not follow the cooling rate [113]. The PEN nanocomposites containing modified CNT exhibit higher peak temperature and lower overall crystallization time at a given cooling rate than those of pure PEN. In general, homogeneous nucleation started spontaneously below the melting temperature and required longer times, whereas heterogeneous nuclei formed as soon as samples reached the crystallization temperature [114]. As the crystallization of PEN nanocomposites proceeds through heterogeneous nucleation, the introduction of modified CNT enhances the crystallization of PEN due to high nucleation induced by modified CNT. For CNT-filled polymer nanocomposites, accelerated crystallizations by the presence of CNT through heterogeneous nucleation have been also reported [69-73,115-117].

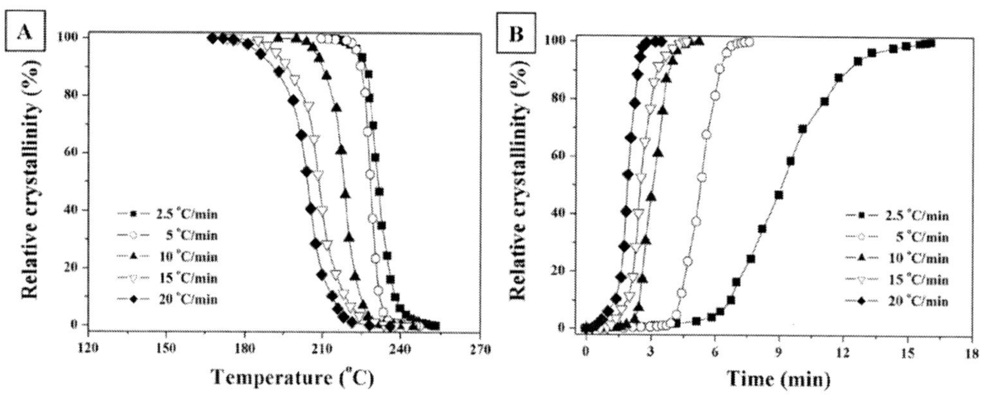

Figure 7. Relative degrees of crystallinity of PEN nanocomposites containing 0.5 wt.% of modified CNT with (A) the temperature and (B) the time at various cooling rates. Reproduced with permission from Ref. [70]. © 2009 Elsevier Ltd.

The non-isothermal crystallization of PEN nanocomposites was performed at various cooling rates to clarify the effects of modified CNT on their crystallization behavior. The relative degrees of crystallinity($X(T)$) with temperature and time for PEN nanocomposites containing modified CNT at various cooling rates are shown in Figure 7. The values of $X(T)$ for PEN nanocomposites at various cooling rates can be calculated from the ratio of the area of the exothermic peak up to temperature (T) divided by that of the total exotherms of the crystallization. The crystallization occurred at lower temperature with increasing the cooling rate, indicating that the crystallization nucleates at higher temperatures with slower cooling rates because at slower cooling rates there is sufficient time to activate nuclei at higher temperatures [118]. The crystallization of PEN nanocomposites containing modified CNT occurred at higher temperature and over a longer time with decreasing cooling rate, suggesting that crystallization may be controlled by the nucleation. As the cooling rate increased, the time for completing crystallization decreased, and the value of $X(T)$ for PEN nanocomposite containing modified CNT was higher than that of pure PEN.

As shown in Figure 8 and Table 4, the peak temperature (T_p) and the crystallization half-time ($t_{1/2}$) for PEN nanocomposites decreased with increasing cooling rate. At a given cooling rate, the T_p values of PEN nanocomposites containing modified CNT were higher than that of pure PEN, while the $t_{1/2}$ values were lower than that of pure PEN. This result indicated that the incorporation of modified CNT increased the crystallization rate of PEN matrix. The lowering of the $t_{1/2}$ values induced by CNT was more pronounced in PEN nanocomposites containing modified CNT than in the case of pristine CNT, suggesting that modified CNT can be more effective for enhancing the crystallization rate of PEN as compared to pristine CNT. This effect of the crystallization rate induced by modified CNT was attributed to the interactions between the functional groups on the surfaces of modified CNT and the PEN matrix as well as physical adsorption of PEN molecules onto the nanotube surfaces, enhancing the rate of crystallization for PEN nanocomposites. Wang and co-workers [119] have reported that for the polyamide (PA)/MWCNT nanocomposites, the introduction of carboxylated MWCNT increased more efficiently the crystallization rate of pure PA matrix. This enhancing effect by modified CNT on non-isothermal crystallization of PEN nanocomposites is elaborated during the following discussion on the nucleation activity and activation energy for non-isothermal crystallization.

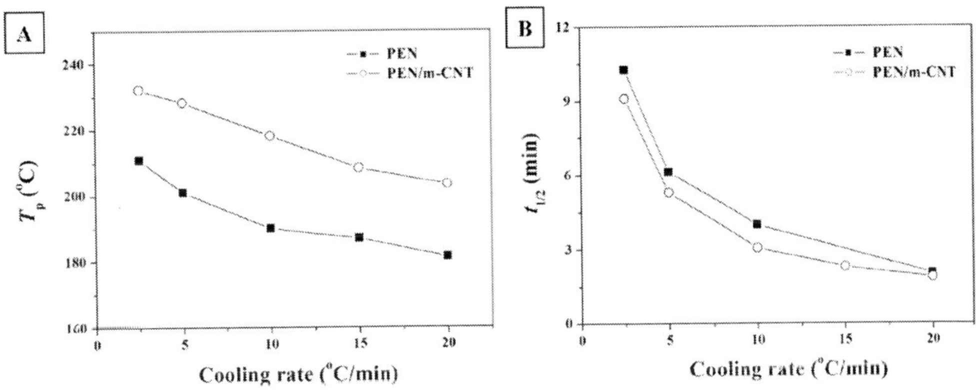

Figure 8. Variations of (A) the peak temperatures (T_p) and (B) the crystallization half-time ($t_{1/2}$) of PEN nanocomposites containing 0.5 wt.% of modified CNT as a function of cooling rate during the non-isothermal crystallization. The crystallization half-time ($t_{1/2}$) can be defined as the time taken to complete half of the non-isothermal crystallization process, i.e., the time required to attain a relative degree of crystallinity of 50%. Reproduced with permission from Ref. [70]. © 2009 Elsevier Ltd.

Table 4. Kinetic parameters of PEN/m-CNT 0.5 nanocomposites during the non-isothermal crystallization.

Cooling rate (°C/min)	T_p (°C)	Z_c	n	$t_{1/2}$ (min)
2.5	232.1	3.72×10^{-3}	6.62	9.10
5	228.2	9.74×10^{-2}	6.88	5.29
10	218.1	4.29×10^{-1}	5.63	3.05
15	208.3	7.18×10^{-1}	4.72	2.29
20	203.3	8.17×10^{-1}	5.69	1.87

The non-isothermal crystallization kinetics can be analyzed using the extension of the Avrami theory [120] proposed by Ozawa [121]. This analysis describes the effect of the cooling rate on the crystallization from the melt by replacing the time variable in the Avrami equation with a variable cooling rate term as follows:

$$1 - X(T) = \exp\left(-Z_t t^n\right) \tag{4}$$

$$1 - X(T) = \exp\left[-\frac{K(T)}{a^m}\right] \tag{5}$$

where $X(T)$ is the relative degree of crystallinity, the exponent n is a mechanism constant depending on the type of nucleation and growth dimension, Z_t is a growth rate constant involving both nucleation and growth rate parameters, a is the cooling rate, m is the Ozawa exponent depending on the dimension of crystal growth, and $K(T)$ is the cooling function related to the overall crystallization rates. The kinetic parameters such as Z_t and n explicit physical meanings for the isothermal crystallization, while in non-isothermal crystallization their physical meaning does not have the same significance due to constant changes in the temperature, influencing the nucleation and crystal growth.

Ozawa plots of PEN nanocomposites containing modified CNT are shown in Figure 9A. Some curvature observed in the Ozawa plots indicated that the Ozawa exponent was not consistent with the temperature during the non-isothermal crystallization, which makes it difficult to estimate the overall crystallization rate because of the inaccurate assumption in the Ozawa's theory, involving the disregard for the secondary crystallization and dependence of the fold length on temperature [122]. Therefore, it can be deduced that Ozawa analysis does not effectively describe the non-isothermal crystallization of PEN nanocomposites, because the presence of modified CNT results in the spontaneous nucleation, the rapid impingement, and the secondary crystallization in the polymer nanocomposites [123].

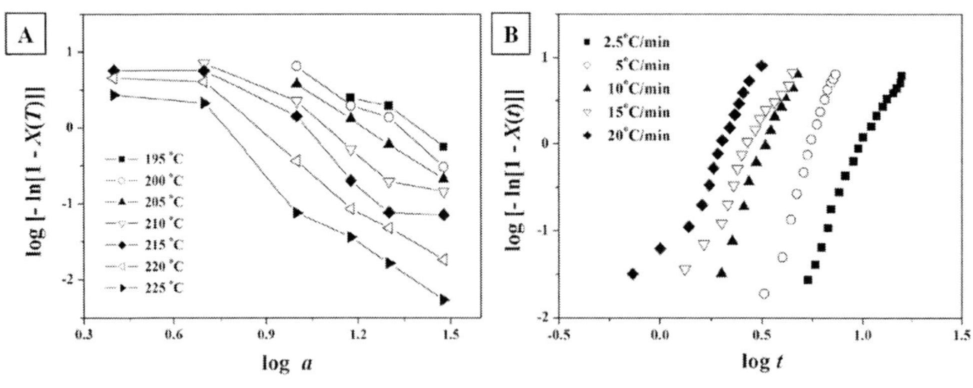

Figure 9. (A) Ozawa plots and (B) Avrami plots of PEN nanocomposites containing modified CNT during the non-isothermal crystallization. Reproduced with permission from Ref. [70]. © 2009 Elsevier Ltd.

Avrami plots of PEN nanocomposites containing modified CNT are shown in Figure 9B. The kinetic parameters such as n and Z_t can be determined from the slope and intercept of the plots of log[-ln{1 - $X(T)$}] versus log t. A poor linear relationship observed in PEN nanocomposites indicates that this analysis does not effectively describe the non-isothermal crystallization of PEN nanocomposites. Based on the non-isothermal character of the process suggested by Jeziorny [124], the rate parameter (Z_t) should be corrected according to the relationships of $\log Z_c = \log Z_t/a$. The kinetic parameters for the non-isothermal crystallization of PEN nanocomposites estimated from the kinetic data selected in the linear region are summarized in Table 4. The n values were in the range of 4.72~6.88 for PEN nanocomposites, depending on the cooling rate. As the dependence of the crystallization kinetics on the temperature is a complex function, and many theoretical models based on the Avrami equation have been developed [123-125]. The PEN nanocomposites containing modified CNT exhibit values of n higher than 4, suggesting that the mechanism of the non-isothermal crystallization of PEN nanocomposites was very complicated and the introduction of modified CNT into PEN significantly influenced the crystallization behavior, leading to the changes of the crystallization kinetics of PEN nanocomposites. The lower $t_{1/2}$ and higher Z_c values of PEN nanocomposites containing modified CNT with increasing the cooling rate as compared to pure PEN revealed that heterogeneous nucleation for PEN nanocomposites was enhanced by modified CNT, suggesting that the introduction of modified CNT resulted in the acceleration of non-isothermal crystallization of PEN nanocomposites.

To describe the non-isothermal crystallization process more effectively for comparison, Mo and co-workers [125] suggested convenient procedures to characterize the non-isothermal crystallization kinetics by combining the Avrami and Ozawa equations based on the assumption that the degree of crystallinity was correlated to the cooling rate and crystallization time. This relationship can be derived by combining the equations (4) and (5) as follows:

$$\log Z_t + n\log t = \log K(T) - m\log a \tag{6}$$

$$\log a = \log F(T) - b\log t \tag{7}$$

where the kinetic parameter, $F(T) = [K(T)/Z_t]^{1/m}$ represents the value of the cooling rate chosen at unit crystallization time when the systems have a defined degree of crystallinity, a is the cooling rate, and b is the ratio of the Avrami exponent (n) to the Ozawa exponent (m). The plots of log a versus log t at a certain degree of crystallinity for PEN nanocomposites containing modified CNT are shown in Figure 10A. The combined Avrami and Ozawa plots exhibit a good linear relationship, suggesting that this analysis can be more effective in describing the non-isothermal crystallization kinetics of PEN nanocomposites. The increase in the $F(T)$ value with the relative degree of crystallinity indicates that higher relative degree of crystallinity can be obtained with higher cooling rate at unit crystallization time. The PEN nanocomposites containing modified CNT exhibit smaller values of $F(T)$ than that of pure PEN. The PEN nanocomposites achieved the values for the fraction crystallinity faster than pure PEN, demonstrating faster crystallization kinetics of PEN nanocomposites containing modified CNT due to the effective function of modified CNT as strong nucleating agent. In addition, the b values were varied from 1.13 to 1.36 for pure PEN, from 1.24 to 1.28 for PEN nanocomposites containing pristine CNT, and from 1.33 to 1.40 for PEN nanocomposites containing modified CNT, respectively [70,73]. Thus, the introduction of CNT can influence significantly the non-isothermal crystallization process involving the nucleation and crystal growth for PEN nanocomposites.

The nucleation activity is a factor by which the work of three dimensional nucleation decreases with the addition of a foreign substrate [126]. The nucleation activity (ϕ) of modified CNT in PEN nanocomposites can be determined from the simple method for calculating the nucleation activity of different substrates using the relationship of $\phi = B^*/B^0$ (where B^0 and B^* are the values of B for homogeneous and heterogeneous nucleation, respectively) [73,76,126]. The ϕ value of 0 means strong nucleation activity and that of 1 means inert nucleation activity. The ϕ values of PEN nanocomposites containing pristine and modified CNTs were calculated as 0.258 and 0.296, respectively. This result demonstrates that pristine and modified CNTs can act as excellent nucleating agents for PEN nanocomposites during non-isothermal crystallization process, which was corresponded with the non-isothermal crystallization kinetics of PEN nanocomposites. The nucleation activities of pristine and modified CNTs in PEN nanocomposites were more excellent than any other nanoscale filler reported to date. Alonso and co-workers [127] reported that for the talc/polypropylene (PP) composite systems the ϕ value of untreated talc was approximately 0.56, while that of the talc modified with silane coupling agent was estimated to be 0.45 in the PP matrix. Mitchell and Krishnamoorti [128] reported that the ϕ values for poly(ε-caprolactone) (PCL) containing 0.35 and 1.8 wt.% CNT were 0.56 and 0.47, respectively. Thus, the incorporated CNT in PEN nanocomposites exhibits much higher nucleation activity than any other nanoscale filler reported to date, with even very small quantity of CNT.

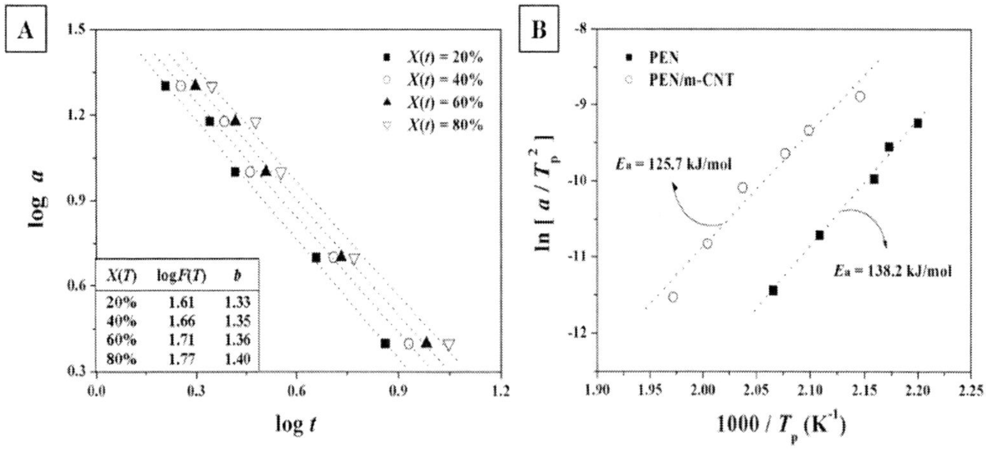

Figure 10. (A) Plots of log a versus log t from combined Avrami and Ozawa equations at different relative degrees of crystallinity for PEN nanocomposites containing modified CNT. (B) Plots of $\ln[a/T_p^2]$ versus $1/T_p$ for PEN nanocomposites containing modified CNT. By the equation (8), the slope provides an estimate of the activation energy for non-isothermal crystallization of PEN nanocomposites. Reproduced with permission from Ref. [70]. © 2009 Elsevier Ltd.

The activation energy for the non-isothermal crystallization can be derived from the combination of the cooling rate and the crystallization peak temperature, suggested by Kissinger [129] as follows:

$$\frac{d\left[\ln\left(a/T_p^2\right)\right]}{d\left(1/T_p\right)} = -\frac{\Delta E_a}{R} \quad (8)$$

where R is the universal gas constant, T_p is the crystallization peak temperature, a is the cooling rate, and ΔE_a is the crystallization activation energy. The ΔE_a values of PEN nanocomposites were obtained from the slope of the plot of $\ln(a/T_p^2)$ versus $1/T_p$. As shown in Figure 10B, the ΔE_a value of PEN nanocomposites containing modified CNT was lower than that of pure PEN and those of pristine CNT (e.g., pristine CNT/PEN nanocomposites, ΔE_a = 136.2 kJ/mol). For PEN nanocomposites, the incorporation of pristine and modified CNTs into PEN matrix probably induced heterogeneous nucleation, leading to lower ΔE_a values. Furthermore, the incorporation of modified CNT well-dispersed in PEN matrix causes more heterogeneous nucleation and decreases the ΔE_a, suggesting that the introduction of modified CNT further induces more nucleation density to enhance the crystallization rate of PEN nanocomposites during non-isothermal crystallization process [130], which was confirmed by the $t_{1/2}$ value that was slightly lower than in the case of pristine CNT. For SWCNT and high-density polyethylene (HDPE) nanocomposites, Winey and co-workers [131] reported that the presence of SWNT dramatically increased the number of nucleation sites and thereby decreased the average crystallite size and given the additional nucleation sites in SWNT/HDPE nanocomposites, the crystallization rate was dramatically faster. Similar effect has been also observed in the case of isotactic PP (iPP)/CNT nanocomposites that the nucleation process of the iPP crystallization was enhanced, with CNT acting as a nucleating agent, promoting faster crystal growth and the creation of a large number of smaller spherulites in heterogeneous nucleation process [132]. Based on the crystallization kinetic analysis on PCL/CNT nanocomposites, Wu and Chen [133] reported that the ΔE_a value of PCL extensively decreased with increasing CNT content, suggesting that the addition of CNT into PCL matrix induced more heterogeneous nucleation during crystallization processes and could lead to lower ΔE_a of PCL/CNT nanocomposites. They also found that the size and the number of spherulites of PCL/CNT nanocomposites were much smaller and higher than those of pure PCL, which was attributed to the increased nucleation density and nucleation rates of PCL/CNT nanocomposites by the presence of CNT. Thus, the ΔE_a value of PEN nanocomposites supported the positive effect on the crystallization of PEN nanocomposites in presence of modified CNT. In addition, as CNT content was relatively low in PEN nanocomposite systems, CNT may not greatly restrict the movement of PEN molecular chains, while modified CNT can act effectively as strong nucleating agents during non-isothermal crystallization process, resulting in the acceleration of the crystallization of PEN nanocomposites.

4. Thermotropic Liquid Crystal Polymer Nanocomposite

4.1. Background

Thermotropic liquid crystal polymer (TLCP) that is a class of the materials to produce the anisotropic melts of relatively low viscosity, has attracted a great deal of scientific interest in fields ranging from the scientific to the industrial, because of their excellent mechanical properties, superior chemical resistance, low gas and liquid permeability, low controllable coefficient of thermal expansion, excellent dimensional stability under temperature or in sever environment, and easy processability with high precision by extrusion and injection molding [134-136]. TLCP with high strength and stiffness due to its rigid-rod like molecules can be preferentially oriented to form fibrils under elongational or shear flow during melt processing, and the oriented fibrous structures are developed in the extruded TLCP, resulting in self-reinforcing characteristics [137-139]. In this regard, TLCP have received considerable attention both in the neat state and as reinforcing fillers for thermoplastic polymers, and much research has been extensively performed to date both to displace conventional thermoplastic polymers and to develop commercial applications of TLCP and its composites such a high performance engineering plastic or fiber [38-42]. Although promising, however, practical applications of TLCP are limited compared to conventional thermoplastic polymers, because TLCP is relatively expensive due to high cost of the monomer. Thus, it is expected that the combination of a very small quantity of CNT with TLCP provides attractive possibilities for improving the overall properties of TLCP nanocomposites using a cost-effective method.

This section focused on the characterization of TLCP nanocomposites reinforced with a very small quantity of modified CNT using simple melt blending in a twin-screw extruder to create high performance composite materials with low processing cost. The effects of unique characteristics of modified CNT on the physical properties of TLCP nanocomposites are discussed. This section will be helpful in providing the preliminary understanding on the properties of TLCP nanocomposites reinforced with CNT, and suggests a simple and cost-effective method that will facilitate the industrial realization of TLCP nanocomposites with enhanced properties, providing a design guide of TLCP nanocomposites with a great potential for industrial uses. The exhaustive description of all LCP composites containing CNT is beyond of the scope of this section, and for this readers are refereed to recent review articles [140-142].

4.2. CNT Modification

The modification of pristine CNT was performed according to procedures in detail elsewhere [55,70,71], and the carboxylic acid groups on the surface of modified CNT were effectively induced via this chemical modification. After chemical modification, modified CNT exhibits less entangled structures as compared to pristine CNT, indicating that the dispersion of modified CNT in the polymer matrix is more effective than that of pristine CNT. It should be noted that the functional groups effectively induced on the surface of the nanotube via chemical modification are helpful for enhancing the interactions between the polymer matrix and the nanotube.

The elemental compositions of TLCP nanocomposites were characterized by XPS. In the XPS survey spectra of TLCP nanocomposites, it can be seen that two characteristic peaks, corresponding to C_{1s} and O_{1s} at approximately 285 and 523 eV, respectively, were observed, which were assigned to the sp^2 carbon and the ester-type groups with oxygen bonded to carbon. In the high resolution C_{1s} XPS spectra, the C_{1s} spectra of pure TLCP exhibited four characteristic peaks at 285.0, 286.5, 287.5, and 289.0 eV, corresponding to the C-C, C-O, C=O, and O=C-O groups, respectively. As compared to pure TLCP, the C_{1s} spectra of TLCP nanocomposites shift to higher binding energy, and also revealed the presence of four peaks, corresponding to the C-C (285.3 eV), C-O (286.8 eV), C=O (287.8 eV), and O=C-O (289.6 eV) groups, respectively (Figure 11A). This slight shifting of characteristic peaks to higher bonding energy for TLCP nanocomposites was attributed to the interfacial interactions of the functional groups effectively induced on the surface of modified CNT via chemical modification with TLCP macromolecular chains as well as good dispersion of modified CNT in TLCP matrix. Possible interactions between modified CNT and TLCP through hydrogen bonding formation are illustrated in Figure 11B.

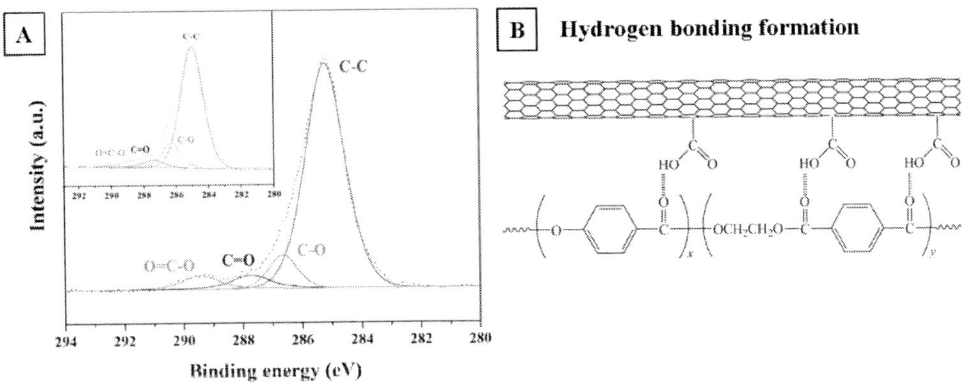

Figure 11. (A) High resolution C_{1s} XPS spectra of TLCP nanocomposites containing 1.5 wt.% of modified CNT. The shit of characteristic peaks to higher bonding energy for TLCP nanocomposites resulted from the interactions of modified CNT with TLCP macromolecular chains as well as good dispersion of modified CNT in TLCP matrix. (B) Schematic showing possible interactions of hydrogen bonding between modified CNT and TLCP matrix. Reproduced with permission from Ref. [67]. © 2009 Elsevier Ltd.

4.3. Rheological Properties

As shown in Figure 12, the complex viscosity ($|\eta^*|$) of TLCP nanocomposites decreased with increasing frequency, indicating a non-Newtonian behavior over the frequency range measured. The shear thinning behavior in TLCP nanocomposites was attributed to random orientation of rigid molecular chains in the nanocomposites during applied shearing force. The TLCP nanocomposites exhibited higher $|\eta^*|$ value than that of pure TLCP at low frequency, indicating that the interconnected or network-like structures were formed in TLCP nanocomposites due to the nanotube–nanotube or the nanotube–polymer interactions. The TLCP nanocomposites exhibited strong shear thinning behavior, resulting from the break-down of these structures with increasing frequency. The $|\eta^*|$ of TLCP nanocomposites increased with increasing CNT content, and this effect was more pronounced at low frequency than that at high frequency. The increased $|\eta^*|$ value of TLCP nanocomposites with increasing CNT content was also attributed to the increase in the nanotube–nanotube and nanotube–polymer interactions. Furthermore, TLCP nanocomposites containing modified CNT show higher $|\eta^*|$ and more distinct shear thinning behavior as compared to pure TLCP, suggesting either better dispersion of modified CNT or stronger nanotube–polymer interactions. Similar observation has been reported that higher viscosity and more distinct shear thinning behavior were indicative of stronger interfacial interactions in CNT/epoxy nanocomposites with more uniform dispersion of CNT [143].

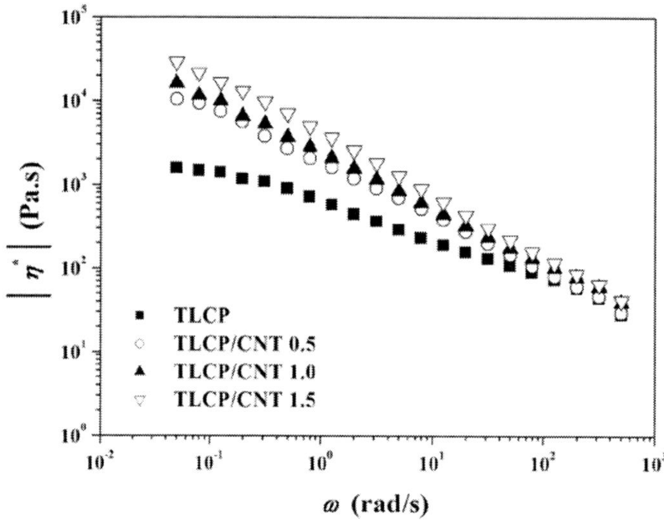

Figure 12. Complex viscosity (|η*|) of TLCP nanocomposites with the modified CNT content measured at 300°C as a function of frequency. Reproduced with permission from Ref. [67]. © 2009 Elsevier Ltd.

Table 5. Variations of low-frequency slopes of $|\eta^*|$, G', and G'' versus ω for TLCP nanocomposites.

Materials	Slope of $|\eta^*|$ versus ω	Slope of G' versus ω	Slope of G'' versus ω
TLCP	-0.43	0.74	0.62
TLCP/CNT 0.5	-0.647	0.45	0.35
TLCP/CNT 1.0	-0.65	0.41	0.30
TLCP/CNT 1.5	-0.72	0.31	0.27

The shear thinning exponent (n) of TLCP nanocomposites can be estimated from the relationship of $|\eta^*| \approx \omega^n$ [98], and the results are summarized in Table 5. There is a significant dependence of the shear thinning behavior of TLCP nanocomposites on the presence of modified CNT. The n values of TLCP nanocomposites decreased with the introduction of modified CNT, which was related to the increase in the interfacial interactions between modified CNT and TLCP as well as the uniform dispersion of modified CNT in TLCP matrix. In addition, TLCP nanocomposites show the relationship between shear thinning behavior and mechanical properties: the more the shear thinning behavior, the better reinforcing effect on the mechanical properties of TLCP nanocomposites. Wagner and Reisinger [100] reported that the tensile modulus of poly(butylene terephthalate) nanocomposites containing modified clay prepared via melt compounding was significantly enhanced with lower values of the shear thinning exponent.

Storage modulus (G') and loss modulus (G'') of TLCP nanocomposites as a function of frequency are shown in Figure 13. The values of G' and G'' of TLCP nanocomposites significantly increased with increasing frequency and CNT content, and this enhancing effect was more pronounced at low frequency. This phenomenon is similar to the relaxation behavior of typical filled-polymer composite system [144]: if polymer chains are fully relaxed and exhibit a characteristic homopolymer-like terminal behavior, the flow curves of polymers can be expressed by a power law of $G' \propto \omega^2$ and $G'' \propto \omega$. Krisnamoorti et al. [101,103] reported that the slopes of $G'(\omega)$ and $G''(\omega)$ for layered silicate and polymer nanocomposite system were much smaller than 2 and 1, respectively, and they suggested that large deviations in the presence of a small quantity of layered silicate resulted from the formation of a network structure in the polymer molten state. As shown in Table 5, the variations of the terminal zone slopes of TLCP nanocomposites indicated the non-terminal behavior with the power law dependence for $G'(\omega)$ and $G''(\omega)$ of TLCP nanocomposites.

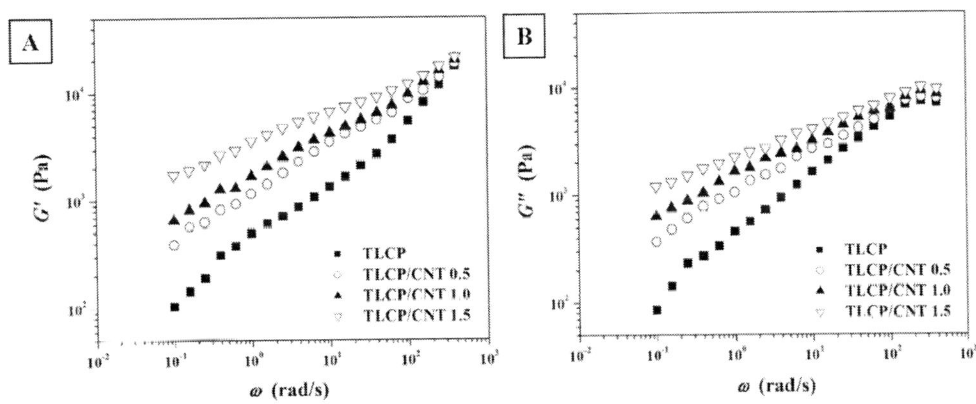

Figure 13. (A) storage modulus (G′) and (B) loss modulus (G″) of TLCP nanocomposites with the modified CNT content as a function of frequency (ω). Reproduced with permission from Ref. [67]. © 2009 Elsevier Ltd.

The decrease in the slope for TLCP nanocomposites with the introduction of modified CNT was explained by the fact that the nanotube–nanotube or the nanotube–polymer interactions can induce the formation of the interconnected or network-like structures in TLCP nanocomposites, leading to the pseudo solid-like behavior of TLCP nanocomposites. Similar observation has been reported in that the difference of the slope of G' and G'' for CNT/epoxy nanocomposites at terminal regime was closely related to the internal structure of the nanocomposites, depending on the particle–particle interaction of CNT in the epoxy matrix

[145]. As shown in Figure 13, higher values of G' and G'' for TLCP nanocomposites at low frequency demonstrates the formation of the interconnected or network-like structures via the nanotube–nanotube and nanotube–polymer with the introduction of modified CNT, resulting in more elasticity than pure TLCP. As the applied frequency increased, the interconnected or network-like structures formed by the nanotube–nanotube and the polymer–nanotube interactions were collapsed by high levels of shearing force and TLCP nanocomposites exhibited almost similar or slightly higher G' and G'' values than that of pure TLCP at high frequency. In addition, the values of G' and G'' of TLCP nanocomposites were higher than that of pure TLCP over the whole frequency range measured suggests the enhanced interactions between modified CNT and TLCP matrix.

4.4. Morphology

The morphologies of TLCP nanocomposites containing modified CNT are shown in Figure 14. In general, the drawbacks related to the uniform dispersion of CNT in the polymer matrix were caused by intrinsic van der Waals attractions between the individual nanotubes in combination with high aspect ratio and large surface area. This feature made it difficult for CNT to disperse in the polymer matrix and could lead to more aggregated bundles of CNT, causing severe stress concentration phenomenon and preventing efficient load transfer to the polymer matrix, thus resulting in some deterioration of the overall performance of CNT/polymer nanocomposites. Chemical modification was performed to achieve both uniform dispersion of CNT in TLCP matrix and strong interfacial adhesion between CNT and TLCP.

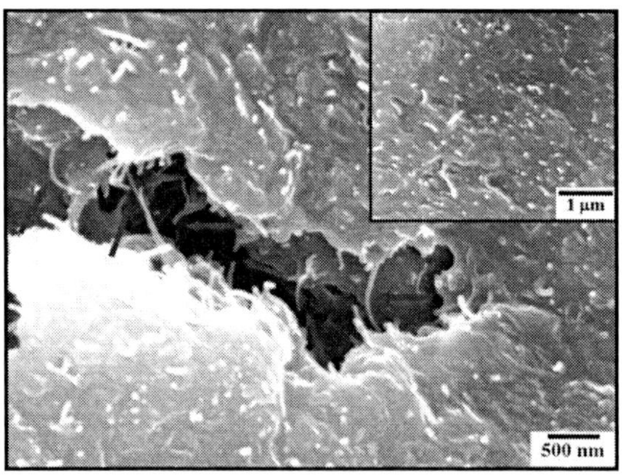

Figure 14. (A) SEM micrograph of TLCP nanocomposites containing 0.5 wt.% of modified CNT. On a large scale, modified CNT was dispersed well in TLCP nanocomposites, and stabilized its dispersion by good interactions with TLCP matrix, resulting from the interracial interactions of modified CNT with TLCP. (B) SEM micrographs of the fracture surfaces for TLCP nanocomposites containing 0.5 wt.% of modified CNT. The arrows indicate that the nanotube were to be broken with their ends embedded in TLCP matrix or they were bridging the local microcracks in the nanocomposites, suggesting good wetting with TLCP matrix and enhanced adhesion between the nanotubes and the polymer matrix, thus being favorable to efficient load transfer from the polymers to the nanotubes. Reproduced with permission from Ref. [67]. © 2009 Elsevier Ltd.

After chemical modification, modified CNT exhibited less entangled organization due to the functional groups formed on the surfaces instead of showing more agglomerated structures or bundles observed in pristine CNT. Modified CNT can stabilize its dispersion by good interactions with TLCP matrix, resulting from the interfacial interactions of the COOH groups of CNT with the C=O groups of TLCP, and lead to the well-dispersion of modified CNT in TLCP nanocomposites. This enhanced interfacial adhesion between modified CNT and TLCP plays a critical role in improving the mechanical properties of TLCP nanocomposites. As shown in Figure 14B, two ends of modified CNT were covered by TLCP despite a few of CNT pulled out from TLCP matrix. Interestingly, some of modified CNT was broken with their two ends still strongly embedded in TLCP matrix, and was bridging the local microcracks, which may delay the failure of the polymer nanocomposites [146]. This result indicates that CNT has a good wetting with TLCP matrix, suggesting the existence of strong interactions between them. Similar observation has been reported that the presence of the fractured nanotubes, along with the polymer matrix still adhered to the fractured tubes matrix in terms of a crack interacting with the nanotube significantly increased the mechanical properties of the polymer nanocomposites [147]. For TLCP nanocomposites, the presence of the functional groups on the surfaces of modified CNT resulted in the interfacial interaction between modified CNT and TLCP: the hydrogen atoms at the -COOH groups of modified CNT form hydrogen boding with the C=O groups of TLCP molecular chains [71], as shown in Figure 11B. This result can be considered as the evidence for efficient load transfer from TLCP to modified CNT during the tensile testing, leading to better reinforcing effect on the overall mechanical properties of TLCP nanocomposites.

4.5. Mechanical Properties

The effects of modified CNT on the mechanical properties of TLCP nanocomposites are shown in Figure 15. There is a significant dependence of the mechanical properties of TLCP nanocomposites on the presence of modified CNT. The incorporation of a very small quantity of modified CNT into TLCP substantially improved the tensile strength and tensile modulus of TLCP nanocomposites due to an effective nanoreinforcing effect of modified CNT with high aspect ratio, and this enhancing effect was more pronounced with lower content of modified CNT. By comparing with pure TLCP, higher tensile strength and tensile modulus of TLCP nanocomposites was attributed to stronger interfacial adhesion between CNT and TLCP matrix as well as better dispersion of CNT in the TLCP matrix. For clarifying the effect of modified CNT on the mechanical properties of the TLCP nanocomposites, it is very instructive to compare the reinforcing efficiencies of pristine and modified CNTs at a given content in TLCP nanocomposites. The reinforcing efficiency of CNT is defined as the normalized mechanical properties of TLCP nanocomposites with respect to those of pure TLCP. As shown in Figure 15B, the enhancing effect of the mechanical properties by incorporating CNT was more significant in TLCP nanocomposites containing modified CNT than in the case of pristine CNT. This result indicated that the incorporation of modified CNT into TLCP matrix was more effective in improving the mechanical properties of TLCP nanocomposites as compared to pristine CNT. The introduction of modified CNT results in good interfacial adhesion between modified CNT and TLCP matrix: the -COOH groups of modified CNT are helpful for enhancing the interactions with the C=O groups of TLCP, being favorable to more efficient load transfer form the polymer matrix to the nanotubes. Thus, the enhanced interfacial adhesion between modified CNT and TLCP as well as uniform dispersion of modified CNT in TLCP matrix can lead to significant improvement in the overall mechanical properties of TLCP nanocomposites.

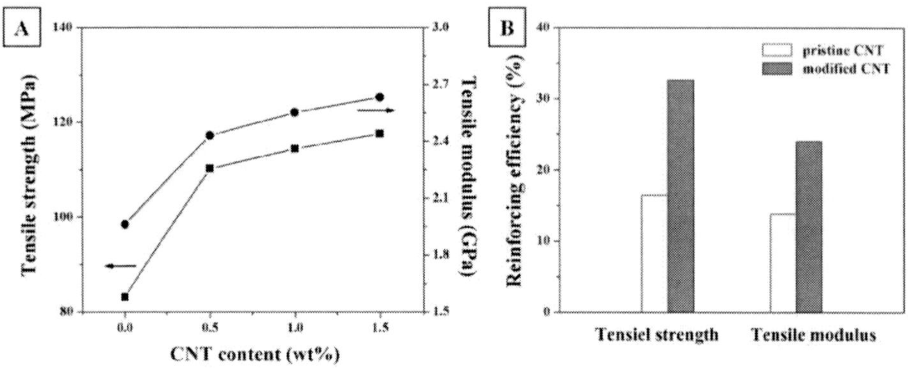

Figure 15. (A) Mechanical properties of TLCP nanocomposites with the modified CNT content. (B) Reinforcing efficiency of pristine and modified CNTs on the mechanical properties of TLCP nanocomposites containing 0.5 wt.% of modified CNT. Reinforcing efficiency (%) = $[(M_c - M_m)/M_m] \times 100$, where Mc and Mm represent the mechanical properties, such as tensile strength and tensile modulus, of TLCP nanocomposites and pure TLCP, respectively. Reproduced with permission from Ref. [67]. © 2009 Elsevier Ltd.

A comparative study of the experimental results with the values predicted from the theoretical models is also very instructive to characterize the effect of modified CNT on the mechanical properties of TLCP nanocomposites. In general, modified Halpin-Tsai equation has been used to predict the modulus of CNT-filled polymer nanocomposites [105-108]. To fit the equation (2) to the experimental results for TLCP nanocomposites, the weight fraction was transformed to the volume fraction, taking the densities of TLCP (1.41 g/cm^3) [148] and perfectly graphitized CNT (2.16 g/cm^3) [105,106]. Theoretical values of TLCP nanocomposite can be estimated assuming the aspect ratio of ~1000 and E_{CNT} of ~450 GPa [110]. In addition, the tensile strength of TLCP nanocomposites can be estimated from the equation (3) [111], and theoretically predicted values of the nanocomposite strength can be determined assuming the σ_{CNT} of ~11 GPa based on the previous literature [25]

Theoretically predicted values and the experimental data for the modulus and strength of the TLCP nanocomposites are compared in Figure 16. At lower content of modified CNT, the experimental results for the mechanical properties of TLCP nanocomposites were fitted well with the theoretical predicted values, while the large deviations were observed at higher content. Nanoreinforcing effect of modified CNT on the mechanical properties of TLCP nanocomposites appeared to be more pronounced at lower content than at higher content. This result suggested that the interfacial interaction was more effective in the enhancement of the mechanical properties of TLCP nanocomposites at lower content of modified CNT. Similar effect has been reported that for modified CNT and PEN nanocomposite, the interfacial interactions were more effective in improving the mechanical properties of modified CNT/PEN nanocomposites at lower loading of modified CNT than at higher loading [71].

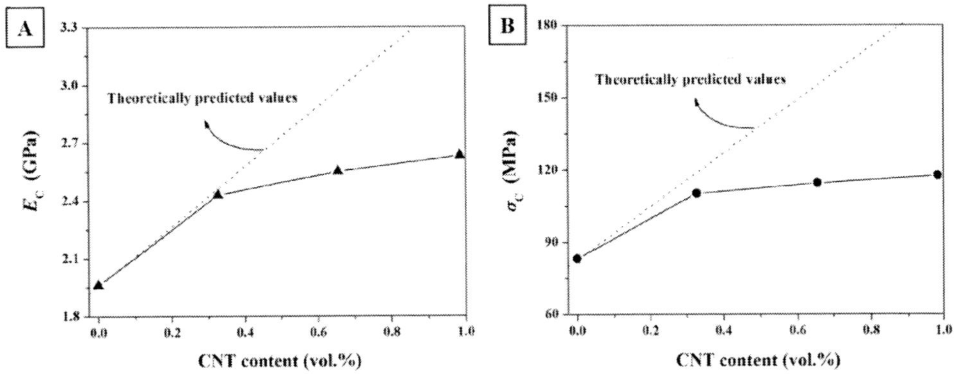

Figure 16. Theoretically predicted values and the experimental results for (A) the nanocomposite modulus (E_c) and (B) the nanocomposites strength (σ_c) of TLCP nanocomposites containing modified CNT. The dotted lines represent the theoretically predicted values based on the equations (2) and (3). Reproduced with permission from Ref. [67]. © 2009 Elsevier Ltd.

Thus, the synergistic effect of the increased interfacial interactions between modified CNT and TLCP in combination with uniform dispersion of modified CNT in TLCP matrixc was more effective at lower content for enhancing the mechanical properties of TLCP nanocomposites. However, the experimental results for the mechanical properties of TLCP nanocomposites were lower than those of theoretically predicted values. This result can be explained by the curvature of the introduced CNT in TLCP nanocomposites: some CNT embedded in TLCP matrix often exhibits curvature morphology, not straight one (Figure 14) and this feature reduces the nanoreinforcing effect of CNT in TLCP nanocomposites in comparison with the theoretical reinforcement provided by straight inclusions. Fisher et al. [149,150] have developed a model combining finite element results and micromechanical methods to determine the effective reinforcing modulus of wavy embedded CNT. They found that even slight curvature of CNT significantly reduced the effective reinforcement as compared to straight nanotubes. This is an additional mechanism to limit the reinforcing effect induced by CNT on the mechanical properties of CNT/polymer nanocomposites, leading to the nanocomposite modulus and strength that were less than predicted by traditional theoretical models [149.150]. In addition, the orientation of CNT can be also important factor in determining the mechanical properties of polymer nanocomposites, and the curvature morphology of CNT was closely related to the orientation of CNT in the polymer nanocomposites. Gorga and Cohen [151] reported that aligned non-entangled nanotubes should disperse and orient more readily in the polymer nanocomposites, resulting in the enhancement of the mechanical properties more significantly with optimal processing conditions. The nanoreinforcing effect of CNT will be more effective in enhancing the mechanical properties of polymer nanocomposites when the introduced nanotubes exhibit straight morphology within polymer matrix and preferentially aligned along their axial direction. Therefore, the overall mechanical properties of TLCP nanocomposites is expected to be further improved by optimizing the unique geometric feature and alignment of CNT in TLCP matrix as well as the combination of the enhanced interfacial adhesion between CNT and TLCP with the good dispersion of CNT in TLCP matrix during melt processing.

4.6. Thermal Stability

Thermal stability of polymer composites plays a crucial role in determining the limit of their working temperature and the environmental conditions for use. The incorporation of modified CNT into TLCP can increase the thermal decomposition temperatures and the residual yields of TLCP nanocomposites, indicating that the presence of modified CNT leads to the stabilization of TLCP matrix, resulting in the enhancement of thermal stability of TLCP nanocomposites (Figure 17A). For TLCP nanocomposites, modified CNT as a protective barrier against thermal decomposition retards thermal decomposition of TLCP nanocomposites, resulting from the effective function of modified CNT acting as a physical barrier to hinder the transport of volatile decomposed products out of TLCP nanocomposites during thermal decomposition [152].

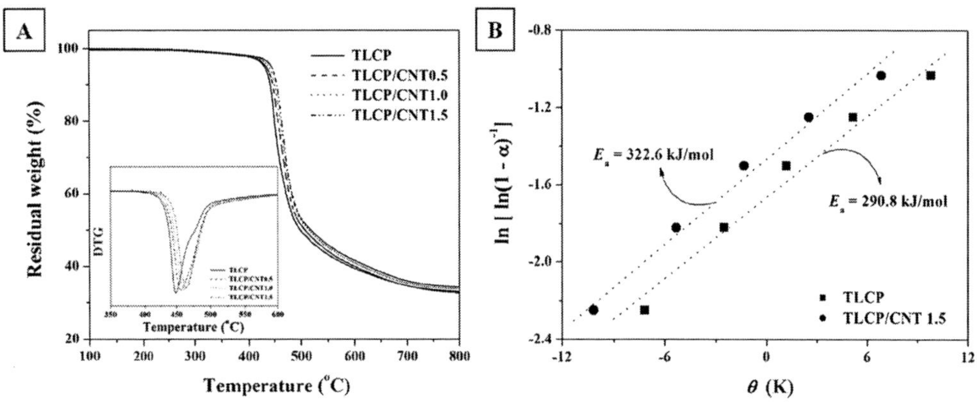

Figure 17. (A) TGA thermograms of TLCP nanocomposites with the modified CNT content. (B) Plots of $\ln[\ln(1 - \alpha)-1]$ versus θ as shown for PEN nanocomposites. By the equation (1), the slope provides an estimate the activation energy for the thermal decomposition (E_a) of TLCP nanocomposites. Reproduced with permission from Ref. [67]. © 2009 Elsevier Ltd.

Good interfacial adhesions between CNT and TLCP matrix also restrict the thermal motion of TLCP macromolecules, resulting in the enhancement of the thermal stability of TLCP nanocomposites. TGA kinetic analysis was conducted on TLCP nanocomposites to clarify the effect of modified CNT on thermal stability of TLCP nanocomposites. The activation energy for thermal decomposition (E_a) of TLCP nanocomposites can be calculated from the TGA thermograms by the Horowits-Metzger integral kinetic method [95] using the equation (1). The E_a value can be determined from the slope of the plot of $\ln[\ln(1-\alpha)^{-1}]$ versus θ as shown in Figure 17B. The E_a values of pure TLCP and TLCP nanocomposites were estimated to be 290.8, 293.7, 303.8, and 322.6 kJ/mol. By comparing with pure TLCP, higher E_a value of TLCP nanocomposites suggested that TLCP nanocomposites were more thermally stable than pure TLCP. This feature was also attributed to the interactions between modified CNT and TLCP matrix, which increase the E_a value of TLCP nanocomposites. Due to excellent thermal conductivity of CNT [97], the enhanced interfacial interactions between modified CNT and TLCP resulted in the increased thermal conductivity of TLCP nanocomposites, leading to the improvement in the thermal stability of TLCP nanocomposites.

5. SUMMARY

This chapter focused on the fabrication and characterization of aromatic polyester nanocomposites containing modified CNT to create high performance polymer nanocomposites with enhanced properties at low processing cost. The chemical modification of CNT appeared to facilitate the dispersion state of CNT in aromatic polyester matrix and the interfacial adhesion between them. The functional groups effectively induced on the surfaces of modified CNT via chemical modification can significantly influence the mechanical and rheological properties, thermal stability, and non-isothermal crystallization behavior of aromatic polyester nanocomposites. The incorporation of a very small quantity of CNT enhanced the thermal stability of aromatic polyester nanocomposites and this enhancing effect was more pronounced in the nanocomposites containing modified CNT than in the case of pristine because modified CNT could effectively act as physical barriers against thermal decomposition of the nanocomposites. Non-terminal behavior of aromatic polyester nanocomposites containing modified CNT was attributed to the nanotube-nanotube and nanotube-polymer interactions. The mechanical properties of aromatic polyester nanocomposites were significantly improved with the introduction of CNT and this enhancing effect was more pronounced in the nanocomposites containing modified CNT at lower content than in the case of pristine CNT and at high content. The uniform dispersion of modified CNT and strong interfacial adhesion or intimate contact between the nanotubes and the polymer matrix can lead to more effective load transfer from the polymers to the nanotubes, resulting in the substantial enhancement of the mechanical properties of aromatic polyester nanocomposites even with a very small quantity of modified CNT. The modified CNT in aromatic polyester nanocomposites exhibited much higher nucleation activity than any other nanoscale filler. The enhanced crystallization of aromatic polyester nanocomposites effectively induced by the modified CNT through heterogeneous nucleation was confirmed by the nucleation activity and the crystallization activation energy. The combined Avrami and Ozawa analysis was found to be more effective in describing the non-isothermal crystallization of aromatic polyester nanocomposites induced by the modified CNT. As compared to pure TLCP, higher complex viscosity and more distinct shear thinning behavior of TLCP nanocomposites containing modified CNT resulted from strong interactions between modified CNT and TLCP, which significantly influenced the relaxation behavior of polymer chains in the nanocomposites. The introduction of modified CNT was attributed to the formation of the interconnected or network-like structures via the nanotube–nanotube and the polymer–nanotube interactions, resulting in the pseudo solid-like behavior of TLCP nanocomposites. The improvement in the mechanical properties of TLCP nanocomposites resulted from the enhanced interfacial adhesion between modified CNT and TLCP as well as the uniform dispersion of modified CNT in TLCP matrix, and their synergistic effect combining good interfacial interaction and uniform dispersion was more effective at lower content of modified CNT than at higher content. The optimization of straight morphology and preferential alignment of CNT during melt processing as well as the enhanced interfacial adhesion and uniform dispersion plays a key role in further improving the mechanical properties of TLCP nanocomposites.

This chapter demonstrates that the overall properties of aromatic polyester nanocomposites are highly dependent on the uniform dispersion of CNT and the interactions between CNT and aromatic polyester, which can be enhanced by modification of CNT, providing a design guide of aromatic nanocomposites containing CNT with a great potential for industrial uses. This chapter also suggests a simple and cost-effective method that will facilitate the industrial realization of aromatic polyester nanocomposites containing modified CNT with enhanced physical properties. Future development of aromatic polyester nanocomposites for advanced composite materials will be performed by balancing higher performance of aromatic polyester nanocomposites containing modified CNT against multi-functionality and manufacturing cost.

REFERENCES

[1] Iijima, S. *Nature* 1991, vol 354, p 56.
[2] Bethune, D. S.; Kiang, C. H.; de Vries, M. S.; Gorman, G.; Savoy, R.; Vazquez, J.; Beyers, R. *Nature* 1993, vol 363, p 605.
[3] Lambert, J. M.; Ajayan, P. M.; Bernier, P.; Planeix, J. M.; Brotons, V.; Coq, B.; Castaing, *J. Chem. Phys. Lett.* 1994, vol 226, p364.
[4] Journet, C.; Maser, W. K.; Bernier, P.; Loiseau, A.; Chapelle, M. L.; Lefrant, S.; Deniard, P.; Lee, R.; Fisher, J. E. *Nature,* 1997, vol 388, p 756.
[5] Thess, A.; Lee, R.; Nikolaev, P.; Dai, H.; Petit, P.; Robert, J.; Xu, C.; Lee, Y. H.; Kim, S. G.; Rinzler, A. G.; Colbert, D. T.; Scuseria, G. E.; Tomanek, D.; Fisher, J. E.; Smalley, R. E. *Science,* 1996, vol 273, p 483.
[6] Guo, T.; Nikolaev, P.; Rinzler, A. G.; Tomanek, D.; Colbert, D. T.; Smalley, R. E. *J. Phys. Chem.*, 1995; vol 99, p 10694.
[7] Yudasaka, M.; Komatsu, T.; Ichihashi, T.; Iijima, S. *Chem. Phys. Lett.,* 1997, vol 278, p 102.
[8] Bronikowski, M. J.; Willis, P. A.; Colbert, D. T.; Smith, K. A.; Smalley, R. E. *J. Vac. Sci. Technol.* A 2001, vol 19, p 1800.
[9] Journet C.; Bernier, P. Appl. Phys. A: *Mater. Sci. Process.* 1998, vol 67, p 1.
[10] Liang, Q.; Gao, L. Z.; Li, Q.; Tang, S. H.; Liu, B. C.; Yu, Z. L. *Carbon* 2001, vol 39, p 897.
[11] Chen, Y.; Shaw, D. T.; Guo, L. *Appl. Phys. Lett. 2000*, vol 76, p 2469.
[12] Sen, R.; Govindaraj, A.; Rao, C. N. R. *Chem. Phys. Lett.* 1997, vol 267, p 276.
[13] Li, D. C.; Dai, L.; Huang, S.; Mau, A. W. H.; Wang, Z. L. *Chem. Phys. Lett.* 2000, vol 316, p 349.
[14] Chen, Q.; Dai, L. *Appl. Phys. Lett.* 2000, vol 76, p 2719.
[15] Hsu, W. K.; Hare, J. P.; Terrones, M.; Kroto, H. W.; Walton, D. R. M. *Nature* 1995, vol 377, p 687.
[16] Laplaze, D.; Bernier, P.; Maser, W. K.; Flamant, G.; Guillard, T.; Loiseau, A. *Carbon* 1998, vol 36, p 685.
[17] Dresselhaus, M. S.; Dresselhaus, G.; Avouris, P. H. *Carbon Nanotubes: Synthesis, Structure, Properties, and Applications*, Springer, Berlin, 2001.
[18] Iijima, S.; Ichihashi, T. *Nature* 1993, vol 363, p 603.
[19] Antonucci, V.; Hsiao, K. T.; Advani, S. G. *Advanced Polymeric Materials*, Shonaike, G. O.; Advani, S. G. Eds., CRC Press, New York, 2003, Chap. 11.

[20] Thostenson, E. T.; Ren, Z.; Chou, T. W. *Compos. Sci. Technol.* 2001, vol 61, p 1899.
[21] Goze, C.; Bernier, P.; Henrard, L.; Vaccarini, L.; Hernandez, E.; Rubio, A. *Synth. Metals* 1999, vol 103, p 2500.
[22] Wong, E. W.; Sheehan, P. E.; Lieber, C. M. *Science* 1997, vol 277, p 1971.
[23] Yao, Z.; Zhu, C. C.; Cheng, M.; Liu, J. *Comput. Mater. Sci.* 2001, vol 22, p 180.
[24] Yu, M. F.; Files, B. S.; Arepalli, S.; Ruoff, R. S. *Phys. Rev. Lett.* 2000, vol 84, p 5552.
[25] Yu, M. F.; Lourie, O.; Dyer, M. J.; Moloni, K.; Kelly, T. F.; Ruoff, R. S. *Science* 2000, vol 287, p 637.
[26] Frank, S.; Poncharal, P.; Wang, Z. L.; Heer, W.A. *Science* 1998, vol 280, p 1744.
[27] Stewart, M. E.; Cox, A. J.; Naylor, D. M. *Polymer* 1993, vol 34, p 4060.
[28] Bedia, E. L.; Murakami, S.; Kitade, T., Kohjiya, S. *Polymer* 2001, vol 42, p 7299.
[29] Kim, J. Y.; Kim, O. S.; Kim, S. H.; Jeon, H. Y. *Polym. Eng. Sci.* 2004, vol 44, p 395.
[30] Fakirov, S. *Handbook of Thermoplastic Polyesters*, Wiley-VCH, Weinheim, 1993.
[31] Rubin, I. I. *Handbook of Plastic Materials and Technology*, John Wiley & Sons Inc., New York, 1990
[32] Zhang, H.; Ward, I. M. *Macromolecules* 1995, vol 28, p 7622.
[33] Ford, T. *Plast. News* 1996, vol 8, p 4.
[34] Uchida Y. *Nikkei Material Technology* 1994, vol 137, p 63.
[35] Kumakawa, S.; Komoriya T. *Japan Patent, 87/156312*, 1987.
[36] Sakamoto S.; Sato, Y. *Japan Patent, 87/115609*, 1987.
[37] Weick, B. L.; Bhushan, B. *Wear* 1995, vol 190, p 28.
[38] Kim, J. Y.; Kim, S. H. 2006, *Polym. Int.*, 55, 449.
[39] Kim, J. Y.; Kim, S. H. 2006, *J. Appl. Polym. Sci.*, 99, 2211.
[40] Kim, J. Y.; Kim, S. H. 2005, *J. Polym. Sci. Part B: Polym. Phys.*, 43, 3600.
[41] Kim, J. Y.; Kim, S. H.; Kikutani, T. 2004, *J. Polym. Sci. Part B: Polym. Phys.*, 42, 395.
[42] Kim, J. Y.; Seo, E. S.; Kim, S. H.; Kikutani, T. 2003, *Macrmol. Res.*, 11, 62.
[43] Kim, J. Y.; Kang, S. W.; Kim, S. H.; Chang, J. H.; Ahn, S. H. *Macrmol. Res.* 2006, vol 11, p 146.
[44] Kim, J. Y.; Kim, D. K.; Kim, S. H. *Polym. Compos.* 2009, vol 37, p 1779.
[45] Paul, D. R.; Robeson, L. M. *Polymer* 2008, vol 49, p 3187.
[46] Bokobza, L. *Polymer* 2007, vol 48, p 5279.
[47] Wong, E. W.; Sheehan, P. E.; Lieber, C. M. *Science* 1997, vol 277, p 1971.
[48] Chen, G. X.; Shimizu, H. *Polymer* 2008, vol 49, p 943.
[49] Choi, J. Y.; Han, S. W.; Huh, W. S.; Tan, L. S.; Baek, J. B. *Polymer* 2007, vol 48, p 4034.
[50] Song, W.; Zheng, Z.; Tang, W.; Wang, X. *Polymer* 2007, vol 48, p 3658.
[51] Ebbesen, T. *Carbon Nanotubes: Preparation and Properties*, CRC Press, New York, 1997.
[52] Lourie, O.; Cox, D. M.; Wagner, H. D. *Phys. Rev. Lett.* 1998, vol 81, p 1638.
[53] Shaffer, M. S. P.; Windle, A. H. *Adv. Mater.* 1999, vol 11, p 937.
[54] Gong, X.; Liu, J.; Baskara, S.; Voise, R. D.; Young, J. S. *Chem. Mater.* 2000, vol 12, p 1049.
[55] Liu, J.; Rinzler, A. G.; Dai, H.; Hafner, J. H.; Bradley, R. K.; Boul, P. J.; Lu, A.; Iverson, T.; Shelimov, K.; Huffman, C. B.; Rodriguez-Macias, F.; Shon, Y. S.; Lee, T. R.; Colbert, D. T.; Smalley, R. E. *Science* 1998, vol 280, p 1253.
[56] Sun, Y. P.; Fu, K.; Lin, Y.; Huang, W. *Acc. Chem. Res.* 2002, vol 35, p 1096.

[57] Hirsch, A. *Angew. Chem. Int. Ed.* 2002, vol 41, p 11.
[58] Cui, S.; Canet, R.; Derre, A.; Couzi, M.; Delhaes, P. *Carbon* 2003, vol 41, p 797.
[59] Haggenmuller, R.; Conmas, H. H.; Rinzler, A. G.; Fischer, J. E.; Winey, K. I. *Chem. Phys. Lett.* 2000, vol 330, p 219.
[60] Jia, Z. J.; Wang, Z. Y.; Liang, J.; Wei, B. Q.; Wu, D. H. *Mater. Sci. Eng.* A, 1999, vol 271, p 395.
[61] Jin, J. X.; Pramoda, K. P.; Goh, S. H.; Xu, G. Q. *Mater. Res. Bull.* 2002, vol 37, p 271.
[62] Pötschke, P.; Fornes, T. D.; Paul, D. R. *Polymer* 2002, vol 43, p 3247.
[63] Jung. R.; Park, W. I.; Kwon, S. M.; Kim, H. S.; Jin, H. J. *Polymer* 2008, vol 49, p 2071.
[64] Mu, M.; Walker, A. M.; Torkelson, J. M.; Winey, K. I. *Polymer* 2008, vol 49, p 1332.
[65] Pegel, S.; Pötschke, P.; Petzold, G.; Alig, I.; Dudkin, S. M.; Lellinger, D. *Polymer* 2008, vol 49, p 974.
[66] Kim, J. Y. *Materials* 2009, vol 2, p 1955.
[67] Kim, J. Y.; Kim, D. K.; Kim, S. H. *Eur. Polym. J.* 2009 vol 45, p 316
[68] Kim, J. Y.; Park, H. S.; Kim, S. H. *J. Appl. Polym. Sci.* 2009 vol 113, p 2008.
[69] Kim, J. Y. *J. Appl. Polym. Sci.* 2009 vol 112, p 2589.
[70] Kim, J. Y.; Han, S. I.; Kim, D. K.; Kim, S. H. *Composites Part A: Appl. Sci. Manuf.* 2009 vol 40, p 45.
[71] Kim, J. Y.; Han, S. I.; Hong, S. *Polymer* 2008, vol 49, p 3335.
[72] Kim, J. Y.; Han, S. I.; Kim, S. H. *Polym. Eng. Sci.* 2007, vol 47, p 1715.
[73] Kim, J. Y.; Park, H. S.; Kim, S. H. *Polymer* 2006, vol 47, p 1379.
[74] Kim, J. Y.; Park, H. S.; Kim, S. H. *J. Appl. Polym. Sci.*, 2007, vol 103, p 1450.
[75] Kim, J. Y.; Kim, S. H. *J. Polym. Sci. Part B: Polym. Phys.* 2006, vol 44, p 1062.
[76] Kim, J. Y.; Choi, H. J.; Kang, C. S.; Kim, S. H. *Polym. Compos.* 2010, vol 31, p 858.
[77] Fornes, T. D.; Yoon, P. J.; Keskkula, H.; Paul, D. R. *Polymer* 2001, vol 42, p 9929.
[78] Kumar, S.; Doshi, H.; Srinivasrao, M.; Park, J.O.; Schiraldi, D. A. *Polymer* 2002, vol 43, p 1701.
[79] Tzavalas, S.; Drakonakis, V.; Mouzakis, D. E.; Fischer, D.; Gregoriou, G. *Macromolecules* 2006, vol 39, p 9150.
[80] Zhu, J.; Peng, H.; Rodriguez-Macias, F.; Margrave, J. L.; Khabashesku, V. N.; Imam, A. M.; Lozano, K.; Barrera, E. V. *Adv. Funct. Mater.* 2004, vol 14, p 643.
[81] Mawhinney, D. B.; Naumenko, V.; Kuznetsova, A. *J. Am. Chem. Soc.* 2000, vol 122, p 2383.
[82] Eitan, A.; Jiang, K.; Dukes, D.; Andrews, R.; Schadler, L. S. *Chem. Mater.* 2003, vol 15, p 3198.
[83] Yuang, Y.; Young, R. J. *J. Mater. Sci.* 1994, vol 29, p 4027.
[84] Ruan, S. L.; Gao, P.; Yang, X. G.; Yu, T. X. *Polymer* 2003, vol 44, p 5643.
[85] Tuinstra, E.; Koening, J. I. *J. Chem. Phys.* 1970, vol 53, p 1126.
[86] Baibarac, M.; Baltog, I.; Lefrant, S.; Gomez-Romero, P. *Polymer* 2007, vol 48, p 5279.
[87] Grossiord, N.; Loss, J.; Regev, O.; Koning, C. E. *Chem. Mater.* 2006, vol 18, p 1089.
[88] Hu, H.; Yu, A.; Kim, E.; Zhao, B.; Itkis, M. E.; Bekyarova, E.; Haddon, R. C. *J. Phys. Chem. B* 2005, vol 109, p 11520.
[89] Cai, H.; Yan, F. Y.; Xue, Q. J. *Mater. Sci. Eng. A* 2004, vol 364, p 94.
[90] Zhao, L. P.; Gao, L. *Carbon* 2004, vol 42, p 423.
[91] Jung, D. D.; Bhattacharyya, D.; Easteal, A. J. *J. Appl. Polym. Sci.* 2001, vol 82, p 1504.
[92] Gupta, M. C.; Umare, S. S. *Macromolecules* 1992, vol 25, p 138.

[93] Watts, P. C. P.; Mureau, N.; Tang, Z.; Miyajima, Y.; Carey, J. D.; Silva, S. R. P. *Nanotechnology* 2007, vol 18, p 175701.
[94] Tseng, C. H.; Wang, C. C.; Chen, C. Y. *Chem. Mater.* 2007, vol 19, p 308.
[95] Horowitz, H. H.; Metzger, G. *Anal. Chem.* 1963, vol 35, p 1464.
[96] Benedict, L. X.; Lourie, S. G.; Cohen, M. L. *Solid State Commun.* 1996, vol 100, p 177.
[97] Berber, S.; Kwon, Y. K.; Tomanek, D. *Phys. Rev. Lett.* 2000, vol 84, p 4614..
[98] Abdel-Goad, M.; Pötschke, P. *J. Non-Newtonian Fluid Mech.* 2005, vol 128, p 2.
[99] Pinnavaia, T. J.; Beall, G. W. *Polymer–Clay Nanocomposites*, Wiley, New York, 2000.
[100] Wagner, R.; Reisinger, T. J. G. *Polymer* 2003, vol 44, p 7513.
[101] Krishnamoorti, R.; Vaia, R. A.; Giannelis, E. P. *Chem. Mater.* 1996, vol 8, p 1728.
[102] Ferry, J. *Viscoelastic Properties of Polymers*, Wiley, New York, 1980.
[103] Krisnamoorti, R.; Giannelis, E. P. *Macromolecules* 1997, vol 30, p 4097.
[104] Han, C. D.; Kim, J.; Kim, J. K. *Macromolecules* 1989, vol 22, p 383.
[105] Qian, D.; Dickey, E. C.; Andrew, R.; Rantall, T. *Appl. Phys. Lett.* 2000, vol 76, p 2868.
[106] Cadek, M.; Coleman, J. N.; Barron, V.; Hedicke, K.; Blau, W. J. *Appl. Phys. Lett.* 2002, vol 81, p 5123.
[107] Mallick, P. K. *Fiber-Reinforced Composites*, Marcel Dekker, New York, 1993.
[108] Zhang, X.; Liu, T.; Sreekumar, T. V.; Kumar, S.; Moore, V. C.; Hauge, R. H.; Smalley, R. E. *Nano Lett* 2003, vol 3, p 1285.
[109] Ito, M.; Kikutani, T. *Handbook of Thermoplastic Polyesters*, Fakirov, S. Ed., Wiley-VCH, Weinheim, 2002, Chap. 10.
[110] Pan, Z. W.; Xie, S. S.; Lu, L.; Chang, B. H.; Sun, L. F.; Zhou, W. Y.; Wang, G.; Zhang, D. L. *Appl. Phys. Lett.* 1999, vol 74, p 3152.
[111] Agarwal, B. D.; Broutman, L. G. *Analysis and Performance of Fiber Composites*, Wiley, New York, 1980.
[112] Wu, C. S.; Liao, H. T. *Polymer* 2007, vol 48, p 4449.
[113] Xu, W.; Ge, M.; He, P. *J. Polym. Sci. Part B: Polym. Phys.* 2002, vol 40, p 408.
[114] Cheng, S. Z. D.; Wunderlich, B. *Macromolecules* 1988, vol 21, p 789.
[115] Xu, D.; Wang, Z. *Polymer* 2008, vol 49, p 330.
[116] Li, L.; Li, C. Y.; Ni, C.; Rong, L.; Hsiao, B. *Polymer* 2007, vol 48, p 3452.
[117] Bikiaris, D.; Vassiliou, A.; Chrissafis, K.; Paraskevopoulos, K. M.; Jannakoudakis, A.; Docoslis, A. *Polym. Degrad. Stab.* 2008, vol 93, p 952.
[118] Lopez, L. C.; Wilkes, G. L. *Polymer* 1989, vol 30, p 882.
[119] Wang, B.; Sun, G.; Liu, J.; He, X.; Li, J. *J. Appl. Polym. Sci.* 2006, vol 100, p 379.
[120] (120a) Avrami, M. *J. Chem. Phys.* 1939, vol 7, p 1103. (120b) Avrami, M. *J. Chem. Phys.* 1940, vol 8, p 212.
[121] Ozawa, T. *Polymer* 1971, vol 12, p 150.
[122] Di Lorenzo, M. L.; Silvestre, C. *Prog. Polym. Sci.* 1999, vol 24, p 917.
[123] Wunderlich, B. *Macromolecular Physics*, Academic Press, New York, 1976.
[124] Jeziorny, A. *Polymer* 1978, vol 19, p 1142.
[125] Liu, T.; Mo, Z.; Wang, S.; Zhang, H. *Polym. Eng. Sci.* 1997, vol 37, p 568.
[126] Dobreva, A.; Gutzow, I. J. *J. Non-Cryst. Solids* 1993, vol 162, p 1.
[127] Alonso, M.; Velasco, J. I.; De Saja, J. A. *Eur. Polym. J.* 1997, vol 33, p 255.
[128] Mitchell, C. A.; Krishnamoorti, R. *Polymer* 2005, vol 46, p 8796.
[129] Kissinger, H. E. *J. Res. Natl. Stand.* 1956, vol 57, p 217.
[130] Chen, E. C.; Wu, T. M. *J. Polym. Sci. Part B: Polym. Phys.* 2008, vol 46, p 158.

[131] Haggenmueller, R.; Fischer, J. E.; Winey, K. I. *Macromolecules* 2006, vol 39, p 2964.
[132] Bikiaris, D.; Vassiliou, A.; Chrissafis, K. *Polym. Degrad. Stab.* 2008, vol 93, p 952.
[133] Wu, T. M.; Chen, E. C. *J. Polym. Sci. Part B: Polym. Phys.* 2006, vol 44, p 598.
[134] Gray, G. W. *Thermotropic Liquid Crystals*, John Wiley & Sons, New York, 1987.
[135] Chung TS. *Thermotropic Liquid Crystal Polymers: Thin Film Polymerization, Characterization, Blends, and Applications*, CRC Press, London, 2001.
[136] Demus, D.; Gray, G. W.; Speiss, H. W.; Goodby, J. W.; Vill, V. *Handbook of Liquid Crystals*, Wiley-VCH, Weinheim, 1998.
[137] Kiss, G. *Polym. Eng. Sci.* 1987, vol 27, p 410.
[138] Dutta, D.; Fruitwala, H.; Kohli, A.; Weiss, R. A. *Polym. Eng. Sci.* 1990, vol 30, p 1005.
[139] He, J.; Bu, W. *Polymer* 1994, vol 35, p 5061.
[140] Lagerwall, J. P. F.; Scalia, G. *J. Mater. Chem.* 2008, vol 18, p 2890.
[141] Zhang, S.; Kumar, S. *Small* 2008, vol 4, p 1270.
[142] Rahman, M.; Lee, W. *J. Phys. D: Appl. Phys.* 2009, vol 42, p 063001.
[143] Abdalla, M.; Derrick, D.; Adibempe, D.; Nyario, E.; Robinson, P.; Thompson, G. *Polymer* 2007, vol 48, p 5662.
[144] Shenoy, A. V. *Rheology of Filled Polymer Systems*, Kluwer Academic Publishers, Dordrecht 1999.
[145] Song, Y. S.; Youn, J. R. *Carbon* 2005, vol 43, p 1378.
[146] Cho, J.; Daniel, I. M. *Scripta Mater.* 2008, vol 58, p 536.
[147] Thostenson, E. T.; Chou, T. W. *J. Phys. D: Appl. Phys.* 2002, vol 35, p L77.
[148] Saengsuwan, S.; Bualek-Limcharoen, S.; Mitchell, G. R.; Olley, R. H. *Polymer* 2003, vol 44, p 3407.
[149] Fisher, F. T.; Bradshow, R. D.; Brinson, L. C. *Appl. Phys. Lett.* 2002, vol 90, p 4647.
[150] Fisher, F. T.; Bradshow, R. D.; Brinson, L. C. *Compos. Sci. Technol.* 2003, vol 63, p 1689.
[151] Gorga, R. E.; Cohen, R. E. *J. Polym. Sci. Part B: Polym. Phys.* 2004, vol 42, p 2690.
[152] Kashiwagi, T.; Grulke, E.; Hilding, J.; Harris, R.; Awad, W.; Douglas, J. F. *Macromol. Rapid Commun.* 2002, vol 23, p 761.

In: Thermoplastic and Thermosetting Polymers and Composites ISBN: 978-1-61209-264-5
Editors: Linda D. Tsai and Matthew R. Hwang

Chapter 2

BIOFIBER REINFORCED STARCH COMPOSITES

Redouan Saiah[1], P.A. Sreekumar[2] and Thomas P. Selvin[2]

Laboratoire de Génie des Matériaux (LGMA)-Pôle de Développement, Ecole d'Ingénieurs en Agriculture Esitpa, 3 rue du Tronquet, BP 40 118, 76 134 Mont Saint Aignan cedex, France[1]

Department of Chemical Engineerng, King Fahd University of Petroleum & Minerals, Dhahran, Saudi Arabia -31261[2]

ABSTRACT

During the last decades, there has been a continuous increase in the production of commodity plastic products because of their low production cost, excellent mechanical properties, as well as chemical, weather and biodegradation resistance. The duration of use of plastic materials is relatively short compared to their life time; it ranges from several months to few years, depending on the conditions of use and environment. Now a day the scientific world is focusing is the development of starch based composites. These materials are available largely from natural resources and are 100% degradable. This chapter presents a brief description about the different biodegradable matrices as well as the different components, structures, transformations and properties of starch-based matrices and biofibers. The reinforcing effect of macro-fibers s well as like nano fibers like whiskers or nanocrystals on starch as matrix will be discussed.

1. INTRODUCTION

Majority of plastic products are made from petroleum-based synthetic polymers. These products have been used in every area of human life. The ease of processing, low cost, high durability, tremendous range of properties increased its usage tremondusly. However these plastics are dangerous to the the environment since it takes months to hundreds of years to degrade. During combustion it produces toxic materials which eventually pollute the surroundings. The landfilling one method to remove the plastic wastages, results in the contamination of water sources, thereby the soil's biological balance. These severe problems created by the plastic wastages to environment triggered the interest in development of biodegradable disposable plastic materials. So that the one time use items can be disposed with the peace of mind that they will not remain for centuries in a landfill, or as litter, which is one of the tenets driving the recent interest in "green" technologies.

Today it appears that it is not practical to use synthetic polymers for certain applications such as bags, agricultural mulch films and food packaging, since these artifacts contain many organic residues and have a less life time. Recently several investigations are focussed to have the biodegradable polymers from renewable resources such as plants, animals and microbes through biochemical reactions. Starch is a natural polymer found in granular form in a variety of plants, such as corn, wheat, rice, cassava and others. They are mainly used in the form of flour and are rarely used to produce thermoplastic materials. Most of the studies were concentrated on the plasticization effects on the structure of the starch during the preparation of starch-based thermoplastic materials, using plasticizers such as water, glycerol, urea etc [1-3]. Another area is the development of biocomposites based on starch as matrix. The term ''biocomposite'' consists of a biodegradable polymer as matrix material and natural fiber such as sisal, coir, jute, kenaf etc as reinforcing element. Moreover other varieties of biodegradable plastics, like Mater Bi®, PHBV, PLA etc were developed, but their cost is considerably greater than that of the conventional commodity thermoplastics.

2. Biodegradable Polymers

2.1 Significance of Biodegradable Polymers

Biodegradation is a natural process and can be defined as the intrinsic ability of a material to be degraded by the action of micro-organisms for progressive simple structures and molecules of water, carbon dioxide and/or methane and new biomass. Each standard organization gives its own definition for biodegradable plastics (Table 1).

Table 1. Definitions of biodegradable plastics according to the standards authorities [4].

Organizations	Definitions of biodegradable plastics
ISO 472-1988	A plastic designed to undergo a significant change in its chemical structure under specific environmental conditions resulting in a loss of some properties that may vary as measured by standard test methods appropriate to the plastic and the application in a period of time that determines its classification.
ASTM sub-committee D20-1996 proposal	A degradable plastic in which the degradation results from the action of naturally occurring microorganisms such as bacteria, fungi and algae.
DIN 103.2-1993 Working groupe on biodegradable polymers	A plastic material is called biodegradable if all its organic compounds undergo a complete biodegradation process. Environmental conditions and rates of biodegradation are to be determined by standardized test methods.
C.E.N.-1993	A degradable material in which the degradation results from the action of microorganisms and ultimately the material are converted to water, carbon dioxide and/or methane and a new cell biomass.
Japanese Biodegradable Plastics Society – 1994	Biodegradable plastics are polymeric materials which are converted into lower molecular weight compounds where at least one step in the degradation process is though metabolism in the presence of naturally occurring organisms.

Biodegradation are of mainly two types according to the fate of polymers [5, 6]:

- Primary biodegradation (or partial biodegradation) is the alteration of the chemical structure. This results in a loss of specific properties of the polymer.
- The ultimate biodegradation consists of a total mineralization. The material is completely degraded by microorganisms with the production of CO2 (under aerobic conditions degradation) or CH4 (under anaerobic conditions degradation), water, minerals and a biomass new non-toxic to humans and the environment. However, sometimes, the mineralization is not complete and final products can be still in the form of oligomers.

Due to the multiplicity of definitions, G. Swift [7] introduced the concept of "biodegradable polymer evironmentally acceptable". A biodegradable polymer may be partially or totally degraded, so that a "biodegradable polymer evironmentally acceptable" must be completely mineralized. If partially digested, it does not produce toxic residues in the environment. Therefore, this definition introduces, in addition to the degree of biodegradation, a concept of environmental impact.

2.2 Different Biodegradable Polymers

A nonexhaustive list of biodegradable polymers with its components, commercial name and the producer are given in Table 2.

Table 2. List of biodegradable polymers with its components, commercial name and the producer.

Name	Composition	Company
	Materials based on synthetic polymers and starch	
	Granular starch	
Ecostar^R	Starch/PE	St. Lawrence Starch Company, UK
Ecopolym^R		Polychim, France
Polyclean^R		Archer Daniels Midland, USA
Amyplast^R		Amylum, Belgium
	Unstructured starch	
Mater-Bi^R		
Class A	Starch / PVA	Novamont, Italy
Class Z and V	Starch / PCL	
Bioflex^R	Starch / PCL	Biotec-Mellita, Germany
Novon^R	Starch / PP	Novon International, China
Ecofoam^R		National Starch, USA
Poly Grade II^R		Ampacet, USA
The polymers of microbial origin		
Biopol^R	PHBV	Monsanto, USA
EcoPLA^R	PLA	Dow Cargill, USA
Heplon^R	PLA	Chronopol, USA
Lacea^R	PLA	Mitsui chemical, Japan
Lacty^R	PLA	Shimadzu, Japan
Pullulan^R	Pullulan	Hayashibara, Japan
Doron^R	Chitosan	Aicelo Chem., Japan
Kytex^R	Chitosan	Marine Commodities Inc. USA

Table 2. (Continued).

The polymers of petrochemical origin		
Tone[R]	PCL	Union Carbide, USA
CAPA[R]	PCL	Solvay, UK
Cel-green[R]	PCL	Daicel Chem, Japan
Bionolle[R]	Polybutylène succinate	Showa Highpolymer, Japan
Bak[R]	Polyester amide	Bayer, Germany
Ecoflex[R]	Copolyester	BASF, Germany
Eastarbio[R]	Copolyester	Eastman Chemical, USA
Skygreen[R]	Copolyester	Sunkyong, Korea
Biomax[R]	PET modifié	Dupont, USA
Vinex[R]	PVA	Air products and chemicals, USA
Aquafilm[R]	PVA	Lin Pac Plastics, UK
Polymers from agricultural		
Paragon[R]	Thermoplastic starch	Avebe, Netherlands
Bioceta[R]	Cellulose diacetate	Tubize plastics, Belgium
Ultraphan[R]	Cellulose acetate	Courtaulds, UK
Paragon[R]	Thermoplastic starch	Avebe, Netherlands
Mater-Bi[R] class Y	Starch / cellulose acetate	Novamont, Italy
Bioplast[R]	Starch / cellulose acetate	Biotec-Mellita, All

A comparison of the characteristics of these polymers with synthetic polymers is given in the Table 3.

Table 3. Comparison of natural polymers with synthetic and blends of synthetic and natural poymers.

Polymer	Synthetic Polymers	Biodegradable polymers (blends of polymer synthetic and natural)	Biodegradable polymers (natural polymers)
Raw materials	Nonrenewable	Only a fraction is renewable	Renewable
Examples	PE, PP, PS, etc.	PE + starch PE + cellulose, etc.	Plastics based cellulose Plastics based starch
Biodegradability	Non biodegradable	Only natural polymers are attacked by microorganisms	Excellent
Photo-degradability	Adding prodégradants favoring low etching	Adding of prodegradant causes the chemical breakdown of the chain to attack natural polymers by microorganisms.	May contribute to or accelerate biodegradability
Price	Very cheap for everyday products	Middle	Very expensive now, but may improve with future production capacity
Physical and mechanical properties	Excellent properties	Variables	Good and variable, depending on the application
During combustion	Toxic pollutants can be produced	Toxic pollutants can be produced	No production of toxic pollutants.
Compostability	No	Slightly compostable	Mostly compostable
Landfills	Stable, possibility of producing pollutants or toxic wastes.	Less stable. Possibility of production of pollutants or toxic effluent	Unstable. Not producing pollutants and toxic effluents
Recycling	Good	Bad	Pretty bad, but can be improved

2.2.1 Natural Polymers

The materials of natural origin are those synthesized by living things: animals, plants and micro-organisms. The natural polymers are biodegradable in the native state with different life spans. Proteins (collagen, gelatin, and casein), carbohydrates, cellulose and lignin produced by animal and plants are good examples for natural polymers. In this most important family is the carbohydrates (polysaccharides). Cellulose and lignin are found in wood, paper, viscose, cellophane and all textile fibers (cotton, flax, hemp, sisal). Proteins such as casein (milk protein) or gluten are widely used as biopolymers. These proteins are involved in formulating adhesives and paints, and are the basis of materials with barrier properties to oxygen and CO2. They are very permeable to moisture. Gelatin is also the basis of many films whose main usage is in the manufacture of tablets and capsules in the pharmaceutical industry. The major drawback of using protein as biopolymer materials reside in the high cost of the resulting materials. Lipids (oil of colza, soya, and sunflower) have a certain future for the manufacture of biodegradable lubricants and greases, biofuels and even rigid products.

2.2.2 Polymers Derived using Biotechnology

These polymers are of bacterial origin. They are produced by biomass from a fixed carbon source, usually glucose, methanol or starch. The choice of the fermentation and microbial strains, determines the composition and production yield of synthesized polymers. Many studies have been conducted on polyhydroxyalkanoates, especially on poly-β-hydroxybutyrate (PHB) (Figure 1), aliphatic polyester accumulated by many micro-organisms as energy reserve, especially when grown on media deficient in nitrogen. Poly-3-hydroxyvalerate (PHV) and the copolymer poly-3- hydroxybutyrate-co-3-hydroxyvalerate (PHBV), called "biopol" are manufactured by Monsanto by fermentation of polysaccharides.

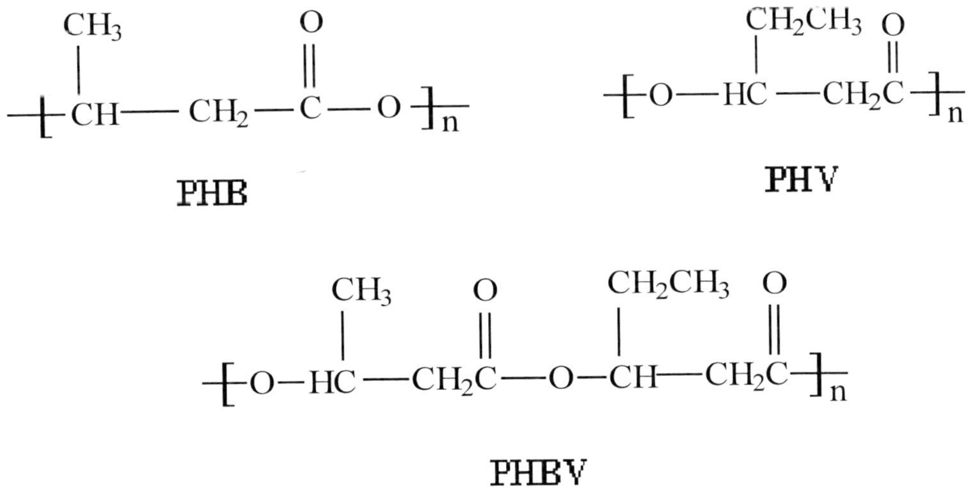

Figure 1. Chemical structure of poly-β-hydroxybutyrate (PHB), poly-3-hydroxyvalerate (PHV) and poly-3- hydroxybutyrate-co-3-hydroxyvalerate (PHBV).

A comparison of physical and mechanical properties for the PHB with PP is given in the Table 4.

Table 4. Mechanical and physical properties for PHB [8] and PP [9].

Properties	PHB	PP
MFI (g/10 min)	43.74	9.0 - 13.0
Density (g/cm^3)	1.23	0.9
Melting temperature (°C)	173.3	160 – 175
Stress at break (MPa)	32.03 ± 0.98	25.0
Strain at break (%)	1.89 ± 0.13	

2.2.3 Synthetic Polymers

The best known biodegradable synthetic polymers are shown in Table 5 with their essential characteristics.

Table 5. Main properties of synthetic biodegradable polymers (PLA, PCL, PEA, and PBAT PBSA) and the polymer derived from biotechnology (PHBV) [10].

Properties	PLA (Naturew orks)	PCL (Tone 787)	PEA (BAK 1095)	PBSA (Bionolle 1000)	PBAT (Eastar bio 14766)	PHBV (Biopol D400G)
Melting temperature (°C)	158	60	125	114	110-115	177
Tensile modulus (MPa)	2000	386	180	500	81	4000
Strain at break (%)	10	800-1000	400	800	600	6
Density	1.25	1.15	1.07	1.20	1.21	1.25
Biodégradability (%)	100	100	100	90	100	100
WVTR (g/m^2/day)	172	177	680	330	550	21

The poly (lactic acid) (PLA), obtained by polycondensation of polylactic acid, is a semi-crystalline polymer in simple linear structure (Figure 2).

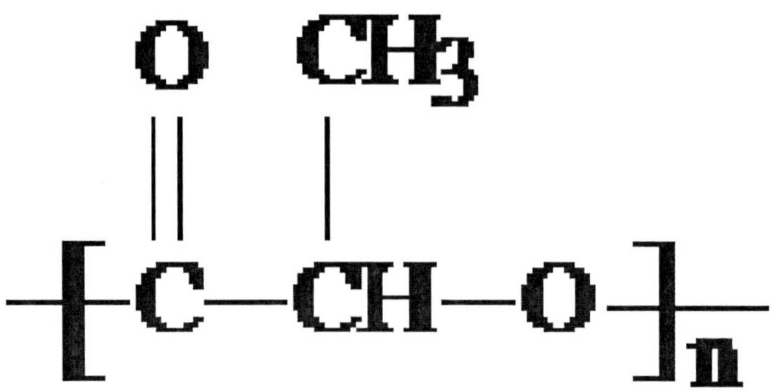

Figure 2. Chemical structure of poly (lactic acid) (PLA).

It is thus used as artificial skin, son of bioresorbable sutures or orthopedic implants. In the field of PLA packaging can be operated easily by conventional methods for formatting.

The polycaprolactone (PCL) is an aliphatic polyester (Figure 3) which is biodegradable, compostable, hydrophobic, rubbery behavior at room temperature (Tg =- 60°C) and whose melting point is the weakest of various polyesters biodegradable (60-80°C). The PCL is now used as a copolymer as the "Ecoflex" produced by BASF.

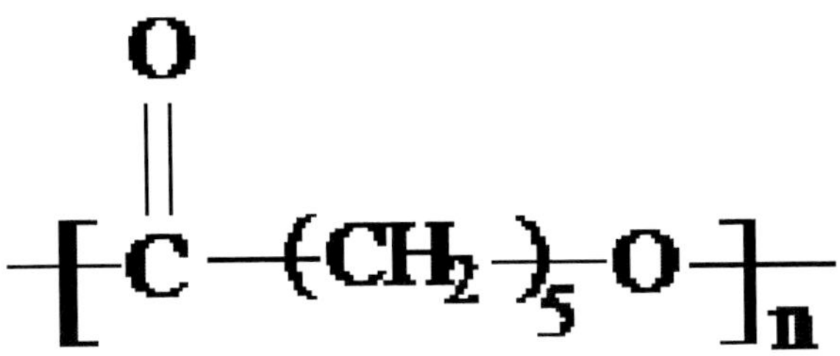

Figure 3. Chemical structure of polycaprolactone (PCL).

The polyesteramide (PEA) is a copolymer of polyamide (Nylon 6 and Nylon 6, 6) and aliphatic ester under the trade name BAKR with properties similar to those of low density polyethylene (Figure 4).

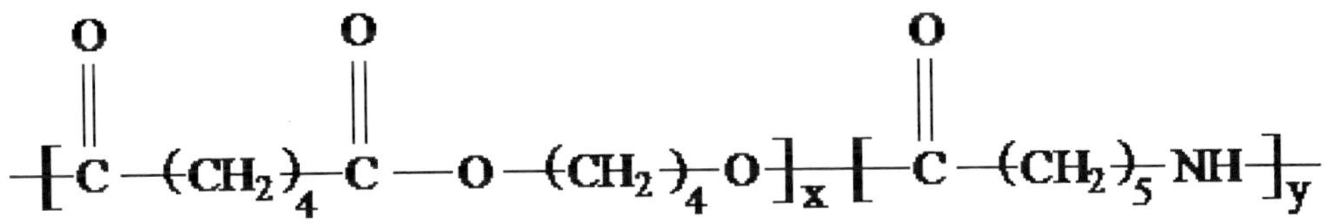

Figure 4. Chemical structure polyesteramide (PEA).

The main applications are for producing agricultural tarps, flower pots and disposable bag for household waste and horticultural.

The polybutylene succinate adipate (PBSA), is an aliphatic polyester (Figure 5), easily transformable and having properties similar to polyethylene. It is marketed under the name Enpol by IreChem (Korean) and under the name of Bionolle by Showa (Japan). Its chemical formula is:

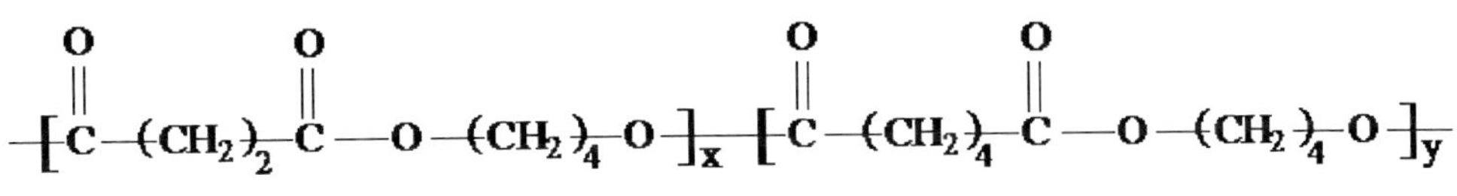

Figure 5. Chemical structure of polybutylene succinate adipate (PBSA).

Polybutylene adipate-co-terephthalate (PBAT), is marketed under the name Ecoflex from BASF (Germany), Eastar Bio from Eastman Chemical (USA) and Biomax by Dupont (USA). The production and applications of this polymer are still marginal. The chemical structure of this material is shown in Figure 6.

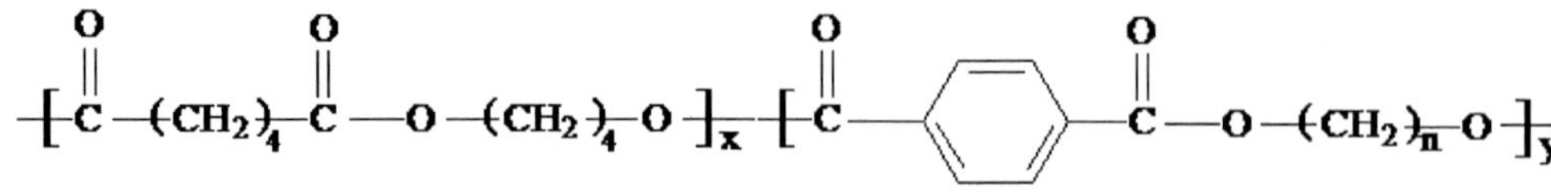

Figure 6. Chemical structure of polybutylene adipate-co-terephthalate (PBAT).

Poly (vinyl alcohol) (PVA), are the only vinyl polymer with carbon-carbon bonds that are completely biodegradable. They are soluble in water and are easily degraded in the presence of activated sludge, but also by soil bacteria. Despite its biodegradability, its applications in the environmental field are limited to the role of carrier of pesticides and herbicides. In the field of packaging, its soluble nature limits its use in blends with other polymers.

2.2.4 Blends of Synthetic and Natural Polymers

The polymer blends or mixtures, obtained by combining a natural polymer with synthetic polymers are more susceptible to microbial flora. Blends of plasticized starch with polycaprolactone are most prevalent. They can provide products with the characteristics of thermoplastics with very good water resistance. This is the case of Mater-Bi, Novamont, whose properties can be compared to that of LDPE. Table 6 combines the properties of these materials. The class A consists of wheat starch mixed with synthetic hydrophilic polymers such as poly(vinyl alcohol) and poly(ethylene-co-vinyl alcohol). For Class Y, wheat starch is mixed with cellulose acetate. Similarly, it is possible to mix natural fibers (flax, hemp) in various biopolymers for making materials "armed" to make such as dashboards of cars. The classes Z and V contain PCL, 50% and 85% respectively. This type of material has been proposed for a wide variety of applications such as packaging products, fast food items and movies. Other types of Mater-Bi are also exists.

Table 6. Physical and mechanical properties of Mater-Bi Novamont (Class Z and Y) compared with PS and LDPE [11].

	Mater-Bi (YI01U)	PS	Mater-Bi (ZF03U/A)	Mater-Bi (ZI01U)	PEBD
MFI (g/10 min)	10 – 15	8 - 12	4 - 5.5	1.5	0.1 – 22
Resistance to stretching (N/mm)	-	-	68	55	60
Stress at beak (MPa)	25 – 30	35 - 64	31	28	8 – 10
Strain at break (%)	2 – 6	1 - 2.5	886	780	150 –600
Tensile modulus (MPa)	2100 - 2500	2800 - 3500	185	180	100 – 200

3. STARCH

3.1 Constituents

Starch is primarily (98-99%) a homopolymer of D-glucose in its most stable chair conformation (4C1), where the hydroxyl groups of C2, C3, C4 and C6 are in equatorial positions (Figure 7).

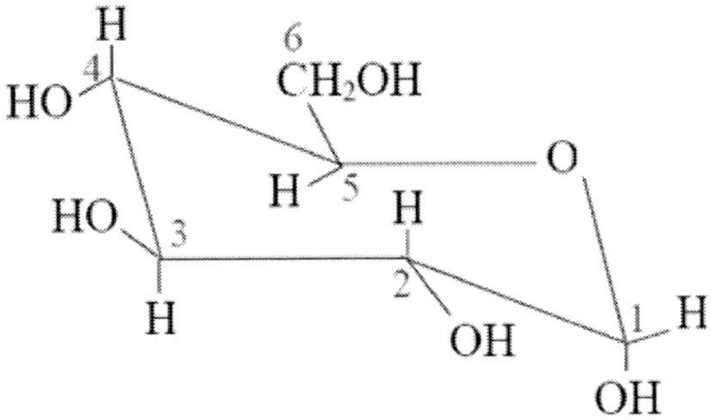

Figure 7. Chemical structure of D-Glucose.

D-glucosyl units linked mainly by 1, 4 bonds (95-96%) and to a lesser extent by 1, 6 type bonds (4-5%). Starch is mainly composed of a mixture of two polymers with very different primary structures: amylose, a molecule that is essentially linear and amylopectin, a branched molecule. The amount of amylose amylopectin in starch varies depending on its origin (Table 7) [12].

Table 7. Composition of different starches (% of dry starch) [12].

Starch	Amylose	Amylopectin
Corn	28	72
Waxy corn	1	99
Wheat	26	74
Potato	23	77
Smooth pea	35	65

The starch contains minor components of lipids, proteins, minerals, whose quantities also depend on the botanical origin of the product and the method of fractionation used. Amylose is a linear polymer (Figure 8) consisting of D-glucosyl units linked by α-(1, 4) type bonds, which is linear fraction of starch.

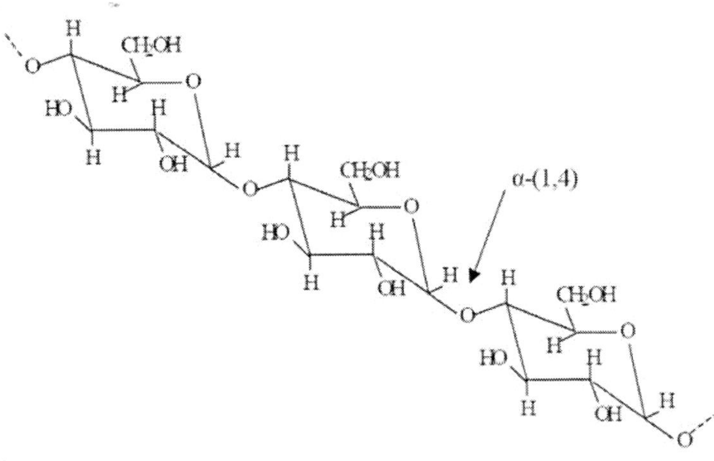

Figure 8. Chemical structure of amylose.

There may be a small amount of α-(1, 6) bonds, which gives slightly a highly branched structure. The proportion of these branches depends on the origin of amylose and its molecular weight. Each molecule contains amylose 500-6000 glucosyl units, divided into several chains (1-20) with average polymerization degree (PD) 500[13]. The molar mass of amylose is around 105-106 g mol-1. Amylopectin, the main carbohydrate constituent of starch (70-80% in normal genotypes) is a branched molecule (Figure 9) where D-glucose units are connected by bonds of type α-(1, 4) and 5-6% of mostly links-type α (1,6).

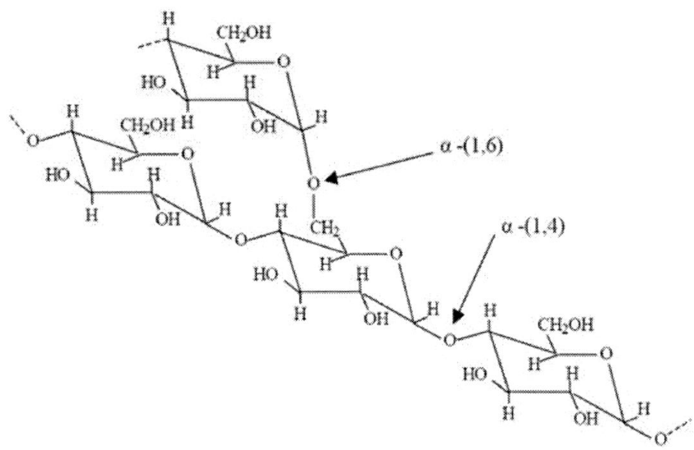

Figure 9. Chemical structure of amylopectin.

It (Figure 10) [14] consists of several clusters of short chains with an average polymerization degree (PD) 15 to 20 (A-chains), linked by longer chains of average polymerization degree (PD) 40-45 or above 60 (B-chains) [15]. Amylopectin is characterized by very high molecular weight 107-108 g mol-1.

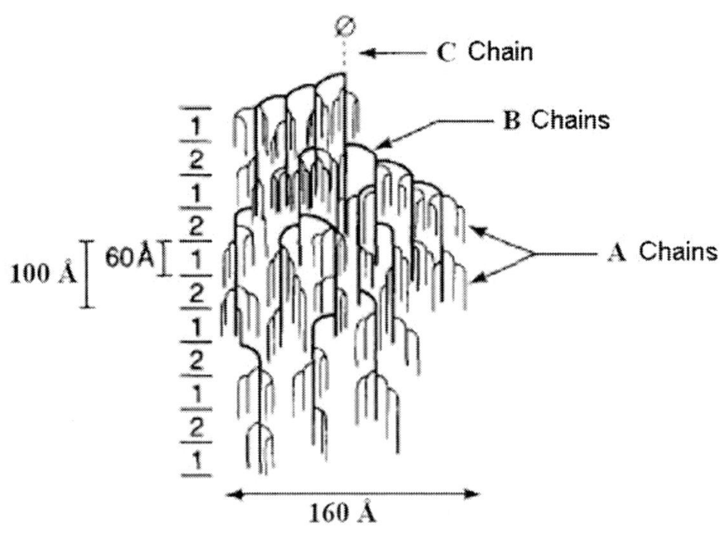

Figure 10. Model of amylopectin cluster structure [8].

The noncarbohydrate fraction (1 to 2%) consists of lipids, proteins, minerals and the amount of these components in various starches is given in Table 8.

Table 8. Composition of different starches (% of dry starch) [11].

Starch	Lipids	Proteins	Minerals	Phosphate
Corn	0.65	0.30	0.10	0.015
Waxy corn	0.23	0.10	0.10	0.003
Wheat	0.24	0.33	0.30	0.050
Potato	0.09	0.05	0.30	0.100
Smooth pea	0.19	0.18	0.05-0.22	0.040

Lipids are classified according to their position within the starch: external lipids, easily extractable and internal lipids that vary with the type of starch. Tuber starches (potato, cassava) and legumes (peas, beans) have very low levels of lipids. The external lipids consist essentially of triglycerides. The internal lipids are fatty acids and lysophospholipids. The presence of internal fat in excess of 0.5% is typical of cereal starches (wheat 0.8 to 1.2%, corn 0.6-0.8%), but their nature differs according to the botanical species. All internal lipids have the ability to form complexes with amylose. Moreover, despite their low quantity, lipids have a significant place on the physicochemical properties of starch. Proteins are the nitrogen fraction which is abundant in starch from wheat and corn. Their presence affects mainly the rheological properties and the susceptibility of starch granules to enzymatic hydrolysis. The minerals are mainly composed of phosphorus. Only the potato starch contains a large fraction of phosphate (0.06 to 0.1% dry matter).

3.2 Structure

The starch granules are semi-crystalline entities whose crystallites diffract X-rays giving two main types of diagrams based on the botanical origin of starch and possibly sustained technological treatments [16]. The native starches can be classified into three groups according to their type of diffraction pattern (Figure 11) [17]: A, B and C. A-type characteristic of cereal starches, and B-type characteristic of starches from tubers, cereals rich in amylose (> 40%) [16-18]. The C-type is a mixture of both crystal types A and B [19], is typical of legume starches [20]. The shape of X-ray diffraction spectrum of starch depends on the moisture content of grain. More starch is hydrated, the more refined lines of the spectrum. Water is an integral part of the crystalline organization of starch. The determination of the starch crystallinity is very difficult because the effect of water content and the lack of standard 100% crystalline.

Table 9 shows the degree of crystallinity for different starches, calculated using three experimental methods: acid hydrolysis, X-ray diffraction and nuclear magnetic resonance (NMR) [21-23].

At the native state, amylose and amylopectin are arranged in a semi-crystalline granular entities, whose size (1-100μm) and morphology (spherical, lenticular, polyhedral ...), the position of hilum (starting point for the granule growth) are under genetic control and therefore according to their botanical origin.

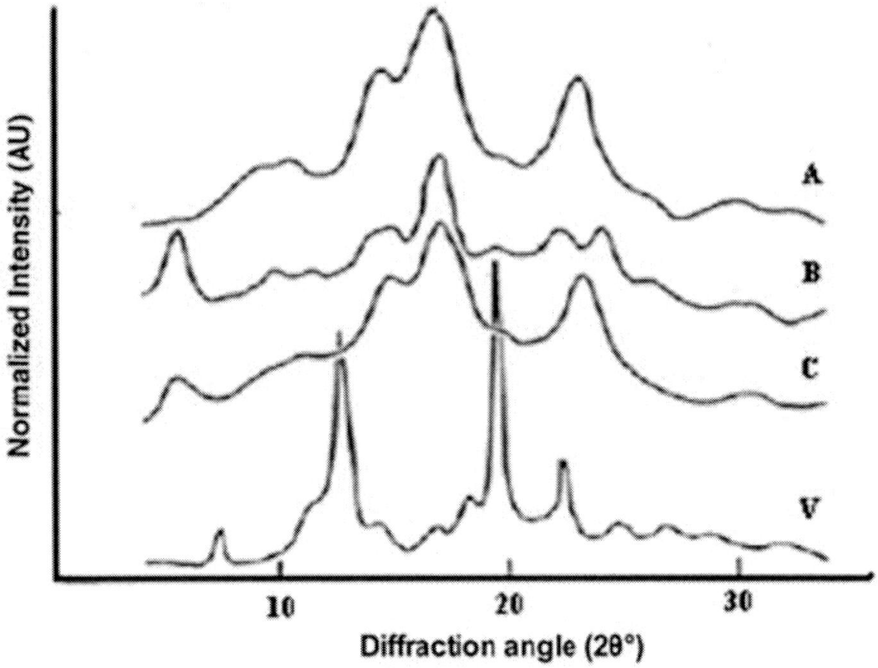

Figure 11. X-ray diffraction patterns of crystal types A, B, C and V [17].

Table 9. Degree of Crystallinity (%) for different starches determined by acid hydrolysis, X-ray diffraction and NMR [21-23].

Starch	Acid hydrolysis	XRD	NMR
A-Type			
Corn	18.1-27.0	38-43	42-43
Waxy corn	19.7-28.0	38-48	48-53
Wheat	20.0-27.4	36-39	39
Rice	-	38-39	49
B-Type			
Potato	18.1-24.0	25-40	40-50
Cassava root	24.0	24	44

The structural models proposed for the crystalline regions of starch A-type and B-type consist of double helices. The two polymorphs differ by the stacking of these double helices and the amount of water present in the crystal lattice [24] (Figure 12). The allomorph A is organized according to a monoclinic lattice (a = 2.124 nm, b = 1.172 nm, c = 1.069 nm and γ = 123 °), B2 space group, where only 8 water molecules are present between the double helixes. The allomorph B would have a hexagonal symmetry (a = b = 1.85 nm, c = 1.04 nm, space group P61) [25], which generates a central channel occupied by 36 water molecules.

Cameron and Donald [26], using X-ray diffraction at small angles, offer a mix of two types of areas, organized and amorphous. The organized areas consisting of alternating of amorphous and crystalline lamellae are repeated in 9-10 nm (Figure 13) [27].

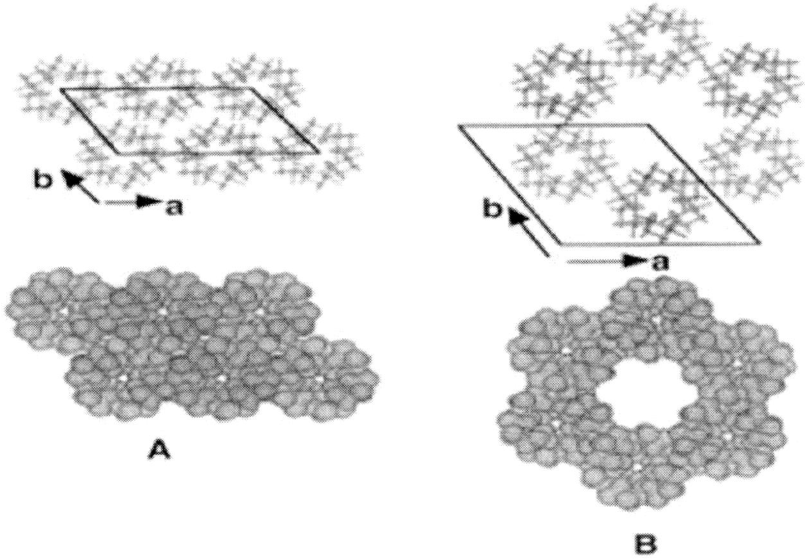

Figure 12. Crystalline arrangement of the double helixes (A) A- type and (B) B-type. Projection according to the plan (a, b).

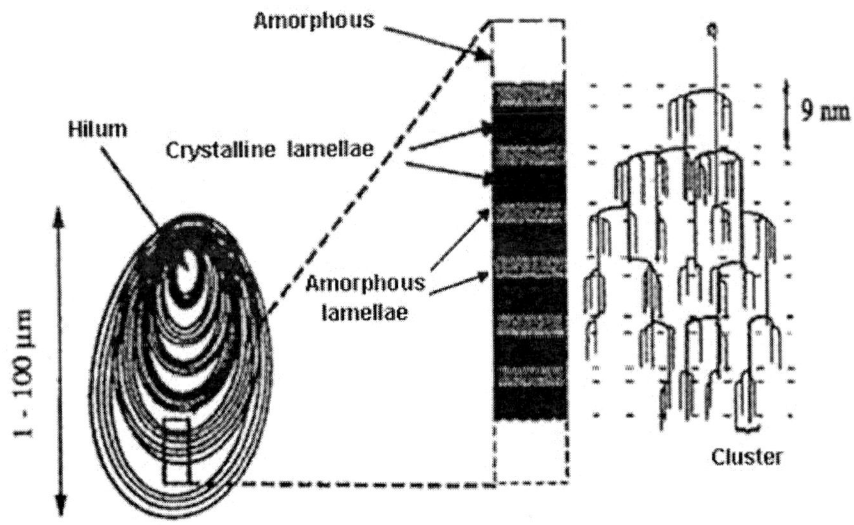

Figure 13. Different organization levels of a grain of starch [27].

The structural model of amylopectin most commonly accepted means of short chains of type S (DP 15-20) involved in double helixes, assembled in clusters. All the clusters form a structure consisting of 10nm thick doublet of crystalline and amorphous lamellae. The double-helix regions of amylopectin are located in crystalline lamellae, while the connections are the bottom of the amorphous lamellae. At the macromolecular level, the starch granules have a radial cohesion. The starch chains are arranged as concentric layers whose center is the hilum of the granule. The granule size may vary from less than 1 mm to more than 100 mm. Generally, granule size refers to the average diameter of the starch granules. Granule size may also be expressed as the average length of the major and minor axes, mean maximum diameter, mean granule volume or mean surface area. No precise categorization of granule size was found in the literature. For this reason, the following classes are defined: large (□25 μm), medium (10–25 μm), small (5–10 μm) and very small (□□5 mm) granules [28]. The table 10 shows the granule sizes of starches from different origins.

Table 10. Granule sizes of starches from different origins [28].

Source	Diameter (μm)
Bimodal	
Barley	2-3 and 12-32 [29]
Immature sweet corn	1-5 and 10-20 [20]
Rye	2-3 and 22-36 [31]
Triticale	5 and 22-36 [31]
Wheat	<10 and 10-35 [32]
Small granule starch	
Buckwheat	2-14 [33]
Cattail	2-15 [31]
Dropwort	0.5-10 [30]
Durian	3-14 [34]
Grain tef	2-6 [35]
Oat	2-14 [33]
Parsnip	1-6 [31]
Rice	2-10 [33]
Small millet	0.8-10 [36]
Wild rice	2-8 [37]
Very small granule starch	
Amaranth	1-2 [30]
Canary grass	1.5-3.5 [38]
Cow cockle	0.3-1.5 [39]
Dasheen	0.5-3 [31]
Pigweed	1.5-4 [40]
Quinoa	0.5-3 [33]
Taro	2-3 [30]

3.3 Morphology

Starch granules from different botanical origins differ in morphology. The starch of maize, waxy maize and rice are polyhedral. The wheat starch is lenticular and polyhedral. The cassava starch is hemispherical and spherical. For the starch of potato, amylo-maize, smooth pea and wrinkled pea, the morphology are respectively ellipsoidal, filamentous, reniform and Rivet washer. Figure 14 presents the SEM images for different types of starches [12].

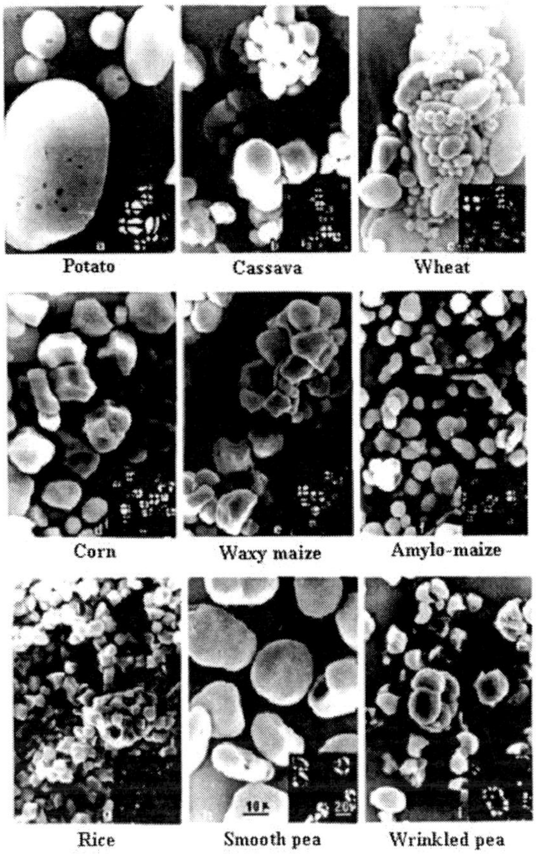

Figure 14. SEM micrographs and light microscopy polarized (inset panels) of different types of starch granules [12].

Figure 15 present the appearance of wheat starch, at a microscopic level, used the SEM photos displayed, obtained for two different magnifications [41].

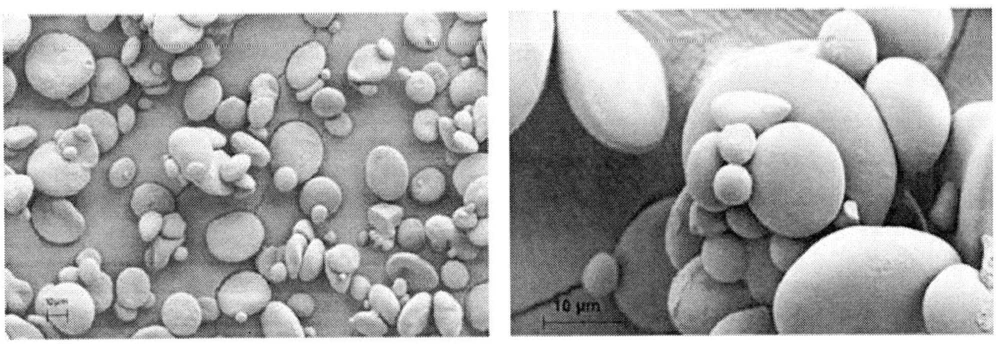

Figure 15. SEM micrographs of native wheat starch [41].

This figure show intact starch granules which are smooth, free from pores, cracks, or fissure around to lenticular and polyhedral shapes and which are relatively thick. For both starch granules there exists a wide distribution of granule size. This wide distribution is a common feature of cereal starches [41].

3.4 Thermal Stability

The high-temperature thermal stability of these materials have been analysed by means of thermogravimetry. The mass loss as a function of the temperature for wheat starch is reported in figures 16 a and b [41]. The TG and DTG curves obtained show that these materials exhibit a first mass loss of 11% w/w between 50 to 200°C. Above 200°C the main degradation process starts and finally at high temperature (T > 500°C) some residues exist (11% w/w). The DTG curves give a better visualization of the degradation kinetics of the composites.

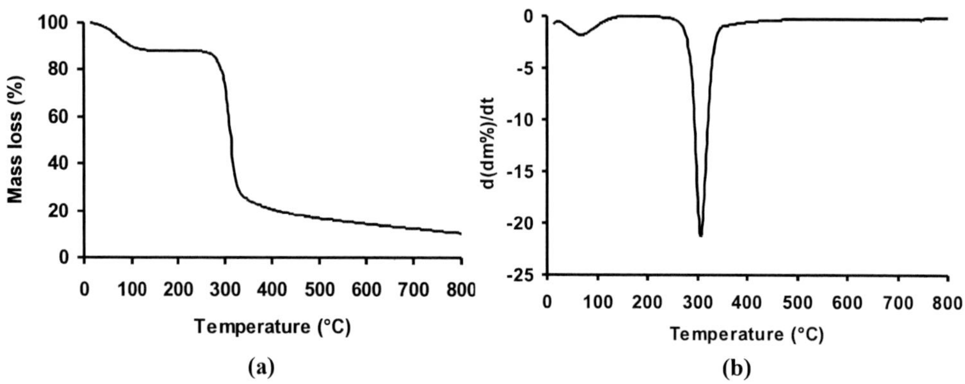

Figures 16 a and b. The thermogravimetric (a) and DTG (b) curves for wheat starch [41].

4. Natural Fibers

Natural fibres have attracted the attention of scientists and technologists mainly because of the environmental advantages when compared to synthetic fibres. They are low cost fibres, with low density and high specific properties. They are biodegradable and less abrasive in nature. Specific properties of some natural fibres are comparable with those of different synthetic fibres used as reinforcement in polymer matrices [42, 43]. Depending upon the source, the plant fibres are grouped into several classes as depicted in figure 17.

Natural fibres consist mainly of hollow cellulose fibrils held together by lignin and pectin in a hemicellulose matrix. Each fibril has a complex layered structure consisting of a primary wall encircling a thick secondary wall. The secondary cell wall S2 is made up of three layers and the thick middle layer determines the mechanical properties of the fibre and thus acts as the main load bearing component as shown in figure 18. The middle layer consists of a series of helically wound cellular microfibrills formed from long chain cellulose molecules. The angle between the fibre axis and the microfibrills is called the microfibrill angle. Mechanical properties of fibres depend on the amount of cellulose, degree of polymerisation of cellulose and on the microfibrill angle [44]. Fibres with higher cellulose content and lower degree of microfibrill angle exhibit higher tensile strength and modulus [45, 46].

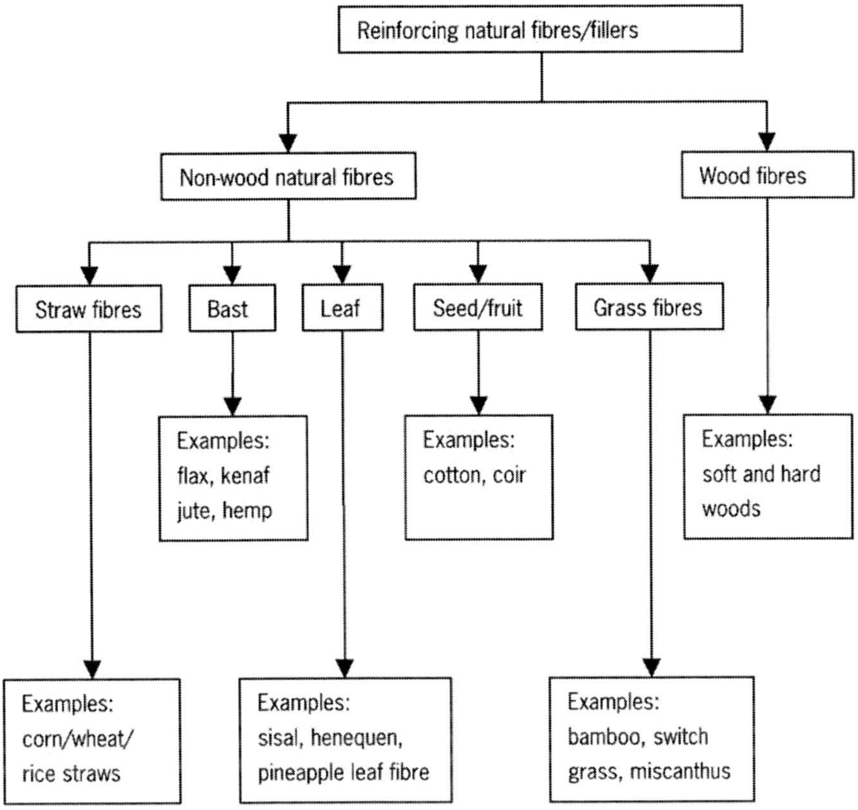

Figure 17. Classification of plant fibres based on origin.

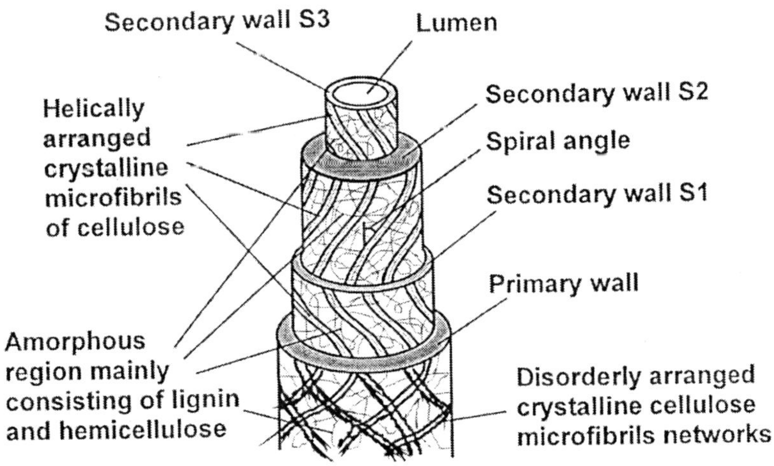

Figure 18. Microstructure of natural fibres [43].

Cellulose is the essential component of all plant-fibres. It is a linear condensation polymer consisting of D-anhydro-glucopyranose units joined together by β-1, 4-glucosidic bonds [47]. The pyranose rings are in the $^4C_1$ conformation, which means that the $-CH_2OH$ and –OH groups, as well as the glycosidic bonds, are equatorial with respect to the mean planes of the rings. The Haworth projection formula for cellulose is given in figure 19 [48].

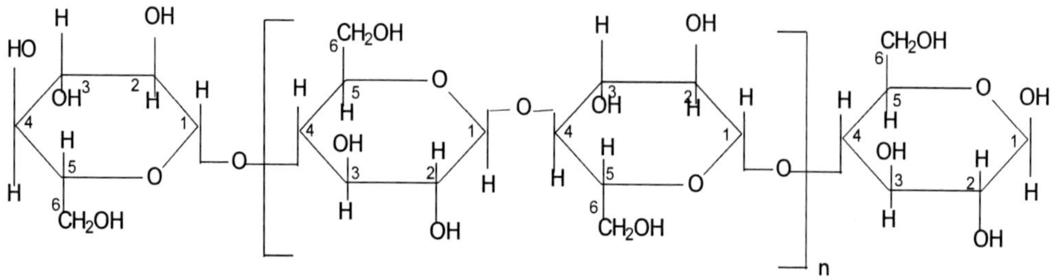

Figure 19. Haworth projection formula for cellulose [48].

The molecular structure of cellulose is responsible for its supramolecular structure and this, in turn, determines many of its chemical and physical properties. In the fully extended molecule, adjacent chain units are orientated by their mean planes at an angle of 180° to each other. Each type of cellulose has its own cell geometry and the geometrical condition determines its mechanical properties. Solid cellulose forms a micro-crystalline structure with regions of high order, i.e. crystalline regions, and regions of lower order, i.e. amorphous regions. Naturally occurring cellulose (cellulose I) crystallises in monoclinic sphenodic structures. The molecular chains are orientated in fibre direction and the geometry of the elementary cell is dependent on the type of cellulose.

Hemicellulose is not a form of cellulose and the name is a misnomer. They comprise a group of polysaccharides that remain associated with cellulose after lignin has been removed. Hemicellulose differs from cellulose in three aspects. Firstly, they contain several different sugar units where as cellulose contains only 1, 4-β-D-glucopyranose units. Secondly, they exhibit a considerable degree of chain branching, where as cellulose is a linear polymer. Thirdly, the degree of polymerisation of native cellulose is 10-100 times higher than that of hemicellulose. Unlike cellulose, the constituents of hemicellulose differ from plant to plant. Lignins are complex hydrocarbon polymers with both aliphatic and aromatic constituents. Their chief monomer units are various ring-substituted phenyl-propanes linked together in ways, which are still not fully understood. Structural details differ from one source to another. The mechanical properties are lower than those of cellulose. At the value of 4 GPa, the mechanical properties of isotropic lignin are distinctly lower than those of cellulose. Pectin is a collective name for heteropolysaccharides, which consist essentially of polygalacturon acid. Pectin is soluble in water only after a partial neutralisation with alkali or ammonium hydroxide. Waxes make up the part of the fibres, which can be extracted with organic solutions. These waxy materials consist of different types of alcohols, which are insoluble in water as well as in several acids.

4.1 Chemical Composition of Natural Fibres

The chemical composition and structure of plant fibres depend to a large extent on the climatic conditions, age and the digestion process of the plant, which they are derived from. The wt% value of each component in some plant fibres are presented in Table 11 [49-51]. The physical properties of fibres mainly depend on cellulose (content and orientation of molecules) hemicellulose and lignin. Hemicellulose and pectin are responsible for the bio-degradation, moisture absorption and thermal degradation of fibre. Individual fibre properties and the structure of fibre can vary widely depending upon the plant, part of the stem, age, extraction technique, moisture content etc.

Table 11. Chemical composition of plant fibers [49-51].

Plant fibres	Banana	63-64	19	-	5	-	10-12	11
	Pine apple leaf	70-82	-	-	5-12	-	11.8	14.0
	Ramie	68.6-76.2	13.1-16.7	1.9	0.6-0.7	0.3	8	7.5
	Coir	36-43	0.15-0.25	3-4	41-45	-	8	41-45
	Hemp	70.2-74.4	17.9-22.4	0.9	3.7-5.7	0.8	10.8	20.0
	Flax	64.1-71	16.7-20.6	1.8-2.3	1.7-2.0	1.5-1.7	10	6.2
	Jute	61-71.5	13.6-20.4	0.2	12-13	0.5	10-12.6	10
	Cotton	82.7	5.7	5.7	-	0.6	10	8
Component		Cellulose (wt %)	Hemicellulose (wt %)	Pectin (wt %)	Lignin (wt %)	Waxes (wt %)	Moisture content (%)	Microfibril angle (°)

4.2 Geometrical and Mechanical Properties

Unlike man-made synthetic fibres, natural fibres extracted from plants have various geometry. The fibre length and diameter are main parameters, which have certain influence on the reinforcing capabilities in composites. This geometrical pattern of the extracted fibres varies not only from one plant to another but vary from one part of the plant to other. Mechanical properties of natural fibres, alike geometrical properties, vary to a large extent. Natural fibres exhibit considerable variation in diameter along with the individual bundles. Quality as well as most of their properties depends on the factors like size, maturity as well as the processing methods adopted for the extraction of fibres. Properties such as tensile strength, modulus, etc. strongly depend on the internal structure and chemical composition of fibres. Table 12 gives us a comparison regarding the mechanical properties of natural fibres and synthetic fibres [52- 54].

According to the data collected from scientific publications, the strength of natural fibres is lower than that of glass fibres, while the stiffness is on the same level. The tensile strength and Young's modulus of natural fibres like jute are lower than those of glass fibres; the specific modulus of jute fibre is superior to that of glass and also on a modulus per cost basis, jute is far superior. The specific strength per unit cost of jute, too, approaches that of glass. However, taking into consideration the low density of natural fibres, which are up to two times lighter than glass, the resulting specific stiffness of natural fibres is substantially higher than the same parameter of glass fibres. Flax, hemp, and jute fibres are particularly standing out due to good specific stiffness. High temperatures can result in physical and chemical changes in their structure. The low temperature degradation is associated with the degradation of hemicellulose where as high temperature degradation is due to lignin. The degradation level depends not only temperature, but also on the duration of the exposure. All natural fibres derived from plants are based on cellulose, which is hydrophilic in nature. Usually, the moisture content in natural fibres varies from 5-15%. In composites, higher moisture content can lead to dimensional variations and can affect the mechanical properties. It can also lead to poor processability and porous products. Therefore, natural fibres must be processed and stored under optimal conditions and must be dried before impregnating them with polymers.

Table 12. Mechanical properties of natural and synthetic fibres [52- 54].

Fiber	Density (g/cm^3)	Diameter (μm)	Tensile strength (MPa)	Young's modulus (GPa)	Elongation at break (%)
Cotton	1.5-1.6	-	287-800	5.5-12.6	7-8
Jute	1.3-1.45	25-200	393-773	13-26.5	1.16-1.5
Flax	1.5	-	345-1100	27.6	2.7-3.2
Hemp	-	-	690	-	1.6
Ramie	1.5	-	400-938	61.4-128	1.2-3.8
Sisal	1.45	50-200	468-640	9.4-22.0	3-7
PALF	-	20-80	413-1627	34.5-82.51	1.6
Coir	1.15	100-450	131-175	4-6	15-40
E-Glass	2.5	-	2000-3500	70	2.5
S-Glass	2.5	-	4570	86	2.8
Aramid	1.4	-	3000-3150	63-67	3.3-3.7
Carbon	1.7	-	4000	230-240	1.4-1.8

5. Thermoplastic Starch

5.1 Fabrication

The mixture of native starch granules with a sufficient amount of plasticizer to allow their fusion, at a lower temperature than their degradation, leads to a starchy material consisting of a continuum of polysaccharide chains tangled. This material is called thermoplastic starch.

Most commonly used methods for the fabrication of starch composites are extrusion and solvent casting method. For that various types of plasticizers such as glycerol, water, urea fromamide etc are used during the formulation. The characteristics of mechanical work and flow (temperature, pressure, residence time, energy) are known and modeled in the case of implementation by extruding one or two-screw [55,56]. A specific mechanical energy (SME) higher than 300kJ/g is necessary to achieve a complete destructuration. The presence of low molecular mass (plasticizers) raises the threshold energy like the action of sugars in the case of cereal products [57]. The SME transmitted as shear, leads to the breakdown of starch grains by fragmentation, and once the melt phase obtained, it accompanied by a moderate depolymerization, in particular the amylopectin [58].

The native starch can be transformed by treatments that destroy its granular structure. The two main types of treatment for the transformation of starch are:

- A heat treatment in excess water, commonly used in industry for a variety of applications (adhesives, paper, food), is the gelatinization.
- Physical treatment, requiring simultaneous action of temperature and shear, lower water content, is extrusion.

The gelatinization corresponds to the phenomena of irreversible swelling and solubilization observed when the starch granules are heated above 60°C in the presence of excess water. During the starching, the starch grains gradually lose their crystallinity. The rupture of hydrogen bonds in crystalline grain areas, allows in a first step the massive absorption of water and in a second step leaching of the constituents of lower molecular weight outside of the grain [59]. After gelatinization, the final state of the starch is a suspension of grains in solution. During cooling, the gelatinized starch forms a gel (Figure 20).

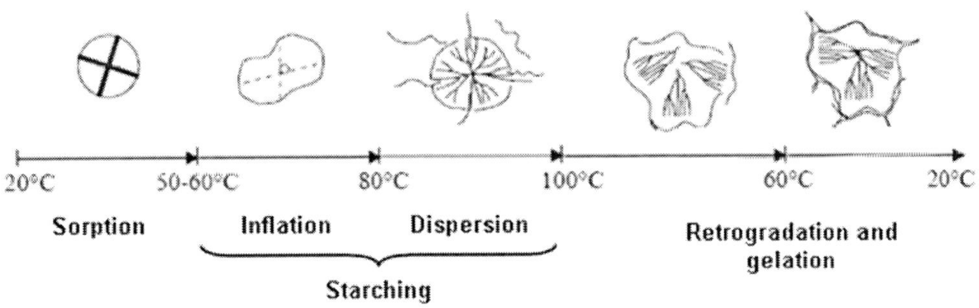

Figure 20. Influence of hydrothermal treatment of different state of starch in excess of water.

These structural changes involve initially amorphous regions more accessible to water and a second time in the crystalline zones. The water should be sufficient to hydrate the starch molecules and allow gelatinization. Obtaining the melt phase from the plasticized starch granules requires a new mechanical power input prior to destroy residual crystalline structures. After mechanical treatment, the crystallinity is reduced.

Figure 21 shows that the viscosities of plasticized starches are of the same order and magnitude as those of common thermoplastics at lower temperatures [60].

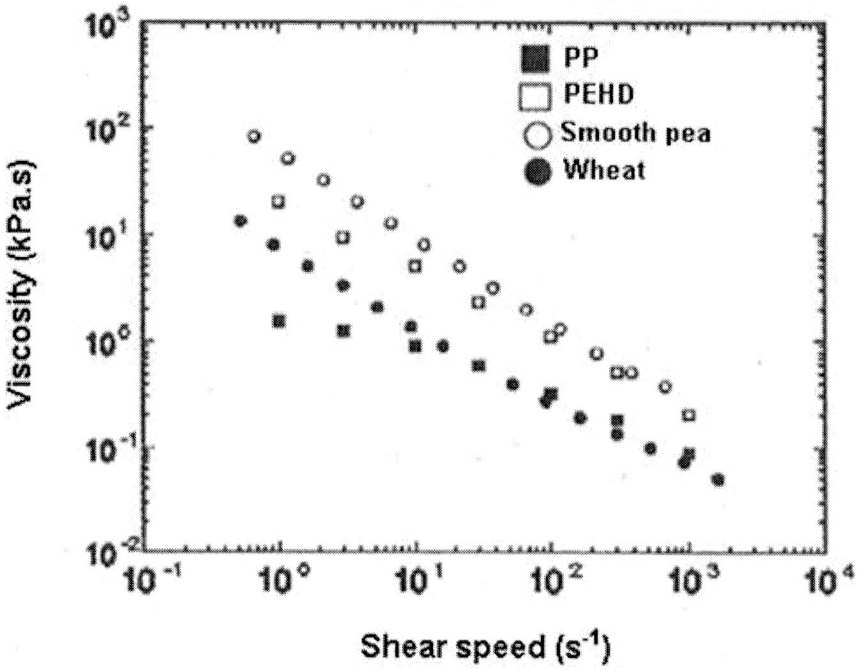

Figure 21. Examples of flow curves of thermoplastic materials at 200°C and plasticized starch from smooth pea and wheat at 125°C.

5.2 Structure

Other crystal types can be obtained during the transformation of starch by complexing of amylose with alcohols and monoacyl lipids (fatty acids and monoglycerides). Such complexes between amylose and lipids are present in the amorphous state in cereal starches and may form crystalline packings generic type V (type V, from the German "Verkleiterung (gelatinization)) whose which depends on the complexing, such as iodine, fatty acids [61], dimethyl sulfoxide [62], lactones [63], acetone [64], minerals and alcohol. The starch-alcohol complex can be obtained with alcohols such as n-butanol [65], methanol, isopropanol, 1-propanol, 1-pentanol [66] and ethanol [67].

The proposed models have in common a single left-handed helix 6 residues per turn and a progress per residue h between 0.132 and 0.136 nm. This low value of h gives the amylose helix with a large diameter feature an inner channel which is most often the complexing (Figure 22).

The allomorph Vh obtained with lipid unicycles and linear alcohols is the most common and most studied for its involvement in many transformations of the starch. It is characterized by an orthorhombic lattice (a = 1.37 nm, b = 2.37 nm and c = 0.805 nm) and P212121 space group type. The Va structure is another form of crystalline amylose-lipid complex. The characteristic angles of XRD peaks are respectively with 2θ = 7.8, 13.5 and 20.9° [68]. It is characterized by an orthorhombic lattice (a = 1.30 nm, b = 2.25 nm, c = 0.79 nm) with a P212121 space group [69-71]. In the Va crystal structure, the helices of amylose are more contracted and there is less water compared to the Vh crystal structure. The transition Vh form to Va is observed when the form Vh is dried to a water activity less than 0.6 [72-74]. This transition is reversible because it is possible to transform the Va type to Vh type by hydration in the vapor phase. It is usually interpreted as a change of form during the hydration with increasing distance between helices of 1.30 to 1.37 nm [75, 76] due to the introduction of water molecules between amylose helix. The Va form is never obtained by direct crystallization solutions of amylose, which is not the case of structures of type A, B and Vh [77].

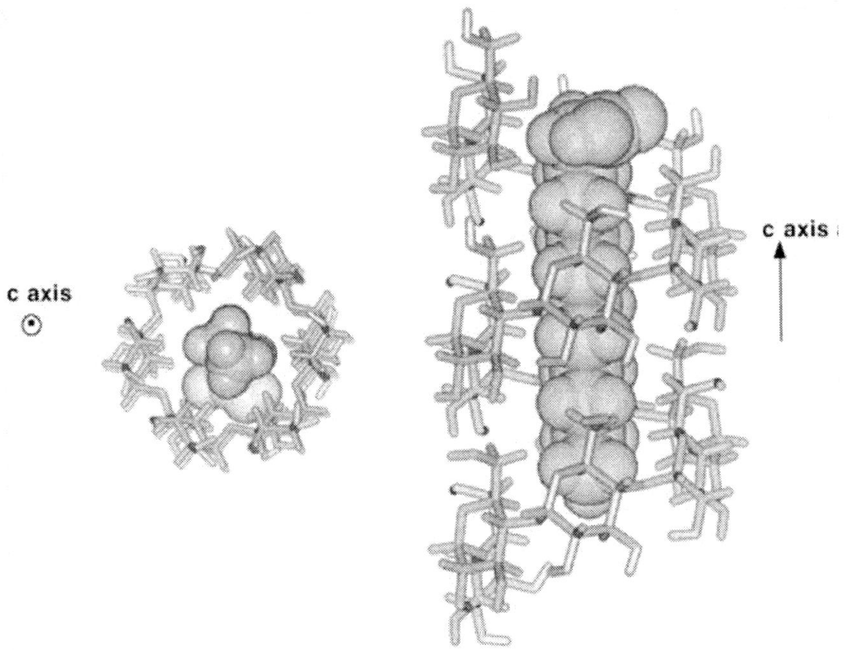

Figure 22. Conformation of inclusion model of an fatty-acid in an amylose helixe (Structure of Vh-type).

Figures 23 (a) and (b) present the isotherm curves at various temperatures (60°C, 90°C, 120°C and 150°C) and non isothermal curves for the wheat flour based polymeric film [78]. The materials which underwent an isotherm of 24 h at temperature 90°C, 120°C and 150°C gives the same diffraction peaks at 2θ = 7.8°, 13.5°, 17.5°, 19.4°, 20.9°, 26.3° and 30.9° which corresponds to Va structure of starch content in wheat flour. But in the case of material kept at an isotherm of 60°C, peaks having value 2θ = 12.9° and 19.8° appears with other peaks having 2θ value of 7.8°, 13.5°, 17.5°, 20.9° and 26.3° which is a characteristics of the Vh and Va structure respectively. This is also observed in the material which is kept at non isothermal temperature.

On the comparison of the intensity of peak having 2θ value at 20.9° with the isotherm to corresponding peaks at temperature of 90°C, 120°C and 150°C have a weaker magnitude. After extrusion, the observed peaks at 2θ = 7.2°, 12.9° 19.8° and 22.6°θ are characteristics of Vh-type structure for the non isothermal. For the materials annealed at 60°C, (during 24h), both peaks of Vh and Va structures are observed. This indicates that the transition from Vh to Va structure is not complete.

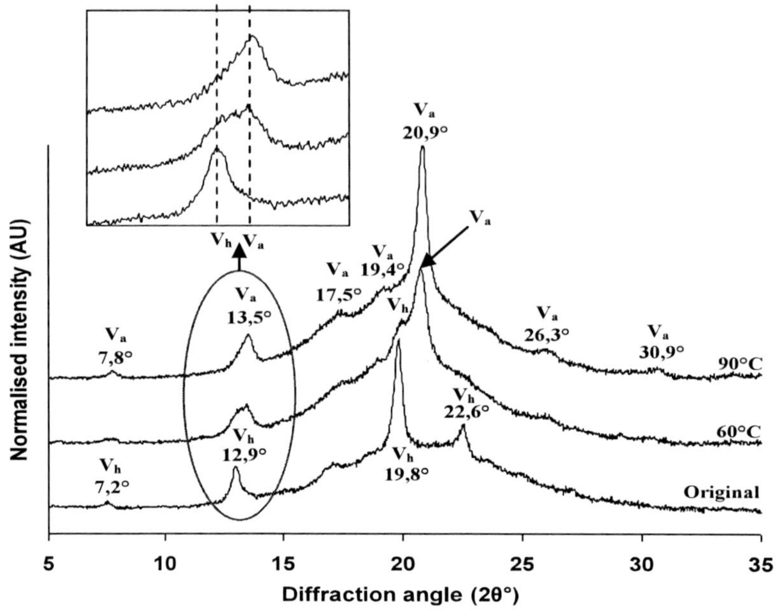

(a)

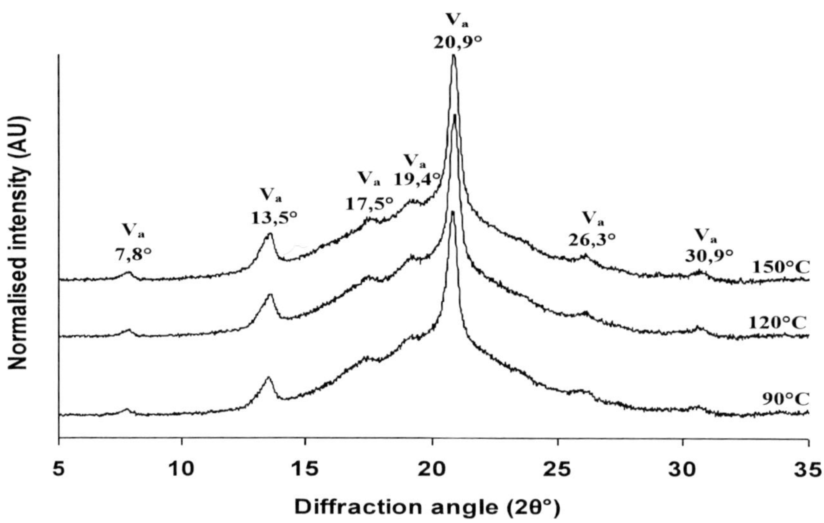

(b)

Figure 23 (a) and (b). The isotherm curves at various temperatures (a) Original, 60, 90°C and (b) 90, 120, 150°C for the wheat flour based polymeric film [78].

Saiah [78] calculated the content of the amorphous and crystalline phases of the wheat flour based TPS for isothermal and nonisothermal annealing at 60°C and 90°C (Figures 24 a, b and c) assuming that the sum of the fitting curves for the crystalline and amorphous phase exactly fits with the experimental data. The change of the Vh towards Va structure results an increase in crystallinity of the polymeric material from 14 to 16% w/w. No change was observed in the percentage of crystallinity for the materials annealed isothermaly and nonisothermal at 60°C.

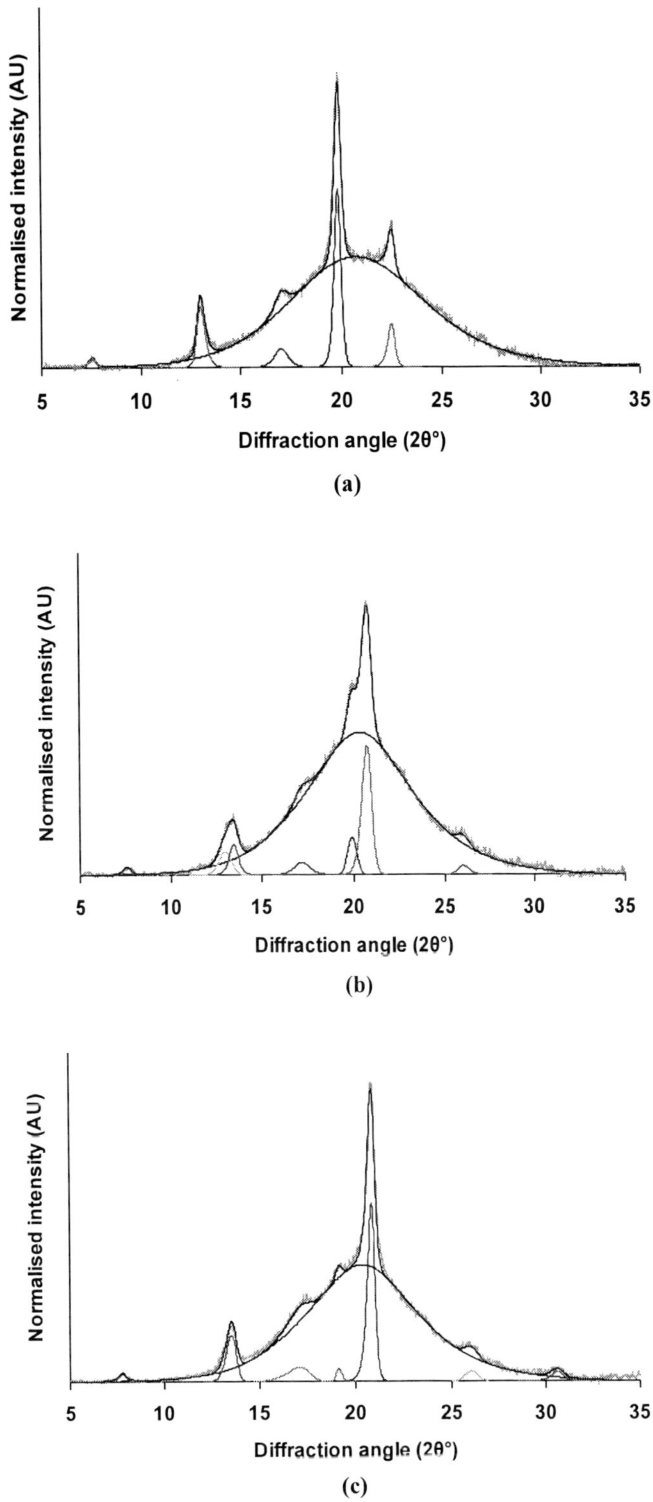

Figure 24 a, b and c. The contributions of amorphous and crystalline phases for materials which undergone nonisothermal (a) and for isotherm curves kept at 60°C (b) and 90°C (c) [78].

5.3 Mechanical and Aging Behvaiour

The mechanical properties are influenced by the botanical origin of starch, more specifically the proportion of amylose and amylopectin. The materials obtained from wheat, corn and potatoes starch, have failure stress above that of a material based on waxy maize starch (rich on amylopectin) (Figure 25 (a)) [79]. Tensile tests on films of pure amylose and amylopectin highlight the difference in behavior between these two materials (Figure 25(b)) [80]. The amylopectin based material has a ductile behavior (high failure strain), while that based amylose has a typical behavior of a brittle material.

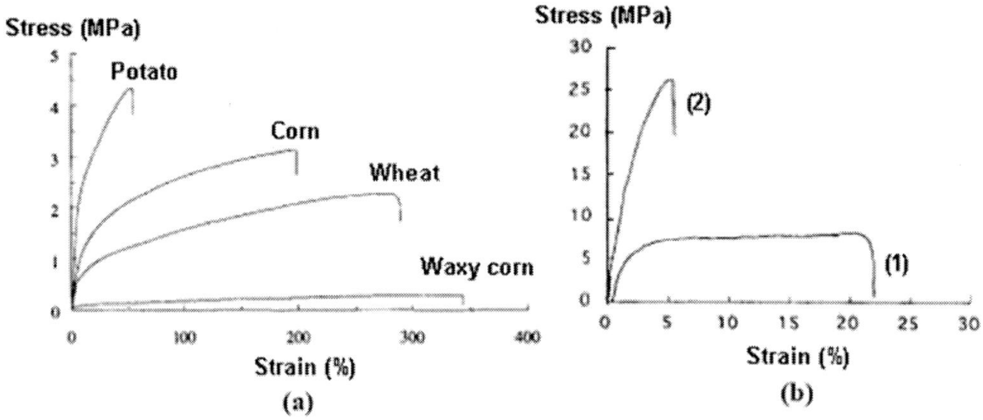

Figures 25 a and b. Stress-strain curves for: (a) based materials starch potato, corn, wheat and waxy maize [79] (b) materials based pure amylopectin (1) and pure amylose (2) [80].

The effect of plasticizers on film resulting from agro-resources, generally leads to a decrease in the modulus and failure stress, and an increase in failure strain (Figure 26) [81-83].

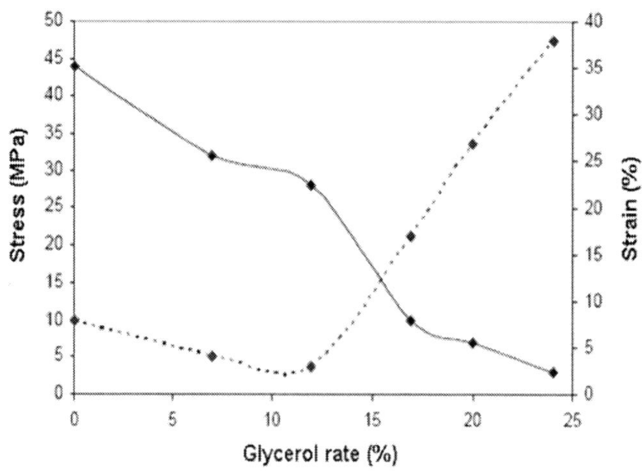

Figure 26. Influence of glycerol rate on the strain at break (—) and stress at break (---) for potato starch films [83].

For glycerol level below 12%, a phenomenon antiplastification is observed, resulting in decreased stress and strain at break. This phenomenon is due to strong interactions between the polymer and the plasticizer, which forms a hydrogen network reinforcing material. From 12%, the strain at break increases quickly, against the stress at break decreases, and that up to 25% glycerol. This change in mechanical behavior is due to displacement of the glass transition temperature (Tg) of the system below the ambient temperature. The maximum stress and strain at break of starch plasticized with different levels of sorbitol are shown in Figure 27 [84]. Similar to glycerol, addition of sorbitol led to a significant change in the mechanical behavior of plasticized starch and antiplastification effect is observed at sorbitol content less than 27%. Beyond this rate, sorbitol acts as a plasticizer. During storage of films, changes in mechanical properties occur [85-87]. The strain at break decreases, while the stress at break increases. The changes observed over time are due to several concurrent factors: reduction in water content [88, 89], increased Tg and increase in crystallinity. Ageing causes reorientation and/or crystallization of molecules of amylose and amylopectin [87]. The crystallites acting as physical nodes, generate as stress concentrations and thus weaken the material [89].

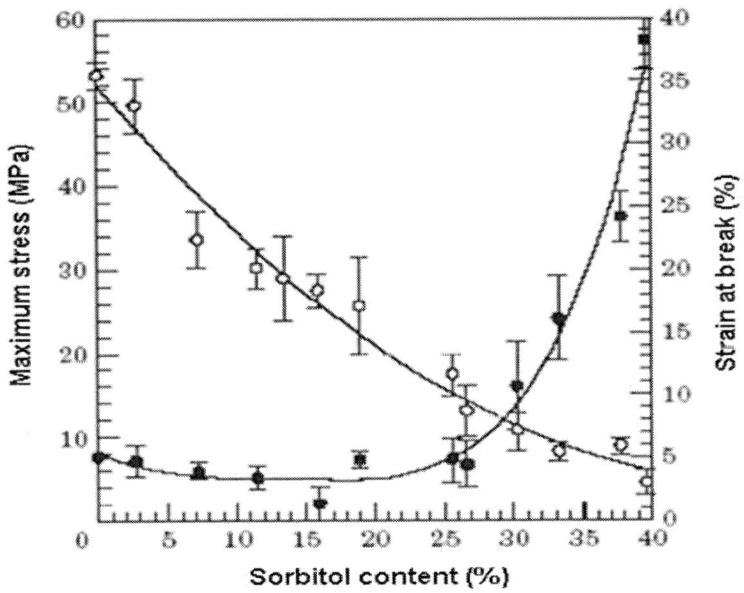

Figure 27. Influence of sorbitol content on stress (○) and strain at failure (●) for films plasticized starch [84].

Aging studies were performed on wheat flour material with 9% water and 12.8% glycerol (at 75% RH for 1 week, 1 month, 6 months, and 12 months) [90]. The XRD patterns for this material are shown in figure 28 [90].

Aging for a long period (12 months) increases the intensity of the peak located at $2\theta = 17.3°$ which corresponds to the A-type crystalline structure. Moreover, the starch macromolecules of wheat flour film can be reorganized and undergo retrogradation (recrystallization). During this period of storage, the V_h structure is not modified. Therefore, the percentage of crystallinity increases from 14 to 18% (Figure 29).

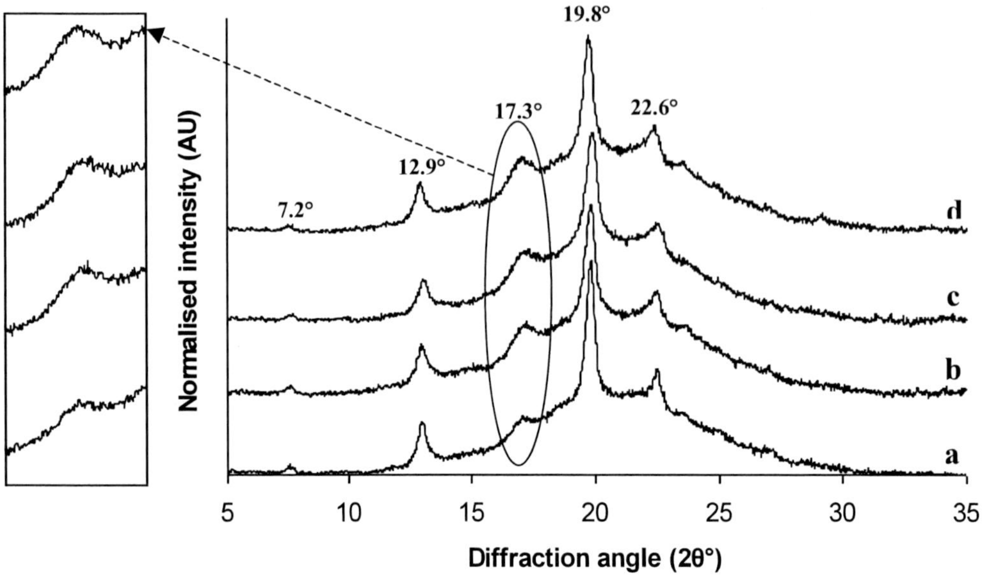

Figure 28. XRD pattern of extruded wheat flour with a composition of a) 12.8% glycerol and 9% water for one week; b) one month; c) six months; and d) 12 months. Zoom of the domain 2θ = 17.3° shows the existence of residual A-type crystalline structure [90].

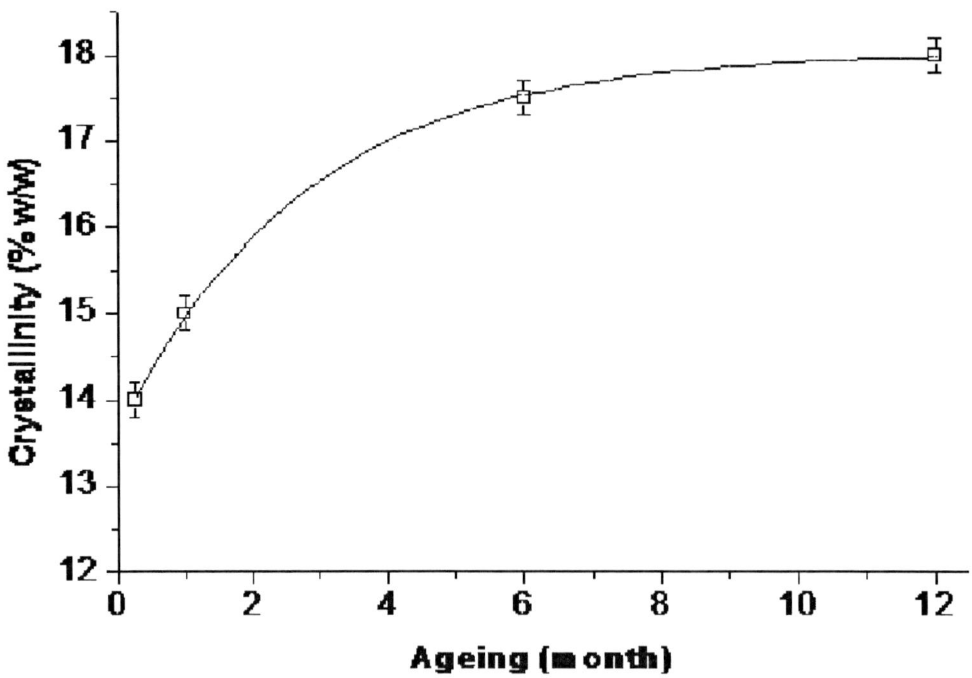

Figure 29. Variations of crystallinity with duration of aging (75% RH) [90].

This retrogradation phenomenon depends on the nature of plasticizer used because smaller plasticizers favor crystallization by mobilizing the starch molecules, while larger particles prevent crystal propagation [91].

5.4 Glass Transition Temperature

The kinetics of crystallization during aging was not linear and 87.5% of possible recrystallization occurred during the first six months. Figure 30 display the DSC curves obtained for the wheat flour based thermoplastic materials [92].

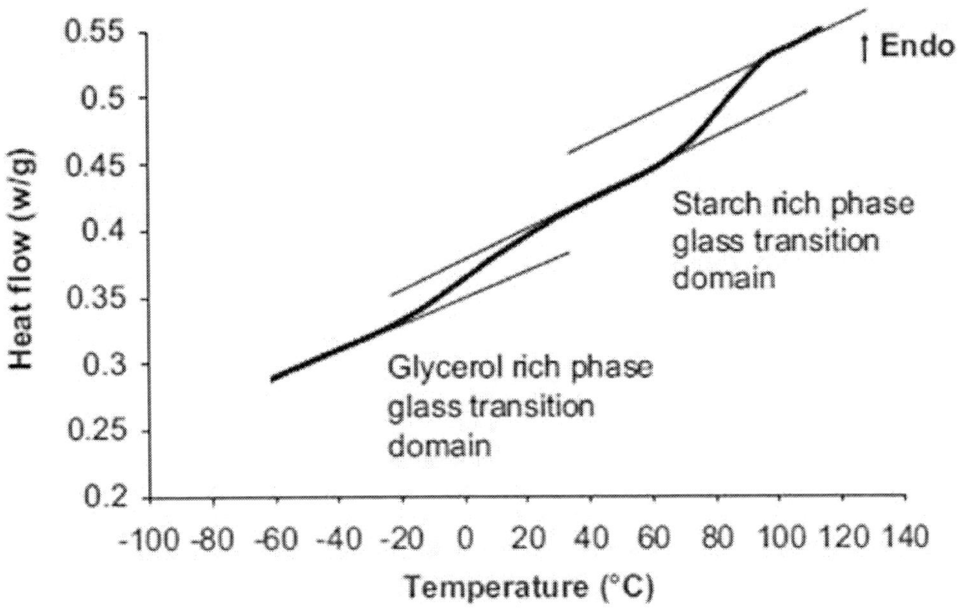

Figure 30. DSC curves obtained for a wheat flour thermoplastic showing the two glass transitions; low temperature (Tg of the glycerol rich phase), higher temperature (Tg of the starch rich phase) [92].

Only two endothermic steps showing glass transitions are observed at -56 and 10°C, no exothermic transitions of crystallisation and no degradation occurs. i.e. vitreous and/or the amorphous fraction of the samples are not modified by heat. When the amount of plasticizer changes the T_g of the starch rich phase is shifted toward the lower temperature [93]. The first transition observed at -56°C is due to the T_g in the glycerol rich phase [94] and the second one at 10°C is due to T_g of the starch rich phase [94]. Finally no melting transition exists because, only the crystalline phase can give a melting signal and, as for many natural polysaccharide based materials, this crystalline phase is destroyed before melting.

The variations of the storage modulus (E') and the damping factor (tan δ) of the film as a function of temperature at a frequency of 5Hz are displayed on figure 31.

As the temperature increases, E' value decreases. For damping curves, two peaks are observed at a temperature of -43°C and +27°C. This is expected according the results obtained previously by means of DSC. These peaks are associated to the α modes of the molecular relaxations of glycerol and starch rich phases, respectively. The α mode can be understood as the mechanical manifestation of the glass transition. The small differences between the values obtained by means of DSC and DMA are linked to the excitation frequencies used, 5Hz for mechanical and 100Hz (equivalence) for the DSC.

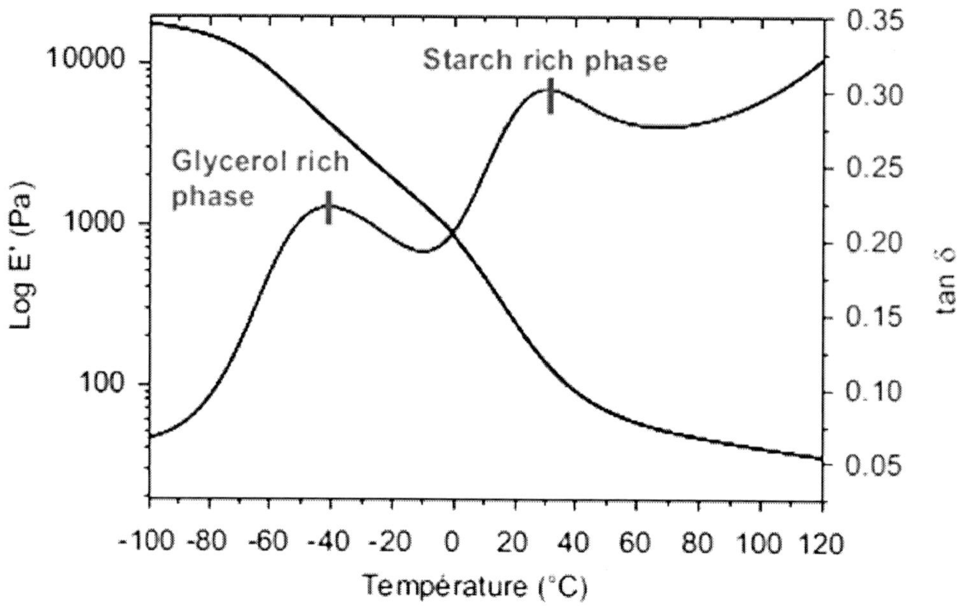

Figure 31. Variation of storage modulus (E') and damping factor (tan δ) for the wheat flour material as a function of temperature at frequency 5Hz [92].

6. Macro Fiber Reinforced Starch Composites

Starch materials are renewable, biodegradable, plently available and low cost when compared to other biodegradable plastics. These materials can be easily molded in the form of thin films. The high moisture sensitivity and poor mechanical properties are the main disadvantages when compared to the commodity plastics. Therefore it's very essential to improve the properties of the biodegradable matrices for the successful applications. Incorporation of natural fibers in these matrices will be an added advantage since both are biodegradable and inexpensive. Also with the growing economic competition and ecological pressure, there is an increasing need to develop more efficient low cost fillers and reinforcements. For the biofiber reinforced composites, the fibers from macro to micro level, cellulose nanocrystalites, and commercial regenerated cellulose fibers were used as reinforcement in the starch matrix. [95-97].

6.1 Mechanical Properties

Sreekumar et al. [98] investigated the reinforcing effect of sisal fiber on the wheat flour based starch materials. The sisal fiber content having fiber length (50-250μm) varied from 0 to 10% w/w keeping the amount of glycerol (plasticizer) as constant (23% w/w). The mechanical properties of these composites are given in the [Table 13].

Table 13. Tensile properties and hardness of the composites having various sisal fiber content (0, 1, 3, 5 and 10%w/w) [98].

Sisal fiber (% w/w)	Tensile strength (MPa)	Young's modulus (MPa)	Elongation (%)	Hardness
0	1.9 ± 0.1	45 ± 3.8	34 ± 2.6	86
1	1.7 ± 0.1	43 ± 3.3	51 ± 8.3	86
3	1.7 ± 0.2	47± 2.6	43 ± 0.4	88
5	2.1 ± 0.1	63± 2.0	38 ± 2.0	92
10	2.0 ± 0.2	93 ± 9.8	21 ± 3.8	90

Rather than tensile strength, prominent improvement was observed for the Young's modulus of the composites. Instead of glycerol alone, Saiah et al. [99] used a mixture of glycerol and water and flax fibers for the fabrication of composites. When the fiber content increased from 0% to 20% w/w, stress at failure and tensile modulus increased from 3.2 MPa to 8.9 MPa and 125 MPa to 465 MPa, respectively [Figure 32].

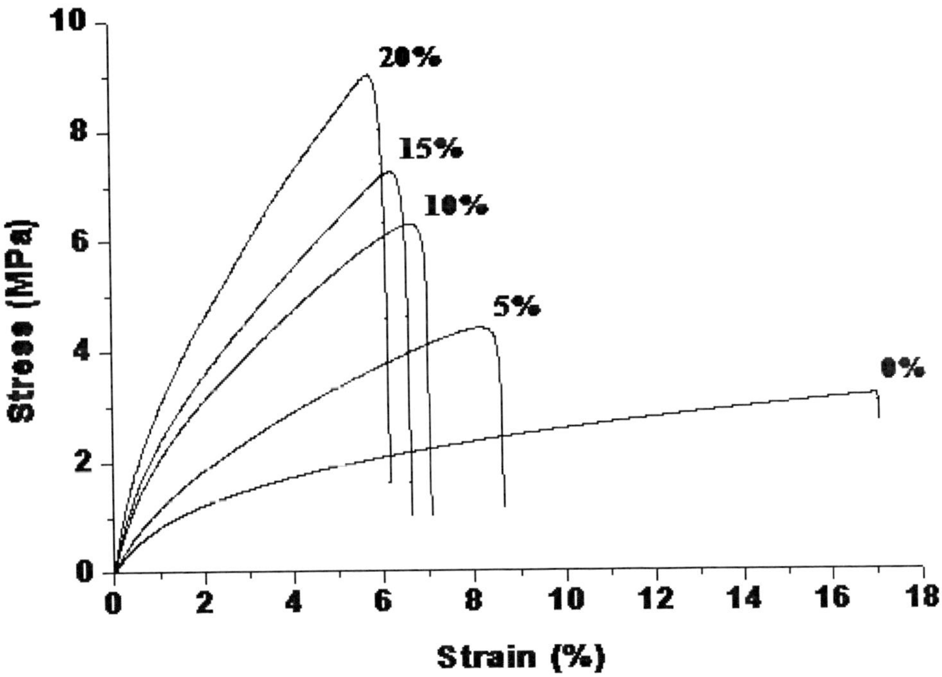

Figure 32. Tensile curves for the flax fibre reinforced thermoplastic composites containing fibre content (0% W/W, 5% W/W, 10% W/W, 15% W/W, 20% W/W) [99].

For the composite having 20% w/w fiber content, an increment of 272% occurred in tensile modulus while for that of the stress at failure is 178%. The chemical similarity of starch and plant fibres, which consist mainly of cellulose, resulted in an increase in mechanical properties [100]. After extrusion, the fibers are oriented along the longitudinal direction of material which corresponds to the direction of extrusion [Figures 33 a and b]. The fibers were very well dispersed in the starch matrix and fiber breakage can be clearly seen.

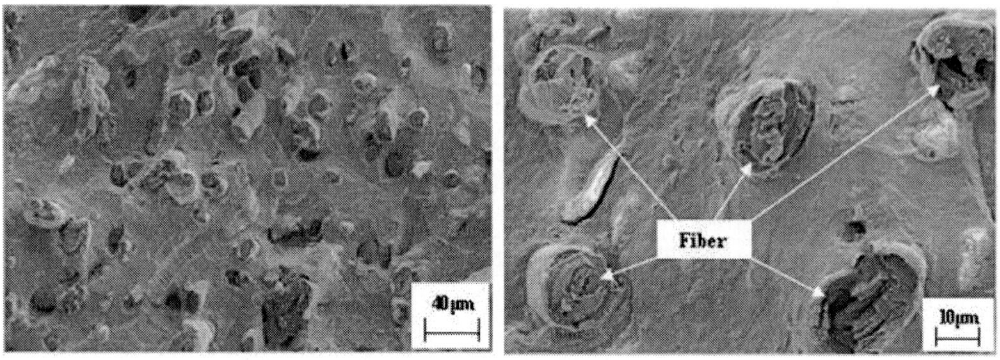

(a) (b)

Figures 33 a and b. Scanning electron micrograph of the cross section of the composite having fiber content of 20% w/w at magnification 40 µm and 10 µm [99].

Similar trend was observed for the cotton/wheat flour starch composites by Larisa et al. [101]. Here the maximum properties were observed for the composite having a fibre loading of 10% w/w, beyond this limit a leveling occurs. In the case of Eucalyptus urograndis pulp derived cellulosic fiber reinforced corn starch composites; Curvelo et al. [102] found that the modulus and tensile strength show 156% and 120% increase, respectively, while elongation was reduced from 31% to 11%. Since the matrix and fiber used in these studies are hydrophilic in nature, fiber-matrix interaction should increase there by the mechanical properties. Also the above mentioned studies convey that the properties of the composites, depends upon the formulation of the matrix as well as the nature of the fiber. The interaction between the fiber and matrix can be increased by the modification of fiber surface. Preetha et al. [103] conducted an investigation regarding the effect of mercerization on the the properties of flax/ starch composites. A comparison on the untreated and treated flax fiber reinforced composites is given in the figure 34.

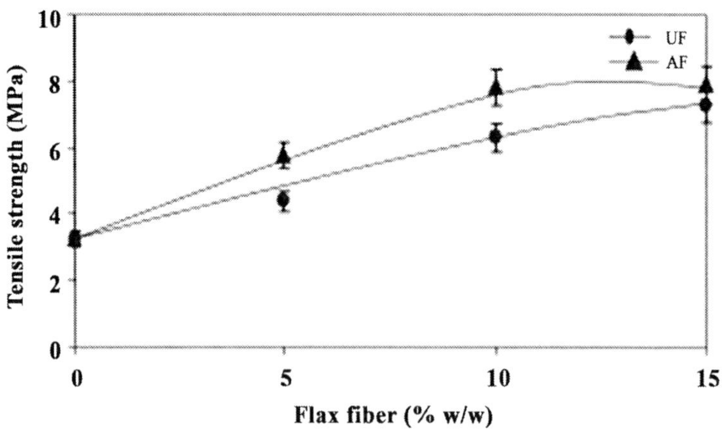

Figure 34. Tensile strength for 5% NaOH-treated flax fiber reinforced composites with alkali-treated fiber (AF) and untreated fiber (UF) [103].

The mechanical properties of composites with treated and untreated fiber are quasi-identical. From all mechanical properties analysis, it appears that it is not necessary to use more than 10% (w/w) of fiber in the wheat flour matrix, since greater contents do not modify significantly the values of any tensile properties. Duanmu et al. [104] modified starch using allylglycidyl-ether with various amounts of wood fiber and ethylene glycol dimethacrylate as crosslinker. The incorporation of bleached softwood fibre with an aspect ratio of up to 100 and fibre loading of 70 wt% significantly increases the Young's modulus and tensile strength. The fibres were well distributed in the high degree substitution-matrix and have excellent wetting between the fibres and the starch. The reinforcing effect banana and sugarcane fibers from Brazil in starch matrix is analysed by Guimarães et al. [105]. The Young's modulus increased by 186%, 294% and 201% over the matrix for banana fiber contents of 20%, 25% and 35%, respectively, while the ultimate tensile strength remained unchanged. The yield strength increased by about 129%, 141% and 133% for 20, 25 and 35 wt.% fiber contents, while the% elongation decreased about fivefold for 20 wt.%, sixfold for 25 wt.% and sevenfold for 35 wt.% fiber content.

6.2 Thermal Stability

Similar to matrix, in the sisal/wheat flour based thermoplastic [98], an initial mass loss (d_{m1}) has been observed for the composites during heating. Initial mass loss (below 200°C) for the composites is slightly higher. This mass loss is associated with the evaporation of both the glycerol and the moisture content associated in the matrix and fibers respectively. The second mass loss (d_{m2}) occurs between 200 and 500°C. This mass loss corresponds to the degradation of the polymeric matrix as well as the sisal fiber and varies between 70 and 74% w/w. The maximum mass loss (T_m) occurs around the temperature region of 303°C which is due to the degradation of matrix as well as fibers. A small peak of mass loss between 400 and 500°C is observed. The magnitude of this peak increases with sisal fiber content. The values of T_{donset}, d_{m1} and $d_{mresidue}$ (% w/w) at 700°C are given in the Table 14.

Table 14. Thermal properties of composites having sisal fiber content 0, 1, 3, 5 and 10%w/w [98].

Sisal fiber (% w/w)	0	1	3	5	10
Temperature at which degradation starts, $T_{d\ onset}$ (°C)	279	280	274	275	274
Maximum degradation temperature, T_m (°C)	304	306	304	303	303
First mass loss, d_{m1} (% w/w)	6	11	12	13	13
Residual mass, $d_{m\ residue}$ (% w/w)	11	14	13	11	12

The thermogravimetric results obtained are relatively closer for each sample which indicates that higher fiber content does not make a significant change in the thermal stability of the composites. Similar trend was observed for the flax/wheat flour starch thermoplastic by Saiah et al. [99] (Figure 35).

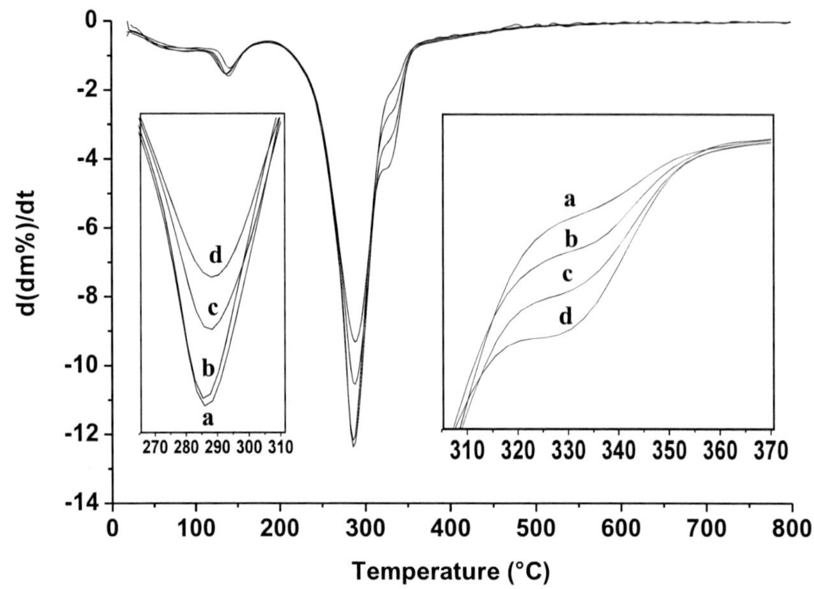

Figure 35. DTG curves of flax fibre reinforced thermoplastic composites having various fibre content (a = 5% W/W, b = 10% W/W, c = 15% W/W, d = 20% W/W) [99].

In the cotton fiber/starch thermoplastic, second mass loss occurred between 250 and 400°C [101]. However, Averous et al. [106] reported that the addition of agro-fillers to wheat starch based films, improves the thermal resistance of the biocomposites. Curvelo et al. [102] showed that mass loss, at the onset temperature, is 30 and 23% for thermoplastic starch and it's composites. This difference is due to the differences in equilibrium moisture content of each sample. Soykeabkaew et al. [107] showed that the flexural strength for all of the tapioca starch based foams prepared exhibited a similar dependence on the moisture content. It increased with increasing moisture content up to around 7–9% where the flexural strength reached a maximum and then decreased with further increase in the moisture content. The reinforcing effect of the fibers was found to increase with increasing fiber content and fiber aspect ratio, with jute fibers providing more improvement to the flexural strength of the composites than flax fibers did. The composites with flax fibers being oriented in the longitudinal direction showed a dramatic improvement in the flexural strength and the flexural modulus of elasticity.

6.3 XRD Analysis

X-ray diffraction analysis gives us more information regarding the variation of the crystallinity of the material during the addition of fiber to the matrix. Saiah et al [99] studied the variation of the crystallinity in the flax/wheat flour based starch composites. Figure 36 represents that XRD pattern of flax fiber reinforced thermoplastic composites having various fiber content

Here the peaks appear at 2θ = 7.28°, 12.98°, 15.18°, 16.88°, 19.88°, 22.78°, and 34.48°. The peak values are associated with the characteristics peaks of V_h structure of the matrix and the crystalline structure of flax fibers (cellulose type I). The V_h structure of the matrix was obtained by complexation of amylose with lipids. The models suggest that the chain conformation consists of six left-handed residues per turn helix with a rise per monomer between 0.132 and 0.136 nm. This structure is characterized by an orthorhombic unit cell (a = 1.37 nm, b = 2.37 nm, c = 0.805 nm) with the space group $P2_12_12_1$ and 16 water molecules within the unit cell. As expected, the intensity of the peaks associated to flax fibers (2θ = 15.18°, 16.88°, 22.78°, and 34.48°) increases with the fiber content. However, the intensity of the peaks at 2θ = 7.28°, 12.98°, and 19.88° associated with the V_h crystalline structure decreases due to the addition of fibers.

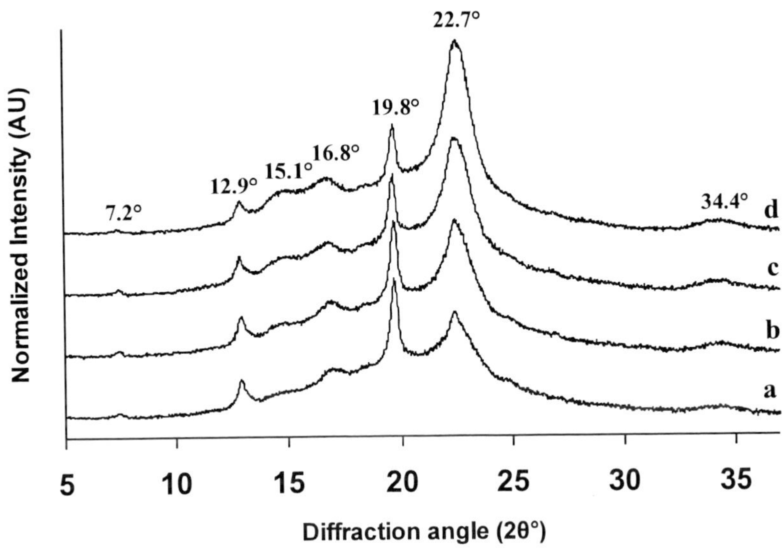

Figure 36. X-ray diffraction diagram of flax fiber reinforced thermoplastic composites having various fiber content (a = 5% w/w, b = 10% w/w, c = 15% w/w, d = 20% w/w [99]).

The banana and sugarcane fiber reinforced starch composites shows lower crystallinity (20–21%) compared to that of banana fibers (39%), due to the plasticization of starch. Also the diffraction peak between 5° and 20° disappears as a consequence of the destruction of starch's crystalline region [105]. In the cotton fiber/starch composites also peaks appeared at 2θ = 7.2°, 12.9°, 15.1°, 16.8°, 19.8°, 22.7° and 34.4° [101]. Except the peak at 2θ = 22.7°, all the other peaks correspond to the V_h structure of the matrix. The intensity of this peak increases with the cotton fiber content. The superimposition of the X-ray diagrams shows that the signal characteristics of the matrix are quasi similar whatever the fiber content and the percentage of crystallinity were around 14%. It is also clear that the incorporation of cotton fiber to the matrix does not influence significantly the crystalline phase of the matrix.

6.4 Contact Angle Analysis

Incorporation of natural fiber to the matrix, which is hydrophilic in nature, can affect the nature of the composites. Contact angle measurements can give a clear picture regarding the nature of the polymer composites. Sreekumar et al. [98] studied the variation of the contact angle for the composites using the three probe liquids (Figure 37).

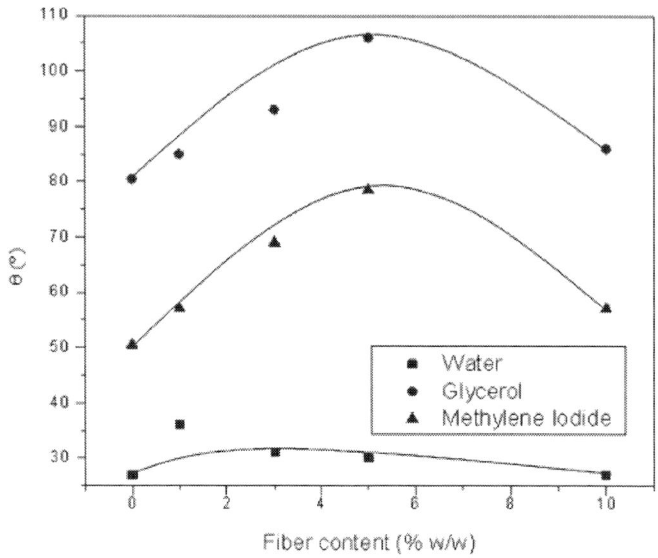

Figure 37. Variation of the contact angle of the composites having sisal fiber content (0, 1, 3, 5, and 10%w/w)[98].

The contact angle of glycerol and methylene iodide increases as a function of fiber content up to 5% w/w and then decreases. Even though glycerol has affinity with the matrix, due to the homogenous dispersion of fiber in the matrix it's not easy for the probe liquid to penetrate into the matrix. Therefore as fiber content increases the contact angle value increases up to 5% w/w. When water is the probe liquid the contact was the lowest and did not vary much with sisal fiber loading. This is because the addition of sisal fiber to the matrix makes its more hydrophilic in nature. The presence of -OH groups of starch macromolecules and sisal fiber forms hydrogen bond between water molecules. Hence the spreading occurs very quickly and gave the lowest contact angle value.

6.5 Viscoelastic Properties

Dynamic mechanical analysis is a good technique to study the variation of viscoelastic properties of the materials for a wide range of temperature. For the sisal fiber reinforced starch composites [108], the storage modulus value (E') value varies as a function of fiber content. At lower temperature i.e., in the glassy state, the molecular chains are in a frozen state and are highly immobile. This imparts good stiffness to the composite, hence higher values for the storage modulus. As temperature increases the polymeric chains become more mobile and lose their close packing arrangement. Therefore in the rubbery region E' values are lower. Above 30°C for the composites containing 10% w/w of fiber the storage modulus value is 390 MPa, i.e. an increment of 42% when compared to the matrix. This is due to the presence of strong interactions between sisal fibers and matrix through hydrogen bonds which leads to the formation of a rigid network governed by the percolation threshold, besides the geometry and stiffness of fibers However when the fiber content of matrix increases from 5 to 10% w/w there is not very much variation in the storage modulus due to the strong tendency for fibre-fibre interaction. In the tan δ curve two peaks were observed, i.e. in the glycerol rich phase and another one in starch rich phases and the peak height of tan δ_{max} decreases when compared to matrix (Figure 38).

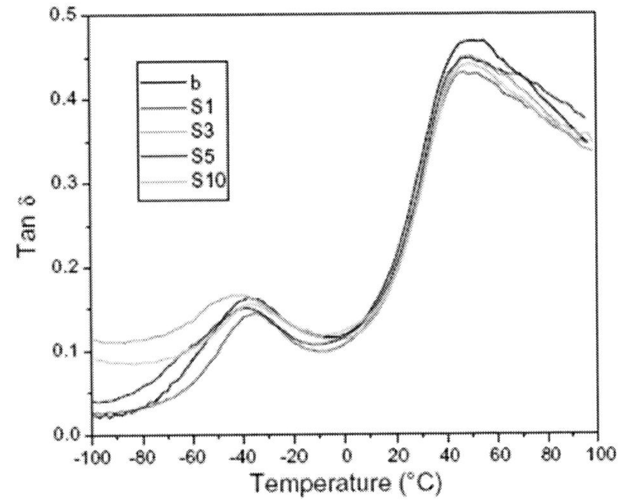

Figure 38. Variation of damping factor of the thermoplastic materials as a function of sisal fiber content (0%, 1%, 3%, 5% and 10% w/w) [108].

Incorporation of sisal fiber is expected to influence the T_g of the matrix by restricting the chain movement of polymer chains, leading to lower flexibility, lower degree of molecular motion, and hence lower damping characteristics. The addition of fibers affects more on the damping nature rather than the glass transition temperature. The activation energy calculated from tan curves at lower temperature region ranges from 142 kJ/mol to 127 kJ/mol while that at higher temperature region from 191 kJ/mol to 234 kJ/mol respectively. Averous et al.[106] fabricated cellulose fiber starch composites by varying the matrix formulation, fibres length (from 60 mm to 1 mm), filler content (till 30 wt%) and fibres nature. The fibres show distinct surface tensions, cellulose fibres are more polar compared to lignocellulose fibres. Compared to the matrix, DMTA analysis shows on these biocomposites some important variations in the main relaxation temperature, which can be linked to interactions resulting in a decrease of starch chain mobility and to a regular reinforcing effect. These results are consistent with the variation of static mechanical (tensile test) behaviour. Comparison between lignocellulose and cellulose fibres seems to show, that for these latter the interface adhesion with the starchy matrix is higher.

6.6 Water Absorption

In order to improve starch-based film characteristics, many researches reported results on the addition of natural fibers as a suitable reinforcing component for thermoplastic materials.

Most of these works focused on films' mechanical properties and have showed that fibers incorporation increases films' tensile strength and elasticity modulus and decreases their elongation capacity. Water sensitivity is another important criterion for many practical applications of starch products. Since cellulose and starch are hydrophilic in nature it's very essential to investigate the barrier properties of starch composites. Concerning to the barrier properties, Carmen et al. [109] studied the water vapor permeability effect of the soft wood short fibers having length of 1.2 mm long with 0.1 mm of diameter on cassava starch. The films water vapor permeability have strong dependency on solubility and water diffusion coefficients and, consequently, on relative humidity gradient range, presenting values up to 2–3 times greater at 33–64% than at 64–90%, depending on film formulation. The reinforced films present lower water vapor permeabilities when compared with starch films without fibers. Also Funke et al. [110] reported that incorporation of fiber reduces the moisture sorption of the composites. Since starch is more hydrophilic than cellulose and the fibers absorb part of the glycerin. This results in a less hydrophilic matrix, since plasticized starch is increasingly sensitive to water uptake the higher the glycerin content. The addition of lignin can reduce the overall water affinity of thermally molded films [111].

7. Nano Fiber Reinforced Starch Composites

Modification of the biodegradable polymers through innovative technology is a challenge for materials scientists. Adding nano-reinforcement to pristine polymers to preparing nanocomposite has already proven to be an effective way to improve these properties concurrently [112]. The pioneer work on nanocomposites that was initiated by researchers at Toyota in the early 1990's created nanoclay reinforced polymers, opening a new research path on composites.

Nanocomposites with starch as one component show unique properties, because of the nanometric size effect, compared to conventional composite even at low filler content [113]. Nanofillers have strong reinforcing effects, and studies have also shown their positive impact in barrier packaging, mechanical properties, flame retardancy and chemical properties. Polysaccharides are good candidates for renewable nanofillers because they have partly crystalline structures conferring interesting properties. Recent reviews have been published on cellulose nanocrystals and cellulose whiskers and their nanocomposites which are by far the most studied polysaccharide for nanoparticles [114-116]. However, as far as we know, nothing similar has been done for starch nanocrystals except for mentioning in book chapters [117, 118]. For these reasons, starch nanoparticles and nanocomposites will be presented in detail with a complete overview of their preparation, characterization, and applications.

One of the most wide sprayed classifications of nanoparticles is made according to particle shape: (i) Particulate nanocomposites, such as metallic nanoparticles or carbon black-reinforced composites, are generally iso-dimensional and show moderate reinforcement due to their low aspect ratio. They are used to enhance resistance to flammability and decrease permeability or costs. (ii) Elongated particles that show better mechanical properties thanks to their high aspect ratio. Such particles include cellulose nanofibrils (also called whiskers or nanocrystals) and carbon nanotubes. (iii) Layered particles, like nanoclays, also referred to as LPN: layered polymer nanocomposites. This latter family is the most used industrially and can show different degrees of dispersion, as shown in Figure 39, namely, intercalated nanocomposites (intercalated polymer chains between layered nanocomposites), exfoliated nanocomposites (separation of individual layers), and flocculated or phase-separated nanocomposites, which are also called microcomposites and consequently show the poorer physical properties (1). Exfoliation is sought by nanocomposite producers as it gives, by far, the best results.

Intercalated Flocculated Exfoliated

Figure 39. Schematic diagram of intercalated, exfoliated and flocculated nanocomposites.

After resolving the incompatibility issues between natural fiber and biodegradable polymer up to a satisfactory extent by adequate modification either in host or filler, it was assumed that dispersion of natural fiber may be enhanced with reduction in the size through introducing nano fillers for positive tailoring of different material properties [119, 120]. Cellulose chains are biosynthesized by enzymes, deposited in continuous fashion and aggregate to form microfibrils and ultimately aggregate into long thread like bundles of molecules stabilized laterally by hydrogen bonds between hydroxyl group and oxygen of adjacent molecules. The extended chain conformation and microfibrilar morphology result in significant load carrying capability. Depending on their origin, microfibrils diameter ranges from about 2 to 30 nm for length that can reach several tens of microns (figure 40) [119,120].

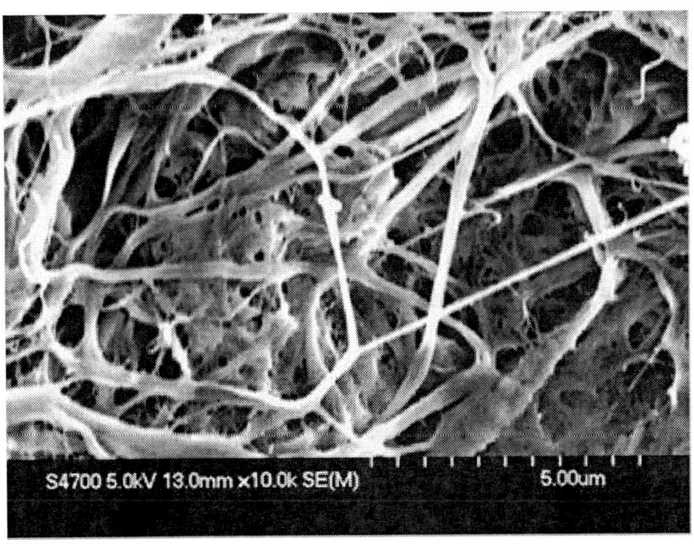

Figure 40. SEM micrograph of cellulose nanofibers [120].

Each microfibril can be considered as string of cellulose crystals linked with amorphous domains. The hydronium ions can penetrate the cellulose chains in these amorphous domains enhancing the hydrolytic cleavage of the glycosidic bonds and releasing individual crystallites [119]. The different treatments of these charged microcrystallites, such as mechanical dispersion or ultrasonification, permit the dispersion of the aggregates and finally produce colloidal suspensions. Because of their thickness, and length, these rod particles are commonly called ''whiskers''. Typical morphology of cellulose nanofiber is given in figure 40. They display a web-like, nano-scale network microstructure [120].

7.1 Cellulose Reinforced Starch Composites

Various types of cellulosic reinforcements have been used and tested in natural polymers such as potato pulp-based microfibrils [121, 122] bleached leaf wood fibers [123], bleached eucalyptus pulp fibers [124], flax fibers and ramie fibers or crystallites [125], tunicin whiskers [126-127] etc. A high compatibility occurs between starch matrix and cellulose nanofibrils with enhanced performance (e.g., mechanical properties and water sensitivity) due to 3D hydrogen bonds network formed between different components.

Nano-biocomposites are organic/inorganic hybrid biomaterials composed of nano-sized fillers (nanofillers) incorporated into a biopolymer matrix. Depending on the nano filler chosen, the nanocomposite materials could exhibit drastic modifications in their properties, like improved mechanical properties, barrier properties, or a change in their thermal and electrical conductivity [128]. Such property enhancements rely both on the nanofiller geometry, on the nanofiller surface area (e.g., 700 m^2/g for the montmorillonite when the nanofiller is fully exfoliated) and nanofiller surface chemistry [129]. According to the literature, three types of nanoparticles have been incorporated into nanocomposites: i) whiskers obtained from cellulose; (ii) nanocrystals from starch; and iii) natural or organo-modified nanoclays. Here we are dealing with the first two only as the third has been reviewed by several authors.

7.2 Whisker-Based Nano-Biocomposites

Through a long pre-treatment, whiskers can be isolated from their original biomass through acid hydrolysis with concentrated mineral acids under strictly controlled conditions of time and temperature [130]. Acid action results in a decrease of the amorphous parts by removing polysaccharide material closely bonded to the crystallite surface and breaks down portions of glucose chains in most accessible, noncrystalline regions and by acid hydrolysis of cellulosic materials. Although chitin can be used, whiskers are typically cellulose monocrystals. Some authors used tunicin (seafood cellulose) whiskers, which are slender, parallelepiped rods of 500 nm to 1–2 μm length and 10 nm width, into polymer matrixes [126, 127, 130, 131]. The whiskers–matrix interactions are important and the high shape ratio of the nanoparticles (50–200) and the high specific area ($\approx$170 m^2/g) increase the interfacial phenomena. Compared to the common cellulose macrocomposites, the global behavior of nanowhisker based material is primarily driven by the matrix/nanofiller interface, which in turn controls the subsequent performance properties (mechanical properties, permeability). For instance, tunicin whiskers favor starch crystallization because of the nucleating effect of the nanofiller [127]. Typical stress strain curve of the starch nanocomposites is given in figure 41. The addition of the cellulose whiskers improves the modulus and tensile strength of the composites.

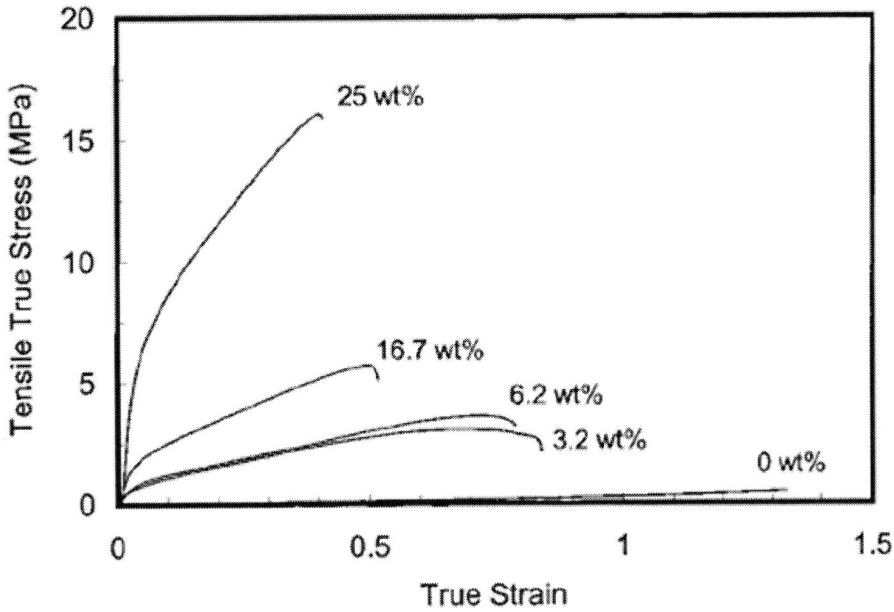

Figure 41. Typical true stress vs true strain curves of glycerol plasticized starch/tunicin whiskers composites conditioned at 35% RH. The tunicin whiskers contents are indicated in the figure [128].

7.3 Starch Nanocrystals-Based Nano-Biocomposites

Waxy maize starch nanocrystals were obtained by acid hydrolysis of native granules by strictly controlling the temperature of the process, the acid and starch concentrations, the hydrolysis duration and the stirring speed. Waxy maize starch nanocrystals consist of 5–7 nm thick, platelet-like particles with a length ranging from 20 to 40 nm and a width in the range 15–30 nm. They were used as a reinforcing agent in a waxy maize starch matrix plasticized with glycerol, where Angellier *at al.* have shown that the reinforcing effect of starch nanocrystals can be attributed to strong filler/filler and filler/matrix interactions due to the establishment of increased hydrogen bonding [122]. The presence of starch nanocrystals leads to a slowing down of the recrystallization of the matrix during aging in high humidity atmospheres [122].

The starch and cellulose nanowhisker based nanocomposites has potential applications and is a challenging research topic now. The table 15 shows the important research works so far carried out in this area [130, 132-137, 140-145].

Table 15. Important research works so far carried out in the cellulose nanowhisker reinforced starch composites [130, 132-137, 140-145].

Matrix	Reinforcement	Processing	Characteristics	Ref
Gelatinized Potato starch	Cellulose microfibrils from potato	Dispersion with aqueous suspension	Strong increase in the thermomechanical stability can be obtained with only a few percent of filler, whereas the water sensitivity linearly decreases with the cellulose microfibril content.	[122]
Sorbitol plasticized waxy maize starch	Tunicin whiskers	Stirring in a preheated autoclave reactor and casting in a Teflon mold	One step decrease of storage modulus and a loss angle tangent peak. For all RH levels, the modulus increased gradually with filler load. The tensile strength and Young's modulus are high at lower RH levels, and elongation at break remains constant.	[130]
Starch	Potato cellulose microfibrills 10%	Casting with gelatinized starch (water/glycer ol) and fibers suspension	Thermal stabilization and decreasing of water uptake.	[132]
Corn starch	Cellulose nano-whiskers	Solution casting, evaporation and application of magnetic field	Oriented nano-whiskers, directional properties exhibited.	[133]
Thermoplastic starch, corn starch–glycerol (30–50 wt%) 1–15 wt%	Cellulose fibers	Intensive batch mixer at 100–150 °C	Characterization by HP size exclusion chromatography; glycerol content reduces starch degradation while increased fiber content increased degradation. Lower polydispersity index for matrix than that of native starch due to shear induced fragmentation of selective breakage of large amylopectin molecules.	[134]
Corn starch	Cellulose fiber (2–45 wt%)	Heating the mixture inside a closed mold	Composite foams formed; tensile strength increased with increasing fiber content up to 15 wt% fiber, then up to 30 wt% fiber no significant change observed; lower strength for still higher fiber content due to non-uniform distribution of fiber.	[135]
Pea starch	Pea hull fiber nanowhiskers	Solution casting and evaporation process	Nanocomposite films showed higher UV absorption, transparency, tensile strength, elongation at break, water-resistance.	[136]
Potato starch, glycerol as plasticizer	Potato cellulose microfibrils	Composite processed from potato cellulose microfibril suspension, blending and casting followed by water evaporation.	The mechanical properties and water absorption behavior of the resulting films were investigated, and differences were observed depending on the glycerol, cellulose microfibrils, and relative humidity content. water sensitivity linearly decreases with the cellulose microfibril content.	[137]

Starch based resin	Cellulose microfibril nanofibers	Stirrer-treatment and hot pressing	Flexural strength and flexural modulus increased with increasing molding pressure. Strength and modulus showed good correlation with their density. Stirrer mixing process is effective, yielding uniform dispersion of nanofibers.	[138]
Thermo plastic starch (TPS)	Cellulose microfibrils from wheat straw	Dispersion using high shear mixer in varying proportions and solution casting to make films	Characterized using AFM, TEM, SEM, TGA, FTIR and WAXRD. XRD and TGA results confirmed the crystalline nature of nanofibrils. TGA depicted an increasing in residue left with increase in cellulose nanofibrils content. Mechanical properties increased with nanofiber concentration. Barrier properties also improved.	[120]
Wheat starch plasticized with glycerol	Bacterial cellulose nanofibers	Solution impregnation method	Possess higher tensile strength and modulus than the unreinforced starch. tensile strength of the bio-composites deteriorates after water absorption. Tensile strength decreases drastically upon exposure to microorganism attacks. They are fully biodegradable.	[139]
Starch	Cellulose producing bacterial suspension	Hot pressing	Morphology revealed the starch as matrix.	[140]
Thermoplastic starch from potato starch	Cellulose microfibrils from wheat straw	Solution casting	Mechanical and thermal properties improved. T_g shifted to higher temperature.	[141]
Thermoplastic starch glyceryl monostearte as surfactant	Cellulose microfibrils	Solution casting	Mechanical and thermal properties improved. Crystallinity increased.	[142]
Thermoplastic starch	Suspension of cellulose microfibrils	Melt mixing	Improvement in tensile strength and stiffness.	[143]
Potato starch	Cellulose nanofiber suspension	Solution casting	A well-dispersed nanofiber network reduced the moisture uptake of the composite to half the value of the pure plasticized starch film.	[144]

From the table it can be easily understood that starch based bio nanocomposites is a challenging research topic and many advances are taking place in the recent years. So far we are reporting the works performed on the processing and behavior of new nanocomposite materials of starch reinforced by polysaccharide microcrystals, nanofibrils, microfibril suspensions and nano whiskers can be considered as an effort aimed at providing further knowledge to a research area presenting yet a variety of pending issues. It was shown that the use of high aspect ratio cellulose whiskers induces a mechanical percolation phenomenon leading to outstanding and unusual mechanical properties through the formation of a rigid filler network. In addition to some practical applications, the study of such model systems can help to understand some physical properties as geometric and mechanical percolation effects. Cellulose can also be used as microfibrillar filler, which is more accessible in terms of available amounts and preparation. The mechanical behavior is then very sensitive to the cellulose purification level and cellulose microfibrils individualization state.

The subject generates substantial interests, and the variety of possible sources for preparing cellulosic nanowhiskers and nanocrystals nanocrystals are available.

Polysaccharide nanocrystals can also be obtained from other abundant renewable resources such as chitin and starch by acid hydrolysis. For the former, they appear as rodlike particles with an aspect ratio related to the origin of the chitin, whereas for the latter, the nanoparticles consist of platelets with nanometer scale dimensions. Practical applications of such fillers and transition into industrial technology require a favorable ratio between the expected performances of the composite material and its cost. There are still significant scientific and technological challenges to take up in this field of green composites based on starch.

References

[1] Ma, X. F.; Yu, J. G.; Ma, Y. B. *Carbohydr. Polym.* 2005, 60, 111-116.
[2] Fishman, M. L.; Coffin, D. R.; Konstance, R. P.; Onwulata, C. I. *Carbohydr. Polym.* 2000, 41, 317-325.
[3] Sreekumar, P. A.; Leblanc, N.; Gattin, R.; Saiter, J. M. *Compos Part A: Appl Sci and Manuf.* 2010, 41, 991- 996.
[4] Pagga, U.; Beimborn, D.B.; Yamamoto, M. *J. Environ. Polym. Degrad.* 1996, 4(3), 173-178.
[5] Battersby, N. S.; Pack, S. E.; Watkinson, R. J. *Chemosphere.* 1992, 24, 1989-2000.
[6] Buchanan, C. M.; Gardner, R. M.; Komarek, R. *J. Appl. Polym. Sci.* 1993, 47, 1709-1719.
[7] Swift G., *"Biodegradable Plastics and polymers"* Doi Y. and Fukuda K. (eds.). Elsevier, Amsterdam. 1994, 228-236.
[8] Abou-Zeid, D.M.; Muller, R. J.; Deckwer, W. D. *J. Biotechnol.* 2001, 86, 113-126.
[9] Kumar, M. S.; Mudliar, S. N.; Reddy, K. M. K.; Chakrabarti, T. *Bioresour. Technol.* 2004, 95, 327-330.
[10] Gontard, N.; Guilbert, S. *"Food packaging and preservation",* Ed. M. Mathlouti, Blackie Academic and Professional, Glasgow. 1994, 159.
[11] Bastioli, C. *Polym. Degrad. Stab.* 1998, 59, 263-272.
[12] Duprat F., Gallant D., Guilbot A., Mercier C., Robin J. P., *"Les polymères végétaux",* Ed. B. Monties, Gauthier-Villars, 1980
[13] Manners, D. *J. Carbohydr. Polym.* 1989, 11, 87-112.
[14] Robin, J. P.; Mercier, C.; Duprat, F.; Charbonnière, R.; Guilbot, A. *Starch/Starke.* 1975, 27, 36-45.
[15] Hizukuri, S. *Carbohydr. Res.* 1986, 147, 342-347.
[16] Katz, J. R. Z. *Phys. Chem.* 1930, 150, 37-59.
[17] Zobel, H. F. *Starch/Stärke.* 1988, 40, 1-7.
[18] Colonna, P. ; Buléon, A. ; Mercier, C.; Lemaguer, M. *Carbohydr. Polym.* 1982, 2, 43-59.
[19] Biliaderis, C. G. *Food Technol.* 1992, 46, 98-145.
[20] Colonna, P.; Buléon, A.; Mercier, C. *J. Food. Sci.* 1981, 46, 88-93.
[21] Robin, J. P.; Mercier, C.; Charbonniere, R.; Guilbot, A. *Cereal Chem.* 1974, 51, 389-406.
[22] Gidley, M. J.; Bociek, S. M. *J. Am. Chem. Soc.* 1985, 107, 7040-7044.
[23] Cooke, D.; Gidley, M. *Carbohydr. Res.* 1992, 227, 103-112.
[24] Imberty, A.; Buléon, A.; Tran, V.; Pérez, S. *Starch/Starke.* 1991, 43, 375-384.

[25] Wu, H. C. H.; Sarko, A. *Carbohydr. Res.* 1978, 61, 7-25.

[26] Cameron, R. E.; Donald, A. M. *J. Polymer. Sci. Part B: Polymer Phys.* 1993, 31, 197-198.

[27] Jenkins, P. J.; Donald, A. M. *Int. J. Biol. Macromol.* 1995, 17, 315-321.

[28] Lindeboom, N.; Chang, P. R.; Tyler, R. T. *Starch/Stärke*, 2004, 56, 89-99.

[29] Tang, H.; Ando, H.; Watanaba, K.; Takeda, Y.; Mitsunaga, T. *Carbohydr. Res.* 2001, 330, 241-248.

[30] Jane, J. L.; Shen, L.; Wang, L.; Maningat, C. C. *Cereal Chem.* 1992, 69, 280-283.

[31] Jane, J. L.; Kasemsuwan, T.; Leas, S.; Ia, A.; Zobel, H.; Il, D.; Robyt, J. F. *Starch/Stärke.* 1994, 46, 121-129.

[32] Evers, A. D. *Starch/Stärke.* 1973, 25, 303-304.

[33] Zheng, G. H.; Sosulski, F. W. *Physicochemical properties of small granule starches,* AACC Annual Meeting, San Diego, 1997.

[34] Oates, C. G.; Powell, A. D. *Food Chem.* 1996, 56, 405-414.

[35] Bultosa, G.; Hall, A. N.; Taylor, J. R. N. *Starch/Stärke,* 2002, 54, 461-468.

[36] Kumari, S. K.; Thayumanavan, B. *Plant Foods for Human Nutr.* 1998, 53, 47-56.

[37] Wang, L.; Wang, Y. J.; Porter, R. *J. Agric. Food Chem.* 2002, 50, 2695-2699.

[38] Abdel Aal, E. S. M.; Hucl, P. J.; Sosulski, F. W. *J. Agric. Food Chem.* 1997, 45, 3049-3055.

[39] Biliaderis, C. G.; Mazza, G.; Przybylski, R. *Starch/Stärke,* 1993, 45, 121-127.

[40] Goering, K. J.; DeHaas, B. *New starches. VIII. Properties of the small granule-starch from Colocasia esculenta.* AACC. 1972, 712-719.

[41] Leblanc, N.; Saiah, R.; Beucher, E.; Gattin, R.; Castandet, M.; Saiter, J. M. *Carbohyd. Polym.* 2008, 73, 548-557.

[42] Wambua, P.; Ivens, J.; Verpoest, I. *Compos. Sci. Technol.* 2003, 63, 1259-1264.

[43] Rowell, R.: *Paper and composites from agro-based resources*, edited by Rowell, R. M.; Young, R. A.; Rowell, J. K.: CRC press/Lewis, Boca Raton, 1996.

[44] Hagstrand, P. O.; Oksman, K. *Polym. Compos.* 2001, 22, 568-578.

[45] Oever van den, M. J. A.; Bos, H. L.; Kemenade Van, M. J. J. M. *Appl. Compos. Mater.* 2000, 7, 387-402.

[46] Bledzki, A. K.; Gassan, J. *Progr. Polym. Sci.* 1999, 24, 221-274.

[47] Nevell, T. P.; Zeronian, S. H. *Cellulose chemistry and its applications,* Wiley, New York, 1985.

[48] Fink, H. P.; Ganster, J.; Fraatz, J.: *Akzo-Nobel viskose chemistry seminar challenges in cellulosic man-made fibres,* Stockholm, 30 May–3 June 1994.

[49] Bledzki, A. K.; Reihmane, S.; Gassan, J. *J. Appl. Polym. Sci.* 1996, 29, 1329-1336.

[50] Pothan, L. A.; Thomas, S. *Compos. Sci. Technol.* 2003, 63, 1231-1240.

[51] Uma Devi, L.; Bhagawan, S. S.; Thomas, S. *J. Appl. Polym. Sci.* 1997, 64, 1739-1748.

[52] Chand, N.; Rohatgi, P. K.: in: *"Natural Fibres and Composites",* Periodical Experts Agency, Delhi, India, 5, 1994.

[53] Saechtling, H.: in *"International Plastics Handbook",* Hanser, Munich, 1987.

[54] Kamath, V. R.; Gallagher, R. B.: in *"Developments in Reinforced Plastics-1"* edited by G. Pritchard, Applied Science Publishers, London. 1980, 362.

[55] Agassant, J. F. ; Avenas, P. ; Sergent, J. P. ; Vergnes, B. ; Vincent, M. *"La mise en forme des matières plastiques",* Tec et Doc, Lavoisier, Paris. 1996, 613.

[56] Vergnes, B.; Della Valle, G.; Delamare, L. *Polym. Eng. Sci.* 1998, 38, 1781-1792.

[57] Fan, J.; Mitchell, J. R.; Blanshard, J. M. V. *J. Food Sci. Technol.* 1996, 31, 67-76.
[58] Barron, C.; Della Valle, G.; Colonna, P.; Vergnes, B. *J. Food Sci.* 2002, 67, 1426-1437.
[59] Leach, H. W.; Schoch, T. V. *Cereal Chem.* 1961, 38, 34-46.
[60] Martin, O.; Avérous, L.; Della Valle, G. *Carbohydr. Polym.* 2003, 53, 169-182.
[61] Godet, M. C.; Buléon, A.; Tran, V.; Colonna, P. *Carbohydr. Polym.* 1993, 21, 91-95.
[62] French, A. D.; Zobel, H. F. *Biopolymers*, 1967, 5, 457-464.
[63] Heinemann, C.; Conde-Petit, B.; Nuessli, J.; Escher, F. *J. Agric. Food. Chem.* 2001, 49, 1370-1376.
[64] Buléon, A.; Delage, M. M.; Brisson, J.; Chanzy, H. *Int. J. Biol. Macromol.* 1990, 12, 25-33.
[65] Hinkle, M. E.; Zobel, H. F. *Biopolymers.* 1968, 6, 1119-1128.
[66] Polaczek, E.; Starzyk, F.; Tomasik, P. *Carbohydr. Polym.* 1999, 39, 37-42.
[67] Le Bail, P.; Bizot, H.; Pontoire, B.; Buléon, A. *Starch/Stärke.* 1995, 47, 229-232.
[68] Paris, M.; Bizot, H.; Emery, J.; Buzaré, J. Y.; Buléon, A. *Int. J. Biol. Macromol.* 2001, 29, 127-136.
[69] Winter, W. T.; Sarko, A. *Biopolymers*, 1974, 13, 1461-1482.
[70] Zaslow, B.; Murphy, K. V. G.; French, A. D. *Biopolymers.* 1974, 13, 779-790.
[71] Winter, W. T.; Sarko, A. *Biopolymers.* 1974, 13, 1447-1460.
[72] Rappenecker, G.; Zugenmaier, P. *Carbohydr. Res.* 1981, 89, 11-19.
[73] Hinkle, M. E.; Zobel, H. *Biopolymers.* 1968, 6, 1119-1128.
[74] Murphy, V. G.; Zaslow, B.; French, A. D. *Biopolymers.* 1975, 14, 1487-1501.
[75] Van Soest, J. J. G.; Hulleman, S. H. D.; De Wit, D.; Vliegenthart, J. F. G. *Indus. Crops and Prod.* 1996, 5, 11-22.
[76] Zaslow, B.; Miller, R. L. *J. Am. Chem. Soc.* 1961, 83, 4378-4381.
[77] Buléon, A.; Duprat, F.; Booy, F. P.; Chanzy, H. *Carbohydr. Polym.* 1984, 4, 161-173.
[78] Saiah.,R. *Properties physiques et comportment dans le temps des agro-materiaux vitreaux issus de la farine de ble*, Ph.D. Thesis, University of Rouen, France. 2007.
[79] Hulleman, S. H. D.; Janssen, F. H. P.; Feil, H. *Polymer.* 1998, 39, 2043-2048.
[80] Lourdin, D.; Della Valle, G.; Colonna P. *Carbohydr. Polym.* 1995, 27, 261-270.
[81] Poutanen, K.; Forssell, P. TRIP, 1996, 4, 128-132.
[82] Gontard, N.; Guilbert, S.; Cuq, J. L. *J. Food. Sci*, 1993, 58, 206-211.
[83] Lourdin, D.; Bizot, H.; Colonna, P. *J. Appl. Polym. Sci.* 1997, 63, 1047-1053.
[84] Gaudin, S.; Lourdin, D.; Le Botlan, D.; Ilari, J. L.; Colonna, P. *J. Cer. Sci.* 1999, 29, 273-284.
[85] Van Soest, J. J. G.; Knooren, N. *J. Appl. Polym. Sci.* 1997, 64, 1411-1422.
[86] Van Soest, J. J. G.; Hulleman, S. H. D.; De Wit, D.; Vliegenthart, J. F. G. *Carbohydr. Polym.* 1996, 29, 225-232.
[87] Forssell, P. M.; Hulleman, S. H. D.; Myllarinen, P. J.; Moates, G. K.; Parker, R. *Carbohydr. Polym.* 1999, 39, 43-51.
[88] Van Soest, J. J. G.; Benes, K.; De Wit, D.; Vliegenthart, J. F. G. *Polymer.* 1996, 37, 3543-3552.
[89] Van Soest, J. J. G.; De Wit, D.; Vliegenthart, J. F. G. *J. Appl. Polym. Sci.* 1996, 61, 1927-.1937
[90] Saiah, R.; Sreekumar, P. A.; Leblanc, N. Castandet, M.; Saiter, J. M. *Cereal Chem.* 2007, 84, 276-281.

[91] Smits, A. L. M.; Kruiskamp, P. H.; van Soest, J. J. G.; Vliegenthart, J. F. G. *Carbohydr. Polym.* 2003, 51, 417-424.
[92] Saiter, J. M.; Dobircau, L.; Saiah, R.; Sreekumar, P. A.; Galandon, A.; Gattin, R.; Leblanc, N.; Adhikari, R. *Physica B.* 2010, 405, 900-905.
[93] Saiter, J. M.; Negahban, M.; dos Santos Claro, P.; Delabarbe, P.; Garda, M. R.; *J. Mater.* Ed., 2008, 30, 51-56.
[94] Lourdin, D.; Bizot, H.; Colonna, P. *J. Appl. Polym. Sci.*, 1997, 63, 1047-1053.
[95] Lu, Y. S.; Weng, L. H.; Cao, X. D. *Carbohydr. Polym.* 2006, 63, 198-204.
[96] Alvarez, V.; Vázquez, A.; Bernal, C. *Polym. Compos.* 2005, 26, 316-323.
[97] Soykeabkaew, N.; Supaphol, P.; Rujiravanit, R. *Carbohydr. Polym.* 2004, 58, 53-63.
[98] Sreekumar, P. A.; Leblanc, N.; Gattin, R.; Saiter J. M. *Polym. Compos.* 2010, 31, 939-945.
[99] Saiah, R.; Sreekumar, P. A.; Gopalakrishnan, P.; Leblanc, N.; Gattin, R.; Saiter J. M. *Polym. Compos.* 2009, 30, 1595-1600.
[100] 100. Wollerdorfer, M.; Bader, H. *Indus. Crops and Prod.* 1998, 8, 105–112.
[101] Dobircau, L.; Sreekumar, P. A.; Saiah, R.; Leblanc, N.; Gattin, R.; Saiter, J. M. *Compos. Part A Appl. Sci. Manuf.* 2009, 40, 329 -334.
[102] Curvelo, A. A. S.; de Carvalho, A. J. F.; Agnelli, J. A. M. *Carbohyd. Polym.* 2001, 45, 183-188.
[103] Gopalakrishnan, P.; Saiah, R.; Gattin, R.; Saiter, J. M. *Compos. Interf.* 2008, 15, 759–770.
[104] Duanmu, J.; Kristofer, G.E.; Rosling, A. *Compos. Sci. Technol.* 2007, 67, 3090–3097.
[105] Guimarães, J. L.; Wypych, F.; Saul, C. K.; Ramos, L. P.; Satyanarayana, K.G. *Carbohyd. Polym.* 2010, 80, 130–138.
[106] Averous, L.; Boquillon, N. *Carbohyd. Polym.* 2004, 56, 111–122.
[107] Soykeabkaew, N.; Supaphol, P.; Rujiravanit, R.; *Carbohyd. Polym.* 2004, 58, 53–63.
[108] Sreekumar, P. A.; Gopalakrishnan, P.; Leblanc, N.; Saiter, J. M. *Compos. Part A: Appl. Sci. Manuf.* 2010, 41, 991-996.
[109] Carmen, M. O. M.; Laurindo, J. B.; Yamashita, F. *Food Hydrocoll.* 2009, 23, 1328 1333.
[110] Funke, U.; Bergthaller, W.; Lindhauer, M. G. *Polym. Degrad. Stab.* 1998, 59, 293-296.
[111] Baumberger, S. ; Lapierre, C. ; Monties, B. ; Della Valle, G. *Polym. Degrad. Stab.* 1998, 59, 273-277.
[112] Ray, S. S.; Okamoto, M. *Prog. in Polym.Sci.* 2003, 28, 1539-1641
[113] Bondeson, D.; Mathew, A.; Oksman, K. *Cellulose* 2006, 13 (2), 171– 180.
[114] Azizi Samir, M. A. S.; Alloin, F.; Dufresne, A. *Biomacromol.* 2005, 6 (2), 612–626.
[115] Siqueira, G.; Bras, J.; Dufresne, A. *Biomacromol.* 2008, 10 (2), 425–432.
[116] Dubief, D.; Samain, E.; Dufresne, A. *Macromol.* 1999, 32 (18), 5765–5771.
[117] Dufresne, A. *Polymer nanocomposites from biological sources.* Biopolymers Technology; Bertolini, A. C., Ed.; Cultura Acade^mica: Sao Paulo. 2007, 59-83.
[118] Dufresne, A.; Meideiros, E. S.; Orts, W. J. *Starch-based nanocomposites. Starch: Characterization, Properties and Applications;* Bertolini, A., Ed.; CRC Press: Rio de Janeiro. 2009, Chapter 9, 205-251.
[119] Miriam, M.; Lima, S.; Vorsali, R. *Macromol. Rapid Commun.* 2004, 25(7), 771–87.
[120] Takagi, H.; Asano, A. *Compos. Part A Appl. Sci. Manuf,* 2008, 39, 685–689.

[121] Noishiki, Y.; Nishiyama, Y.; Wada, M.; Kuga, S.; Magoshi, J. *J. Appl. Polym. Sci.* 2001, 86(13), 3425–3429.
[122] Alain, D.; Danièle, D.; &Michel, R. V. *J. Appl. Polym. Sci.* 2000, 76(14), 2080–2092.
[123] Dufresne, A.; Vignon, M. R. *Macromol.* 1998, 31(8), 2693–2696.
[124] Funke, U.; Bergthaller, W.; Lindhauer, M. G. *Polym. Degrad. Stab.* 1998, 59, 293–296.
[125] Curvelo, A. A. S.; de Carvalho, A. J. F.; Agnelli, J. A. M. *Carbohyd. Polym.* 2001, 45(2), 183–188.
[126] Nattakan, S.; Pitt, S.; Rujiravanit, R. *Carbohydr. Polym.* 2004, 58(1), 53–63.
[127] Anglès, M. N.; Dufresne, A. *Macromol.* 2000, 33(22), 8344–8353.
[128] Anglès, M. N.; Dufresne, A. *Macromol.* 2001, 34(9), 2921–2931.
[129] Alexandre, M.; Dubois, P. *Mater. Sci. Eng. R-Rep* 2000, 28(1), 1–63.
[130] Azizi Samir, M. A. S.; Alloin, F.; Sanchez, J. Y.; El Kissi, N.; Dufresne, A.; *Macromol.* 2004, 37(4), 1386–1393.
[131] Mathew AP, Thielemans W and Dufresne A, *J Appl Polym Sci* 109(6):4065–4074 (2008).
[132] Mathew, A. P.; Dufresne, A. *Biomacromol.* 2002, 3(3), 609–617.
[133] Angellier, H.; Molina-Boisseau, S.; Dole, P.; Dufresne, A. *Biomacromol.* 2006, 7(2), 531–539.
[134] Kvien, I.; Oksman, K. *Appl. Phys. A: Mater. Sci. Process.* 2007, 87, 641–643.
[135] Carvalho, A.; Zambo, M.; Agnelli, *J. Polym. Degrad. Stab.* 2003, 79, 133–138.
[136] Lawton, J. W.; Shogren, R. L.; Tiefenbacher, K. F. *Indus. Crops Prod.* 2004, 19, 1–7.
[137] Yun, C.; Liu, C.; Chang, P. R.; Cao, X.; Anderson, D. P. *Carbohyd. Polym.* 2009, 76, 607–615
[138] Dufresne, A.; Dupeyre, D.; Vignon, M. R. *J. Appl. Polym. Sci.* 2000, 76, 2080–2092.
[139] Anupama, K.; Singh, M.; Verma, G. *Carbohyd. Polym.* 2010, 82, 337–345.
[140] Wan, Y. Z.; Luo, H.; He, F.; Liang, H.; Huang, Y.; Li, X. L. *Compos. Sci. Technol.* 2009, 69, 1212–1217.
[141] Grande, C. J.; Torres, F. G.; Gomez, C. M.; Troncoso, O. P.; Canet-Ferrer, J.; Martinez-Pastor, *J. Polym. Polym. Compos.* 2008, 16, 181–185.
[142] Alemdar, A.; Sain, M. *Compos. Sci. Technol.* 2008, 68, 557–565.
[143] Mondragon, M.; Arroyo, K.; Romero-Garcıa, *J. Carbohydr. Polym.* 2008, 74, 201–208.
[144] Chakraborty, A.; Sain, M.; Kortschot, M.; Cutler, S. *J. Biobased Mater. Bioenergy* 2007, 1, 71–77.
[145] Svagan, A. J.; Hedenqvist, M. S.; Berglund, L. *Compos. Sci. Technol.* 2009, 69, 500–506.

In: Thermoplastic and Thermosetting Polymers and Composites ISBN: 978-1-61209-264-5
Editors: Linda D. Tsai and Matthew R. Hwang

Chapter 3

THERMOPLASTIC AND THERMOSETTING COMPOSITES WITH NATURAL FIBERS

***Daniella R. Mulinari*[1]*, Clodoaldo Saron*[1]*, Kelly C. C. Carvalho*[2] *and Herman J. C. Voorwald*[2]**

[1]Volta Redonda Center University, Department of Engenering, UNiFoa, Volta Redonda/RJ, Brazil

[2]Fatigue and Aeronautic Materials Research Group, Department of Materials and Technology, UNESP/FEG, Guaratinguetá/SP, Brazil

ABSTRACT

Nowadays, great attention has been dedicated to the development of natural fiber reinforced composites. Natural fibers provide with interesting properties the final composite, especially those related to environment the protection such as their capacity to be recyclable, renewable raw material, and less abrasive and harmful behavior. Some advantages associated to the use of natural fibers as reinforcement in plastics are their non-abrasive nature, biodegradability, low energy consumption, low cost, low density and high specific properties. The specific mechanical properties of natural fibers are comparable to those of traditional reinforcements.

However, certain drawbacks such as incompatibility with a hydrophobic polymer matrix, the tendency to form aggregates during processing and poor resistance to moisture greatly reduce the potential of natural fibers to be used as reinforcement in polymers. On the other hand, various treatments are being used to improve fiber-matrix compatibility. This process is considered critical as development phase of these materials due to strong interfiber hydrogen bonding, which holds the fibers together. Methods for surface modification can be physical or chemical according to superficial modification approach of the fiber. Others frequently used treatments are bleaching, acetylation and alkali treatment.

In this chapter, the main results presented in the literature are summarized, focusing attention on the properties in terms of physical and chemical structure of natural fibers, thermal and mechanical properties, processing behavior and final properties of natural fibers with thermoplastics and thermosetting matrixes paying attention to the use of physical and chemical treatments for the improvement of fiber-matrix interaction.

1. INTRODUCTION

Natural fibers can be classified according to the vegetal, animal or mineral obtainment source. Vegetal fibers are constituted mainly by cellulose, lignin and hemicellulose and are the most commercially important fibers with potencial to be applied in composite materials. The use o vegetal fibers as reinforcement in polimeric matrices have been studied for many years. Nowadays due to economical and environmental considerations, research projects are related to the properties and characteristics of natural fibers in order to use them in a large number of applications. Some of them are: as reinforcement in composite materials, in automobile industry, in civil construction and, as filter to retain heavy metals [1,2,3,4,5].

The chemical composition of vegetable fibers, also called as lignocellulosic fibers, as well as morphology and properties depend upon factors as: the source from which they were extracted, the ripening of the plant and soil conditions where they were growing [1,2].

In Table 1 the chemical composition of some vegetable fibers used among other applications, as reinforcement in polymer composites are represented. Values shown in Table 1 were obtained from various scientific studies conducted by different researchers in recent years.

Components of lignocellulosic fibers (cellulose, lignin and hemicellulose) are arranged in fibers in a complex physical structure. The knowledge of such structure is important to develop the use of fibers as reinforcement in composite materials.

Table 1. Chemical Composition of vegetal fibers.

Fiber	Cellulose (%)	Hemicellulose (%)	Lignin (%)	Ash (%)
Sugarcane bagasse*	32-48	27-32	19-24	1.5-5
Coconut	32-43	0.15-0.25	40-45	-
Banana	63-64	10	5	-
Curaua **	73	20	1.5	1.0
Sisal*	47-62	21-24	7-9	-
Jute*	45-63	18-21	21-26	0.5-2

Source: [1], *[2], ** [6], [7],[8]

According to Figure 1, the plant fiber is composed by a complex layer structure consisting of fibrils and microfibrils. Around each fiber, can be broken down into elementary fibrils and microfibrils of cellulose. Each vegetable fiber is formed by a central lumen, which is responsible for water and nutrients transportion into the plant. This lumen is surrounded by three secondary walls (S1,S2 and S3),a primary wall and a middle lamella [9,10].

The primary wall is formed by a network of disordered crystalline cellulose microfibrils, while the secondary walls is composed by crystalline cellulose microfibrils arranged in a spiral arrangement.

These microfibrils are arranged in an amorphous region made up primarily of lignin and hemicellulose and have a diameter of 10-30 nm, and the result packing 30-100 cellulose chain extended [9,10].

The middle lamela is composed predominantly of pectin that acts to cement fibers together into a bundle [10].

1.1 Cellulose

Cellulose is the most abundant organic material on the earth and, together with the lignin, hemicellulose and pectin is the major component of plant cell walls,with an annual production of 50 billion tons. It is formed by a long chain linear homopolymer with an empirical chemical formula $(C_6H_{10}O_5)n$, where n varies from a minimum of 200 to values greater than 7000 [9].

Its repeating unit, called cellobiose (Figure 1), is composed by two glucose molecules etherified by β-(1→4) glycosidic bonds, containing six hydroxyl groups that establish one hundred interactions of type hydrogen bonds intra and intermolecular [12].

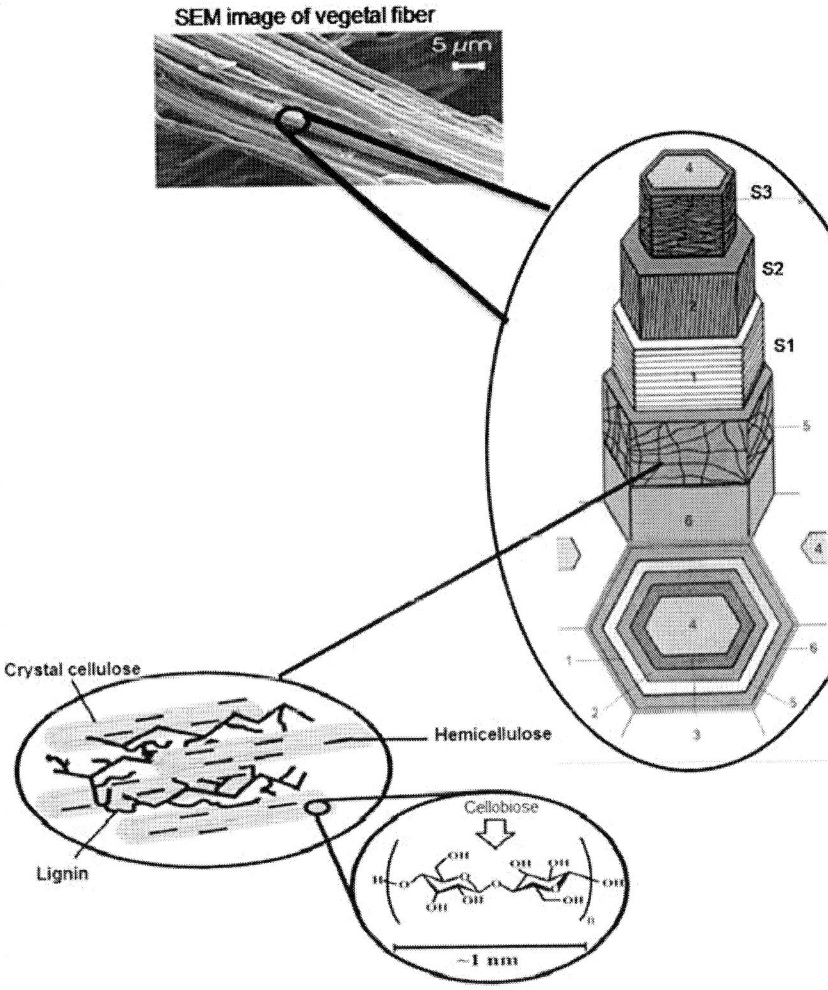

Figure 1. Wall structure of a vegetal fiber seen in tranverse (bottom) and three-dimensional view (top).(1-3) secondary walls (S1,S2 and S3); (4) lumen; (5) primary wall and (6) middle lamella. Adapted from [9-11].

Due to this linear structure (Figure 2), fibrous and moist, it becomes impervious to water and therefore insoluble fibers which leads compact form the cell wall of plants [9,12].

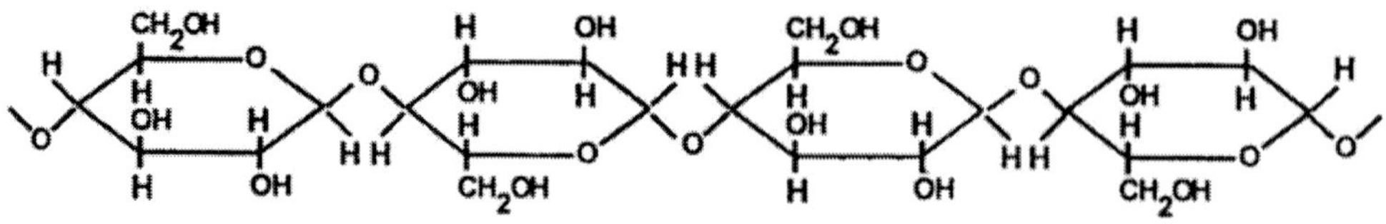

Figure 2. Molecular structure of cellulose [13].

It has been reported [9] that the cellulose crystaliinity degree varies according to their origin and processing. The cellulose of cotton, for example, has more ordered chains, with crystallinity of about 70%, while the cellulose crystallinity index of tree is around 40%.

1.2 Hemicellulose

Following cellulose, the hemicelluloses is the organic complex of major occurrence in the biosphere. It is a polysacharide with low molecular weight, which is interspersed with cellulose microfibrils resulting in elasticity and preventing them from touching each other. Along with cellulose, pectin and glycoproteins, form the cell wall of plant cells.Hemicelluloses are mainly divided into pentosans (as xylose and arabinose) and hexosanas (as galactose, mannose and glucose), with general formula $C_5H_8O_4$ and $C_6H_{10}O_5$, respectively, where n is the degree of polymerization [9,12].

Because of its open structure containing many hydroxyl groups, hemicellulose is partly soluble in water and is hygroscopic [14].

1.3 Lignin

Lignin is a macromolecule complex polyphenolic structure and not converted into fermentable sugars. This resin is an amorphous material with hydrophobic three-dimensional structure, highly branched which acts as a cement between the fibrils and is present in all layers of the cell wall of the plant. However concentrates on primary and secondary layers, and occurance is in association with cellulose and hemicellulose.This biopolymer is formed by three different units of the family of phenyl ether and the proportion of these compounds results in different types of lignin. It is insoluble in water and its architecture as well as the chemical complexity of lignin not only hamper their isolation but also their plasticization by economic processes [9,15].

In the use of lignocellulose fibers as reinforcement in polymers, the lignin increase fiber stiffness, since it is inflexible and prevents fiber reorientation required for proper load transfer and mechanical strength mobilization [16].

2. Lignocellulosic Fibers as Reinforcement for Composites

All vegetals fibers are composed mainly of cellulose, hemicellulose and lignin, while animal fibers consist basic of proteins (hair, silk, and wool). Vegetals fibers are subdivided in classes such as: bast fibers (as jute, flax, hemp, kenaf), leaf fibers (pineapple, sisal, henequen, cuaraua, banana pseudo stem), seed or fruit fibers (as coir, cotton and oil palm) and, grasses and reeds (as bagasse and sape) [1,2,17].

2.1 Jute Fibers

Around the world, about 40 species of jute plant are found, however, commercial jute fibers are extracted from only two types of plants known as *Corchorus Capsularies* (white jute) and *Corchorus Olitorius* (both tossa-daisee) which are annual plants grown from seeds [18].

The length of white jute is around 6 to 12 feet long and ½ to ¾ in diameter while *Corchorus Olitorius* jute is about 5 to 10 feet long and ½ to ¾ in diameter [18].

The main producers of jute in the world are Bangladesh, West Bengal, Bihar, Orissa, Assam, Bombay of India, Pakistan, Nepa, Nigeria, Brazil, Burma, and Japan respectively. Together these countries hold approximately 95% of jute world production [5,18].

It has been reported [18] that jute fiber is totally dependant on the weather and area to be grown. The most suitable environment for fiber jute is Bangladesh, and this is why Bangladesh is the largest jute growing country in all over the world.

Due to toughness and high aspect ratio in comparison with other natural fibers, jute fibers are considered a great promise in the reinforcement of composites, which can be used in the form of fiber or fabric, as can be seen in Figure 3 [19].

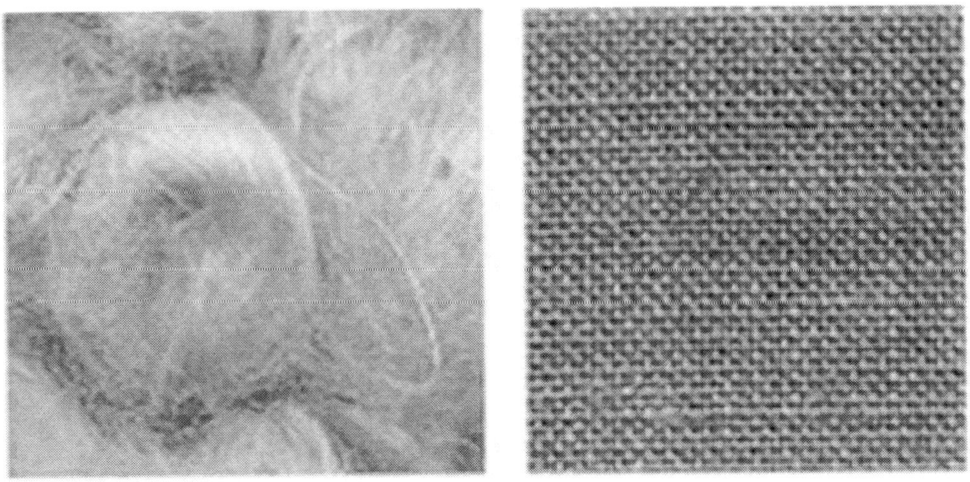

Figure 3. Jute fiber and jute fabric [20,21].

Alves *et al* [5] studied the LCA analysis in order to replace glass fibers by jute fibers as reinforcement of composite materials to produce automotive structural components. According to C. Alves "buggy" case study demonstrated that jute fibers composites present the best solution enhancing the environmental performance of the buggy´s enclosures, hence improving the environmental performance of the whole vehicle.

2.2 Sisal Fibers

Sisal (*Agave Sisalana Perrine*) belongs to the class of monocotyledons, *Agavaceae* family, and its main product is the fiber type hard, with high levels of cellulose and lignin, which offer numerous applications. There are two species in the genus Agave, important economics: *A. Sisalana*, which has Bahia as the world's largest producer (Figure 4) and *A. fourcroydes*, operated in Mexico under the name of henequen and whose fiber is used for the manufacture of wires and cords [22-24].

Sisal is the main hard fiber produced worldwide, corresponding to approximately 70% of commercial production of all fibers of that type. In Brazil, world's largest producer and exporter of sisal fiber and manufactured, the annual production of fiber in the interval between 1995 and 2005 changed in the range 110000 to 140000 tonnes. Currently, about 87% of domestic production is for export, generating annual foreign exchange of about 80 million dolars [22-24].

Figure 4. Sisal plantation [25].

2.3 Curaua Fibers

Curauá fibers are extracted from *Ananas erectifolius* plant, which belongs to the family of pineapple (*Ananas comosus*) grown in soils of semi-arid conditions. In Brazil the plant is cultivated predominantly in the Amazon region. Its leaves rigid and with flat faces (Figure 5) can reach up between 1-5 m in length and 4 cm in width, with great potential for use as reinforcement in polymeric matrices. They are relatively soft with higher mechanical strength compared with other lignocellulosic fibers like sisal, jute and flax [4,27,28,29].

In Brazil, the curauá used in industrial manufacture is obtained from farms in the Amazon. The fibres are processed (extracted, washed and dried in open air) and milled to obtain workable forms of fibres.

Figure 5. Curaua plantation and fruit of curaua plant [30,31].

Curaua fibers have low-cost of production and offer a relatively high tensile strength level wich is necessary for practical applications. If specific modulus (relative to density) is considered, curauá stands out relative to other fibres, which enable composites with curauá fibre up to a 15% weight reduction [4,29].

Curaua fiber is a promising material to reinforce thermosets and thermoplastics [14]. Composites reinforced with curaua fibers have mechanical properties comparable to commercially produced composites HPDE reinforced with fiber glass [6].According to Araujo *et al* [6] curauá fiber is a promising candidate to reinforce polimeric composites because it shows specific mechanical properties similar to fiber glass and superior to other vegetal fibers.

Gomes *et al* [4] studied the development and improvement of green composites reinforced with curauá fibers and verified that alkaline treated fiber composites increased in fracture strain twice to three times more than untreated fiber composites.

2.4 Banana Pseudo-Stem or Husk Fibers

Banana plant is a fruitful plant belonging to the *Musacae* family and produced in countries like Brasil, India and China. Banana plant cultivation generates a considerable amount of cellulosic-based waste. The comestible part, that constitute only 12% in weight of the plant, is a succulent fruit, with many nutrients as potassium, magnesium, phosphorus, carbohydrates, vitamin A, C, B1, B2, B3. The remaining parts of the banana plant becomes agricultural residue[32,33].

From the banana plant it is possible to obtain pseudo stem fiber and from the husks of banana fruit are extracted peel fibers. This residual resource rich in cellulose attracted much interest as a reinforcing component in composite materials [33-37].

The banana peel is a waste of agribusiness and home discarded in large quantities in nature. The husks represents 47-50% of the total weight of mature fruit, and has applications in industrial order and are occasionally used, directly, for animal feed [38].

It has been reported [35] that banana fiber extracted from the waste product of banana cultivation has superior mechanical properties , especially tensile strenght and modulus, due to high cellulose content and comparatively low microfibrilar angle.

Banana pseudo-stem fiber has been used as a reinforcement of polymeric matrices such as polyethylene [38], high density polyethylene (HDPE)/Nylon-6 blends [34] and, polypropylene [39].

2.5 Coconut Fibers

The coconut (*Cocus nucifera L*) is a tropical palm tree originated probably in Southeast Asia. It is a culture of a large expansion that plays an important economic role in over 90 countries.Indonesia is the world's largest producer, with a production in 2005 of approximately 16 300 000 tonnes, followed by the Philippines and India. Brazil is the fourth largest producer with an output of just over 3 million tonnes harvested in an area of 280 800 ha [40].

According to Embrapa, between 1985 and 2001 the coconut planted area in Brazil increased from 166 000 hectares in 1985 to 266 000 hectares in 2001.The Northeast is the largest producer of coconut in Brazil, accounting for 65% of Brazilian production. Data from the Brazilian Institute of Geography and Statistics (IBGE) shows that 2007 production was approximately 1.9 billion fruits [40,41].

Figure 6. Coconut [42].

Coconut husks fibers are lignocellulosic materials that can be extracted from mesocarp (thick fibrous part) as well as exocarp (outer shell) from coconut fruits (Figure 6), as represented in Figure 7.

The use of coconut fiber as reinforcement in a polymer matrix is important, because it is an inexpensive material when compared with glass fiber, decreases the amount of waste accumulated in landfills, and yet improves the mechanical properties of composites materials. The green coconut fibers have been studied for reinforcement application in polymers such as polyester [43], polypropylene [44], polyethylene [45] and biodegradable polymers [46] by changing mechanical properties of these compounds such as tensile strenght and elongation at break [47].

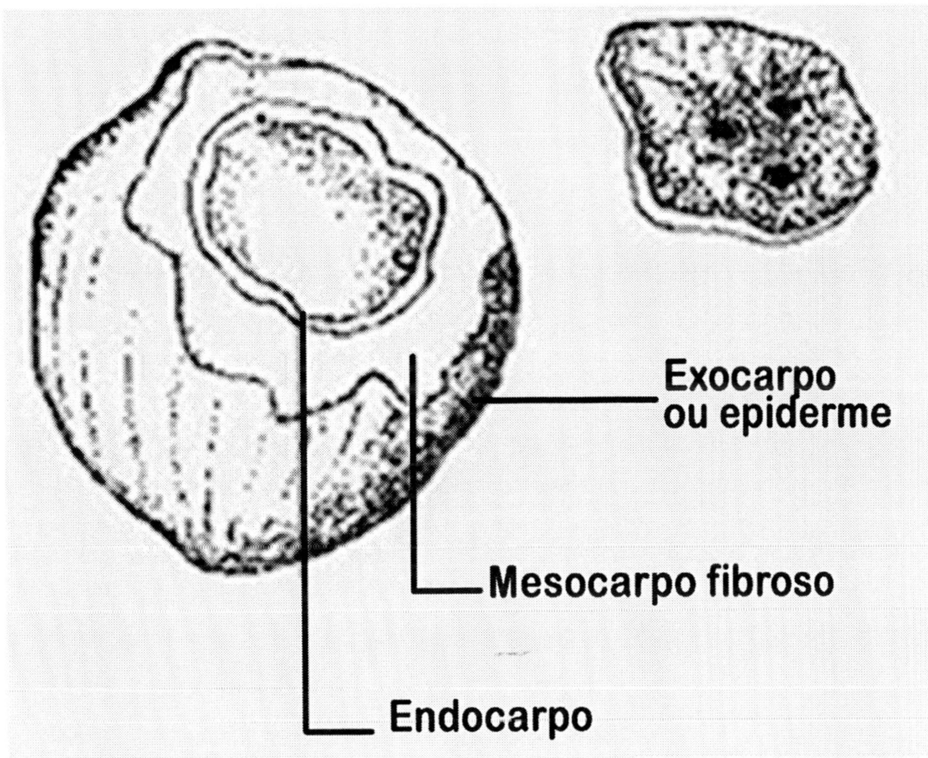

Figure 7. Cross section of coconut fruit [48].

Compared to other vegetable fibers, coconut fiber has a lower percentage of cellulose (32-43%); however, the amount of lignin (40-45%) represents two to four times the existing values for jute and sisal, resulting in greater strength and hardness compared to other fibers [48].

2.6 Sugar-Cane Bagasse Fibers

The sugar cane (Figure 8) is historically a major agricultural product in Brazil, being cultivated since the time of colonization. According to its industrialization processes products as sugar in its various forms and types, alcohol (anhydrous and hydrated) and, bagasse are produced [8].

The sugar cane bagasse is the residue generated after the process of extracting the juice from the sugarcane industry in the manufacturing of sugar and alcohol, which is used also as fuel for the boilers [49].

The large amount of bagasse generated has caused serious problems of storage, and of course, impact the environment, therefore the crushed sugar-cane, in addition to being used for power generation, has been provided for several other applications, such as reinforcement for polymer composites [8,45, 50,50-52], adsorbing materials [53-55] and components for industries of construction [56,57]. From the chemical point of view, the crushed sugar-cane is composed of cellulose (46%), hemicellulose (24.5%), lignin (19.95%), fat and waxes (3.5%), ash (2.4%), silica (2.0%) and other elements (1.7%) [8].

Sugarcane bagasse fiber have been used as a reinforcement in cement composites, used in civil construction to reduce electricity consumption in houses [60], and in high polyethylene (HDPE) have shown special interest in interior components for automotives such as seat frames, side panel and central consoles [8].

Figure 8. Sugar cane plantation [58].

3. Surface Treatments of Fibers

Lignocellulosic fibers are obtained from the direct extraction of plants or from the use of industrial wastes, as for example, sugar cane bagasse, a residue from sugar mills and ethanol and the husks of the coconut fruit and banana fruit.

After obtaining lignocellulosic fibers, normally some type of surface treatment takes place. The aim is to modify the fiber surface changing its morphological characteristics in order to improve compatibility with polymer matrices in the use of fibers as reinforcement in composites or to provide some specific feature such as the use of fibers to heavy metals adsorbent [8,59].

The chemical surface treatments of fibers are needed due to disadvantages as the lack of compatibility with hydrophobic polymer matrix and the tendency to form aggregates during processing. The surface treatments are conducted with the aim of improving the conditions of accession fiber / matrix or change the characteristics such as hydrophilicity and roughness. It is known that natural fibers have several hydroxyl groups along their chains, which gives a large hydrophilicity fiber. According to Mohanty et al. [60], fiber treatments such as degreasing, graphitization, bleaching, acetylation or reaction with alkalis, peroxides, silanes or isocyanates are essential to obtain materials with better performance.

3.1 Alkaline Treatment: Effect on the Vegetal Fiber Surface

The mercerization treatment aims to clean the fiber surface by removing partially amorphous constituents such as hemicellulose, lignin, waxes and greases soluble in alkaline solution. Thus, decreases the degree of aggregation of the fibers and makes the surface rougher [4,48, 61,62].

In the mercerization treatment fiber is washed with NaOH solution with different concentrations in order to remove dirt, waxes, oils and much of the amorphous material, reducing the its hydrophilicity carater. This treatment is commonly used for initial cleaning of fiber and in most of the cases, after mercerization, it is applied other kind of surface treatment.

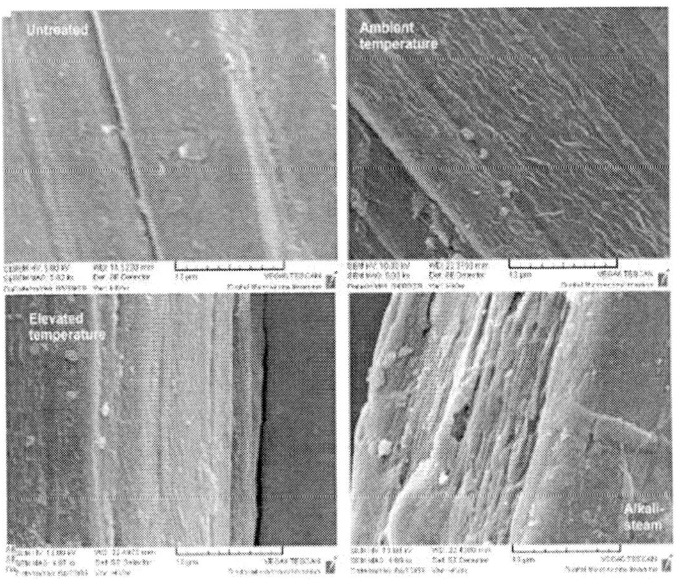

Figure 9. SEM photoghraphs of untreated and alkali treated jute fibers [16].

For use as reinforcement in composites, processing of mercerization is important because it improves compatibility with nonpolar and hydrophobic polymer matrices, improving the mechanical properties of composite material [63].

Saha et al [16] investigated the physico-chemical properties of jute fibers treated with alkali (NaOH) solution under ambient (30 ± 2 °C over 30 min to 24 h duration),elevated temperature (90 ± 2 °C over 30 min to 24 h duration) and, high pressure steaming conditions. In the conditions of high pressure, jute fibers were kept immersed in 0.5-8% NaOH solutions by maintaining a fiber weight to alkali volume ratio of 1:50 at 30°C for 30 min to 8h. The results from this investigation indicate significante changes in surface morphology after the differents alkali treatment conditions, as observed in Figure 9.The removal of surface impurities, non-cellulosic materials, inorganic substances and waxes was found to result in rougher surfaces and better fiber separation due to alkali treatments with the alkali-steam treated fibers leading to the development of the roughest surface.

Kim *et al* [24] studied the effect of mercerization treatment on the mechanical properties of sisal fibers. Fibers were mercerized under tension and no tension, to improve their tensile properties and interfacial adhesion with soy protein resin. Through SEM photographs it was observed that sisal fibers presents a quadrate grid pattern on its longitudinal surface with the presence of some impurities, Figure 10a. Mercerization of sisal fibers improves the surface properties by removing hemicellulose and lignin present between microfibrils resulting in increased tensile properties. By removing the hemicellulose and lignin, this also results in the separation of fibrils as seen in Figures 10b and 10c.

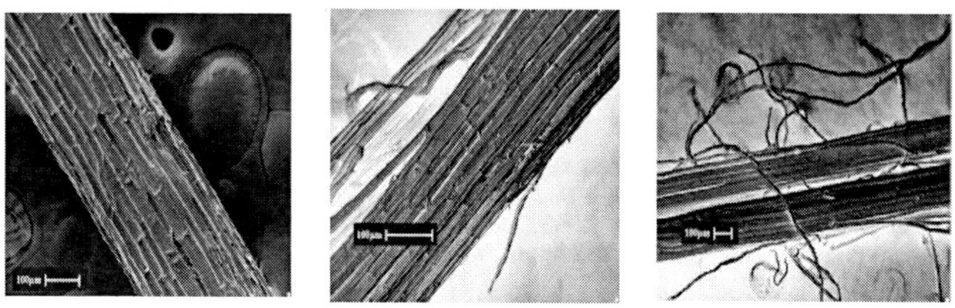

(a) (b) (c)

Figure 10. SEM surface of : (a) control; (b) slack-mercerization and (d) tension –mercerization sisal fibers. [24].

Carvalho et al [59] used coconut fibers from mesocarp of green coconut fruit as a reinfoecement in high impact polystyrene matrix and by SEM photographs (Figure 11) highlighted significant changes in the fiber surface after chemical treatments of mercerization with NaOH 10% m/v solution under room temperature and bleaching with sodium chlorite $NaClO_2$ and acetic acid CH_3COOH.

SEM micrographs of untreated fiber in Figure 11 (a) indicates that green coconut fiber surface is covered with a layer of substances such as oils, waxes, and extractives, part of the natural constitution of lignocellulosic fibers. Micrographs in frames (b) of the figure show a rough surface, which is a consequence of the alkaline solution treatment effect, which removes extractives, waxes, and oil from fiber surfaces and thus increases the overall roughness of surface. With the removal of these substances, it was possible to verify the presence of parenchyma cells that are the natural constituents of lignocellulosic fibers, as well as the presence of globular protusions, which are fatty deposits called "tyloses". According Carvalho et al [59] these globular protusions, are arranged on the fiber surface at regular intervals and their presence on the surface of coconut fibers was also observed by Brigda et al. [63], Bismarck et al. [7], and Calado et al. [64].

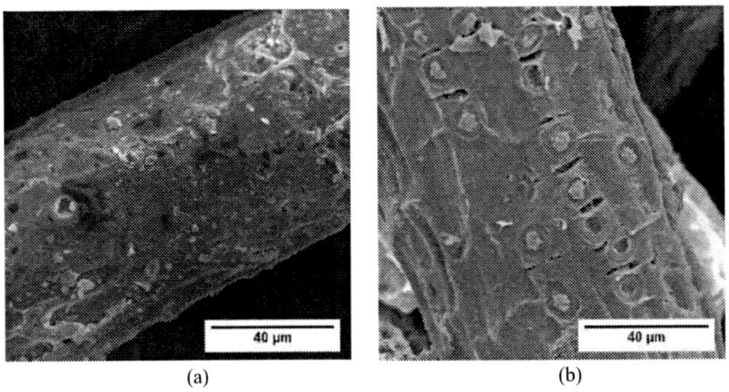

Figure 11. SEM micrographs of fiber surface of: (a) untreated fiber, (b) alkali-treated fiber [59].

3.2 Effect of Differents Treatments on the Vegetal Fiber Surface

Ibrahim et al [38] treated banana plant waste with alkaline pulping and stem explosion to produce banana fibers and banana microfibrils. From the SEM observation, Figure 12, it is possible to notice the degradation of the amorphous part of cellulose during the steam explosion process.A closer look at the microfibril surface of the steam –explode banana fibers at higher magnification show that many terraces, steps, and holes form by the explosion technology. It was proposed [38] that these changes result from the removal of very reactive amorphous cellulose from the surface. In addition, the steam explosion process induces a recondensation of the lignin onto the fibers.

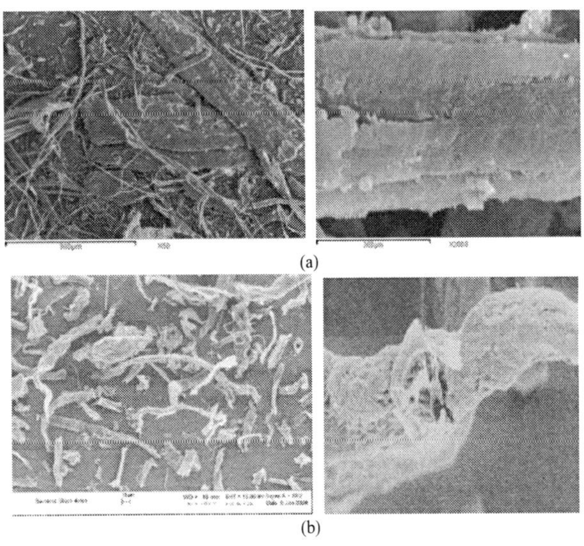

Figure 12. SEM of (a) alkaline pulping banana and (b) steam-exploded banana fiber [38].

Spinacé et al [14] studied the morphology of milled curaua fibers, submitted to different treatments by scaning electron microscopy (SEM). From Figure 13a it is possible to observe a smooth and compact surface with no fibrilation. For treated curaua fibers Figures 13b, c and d, an increased surface rougness and fibrils could be distinguished on the surface for fibers treated with sodium hypochlorite solution (CH) and cold oxygen plasma (CP).

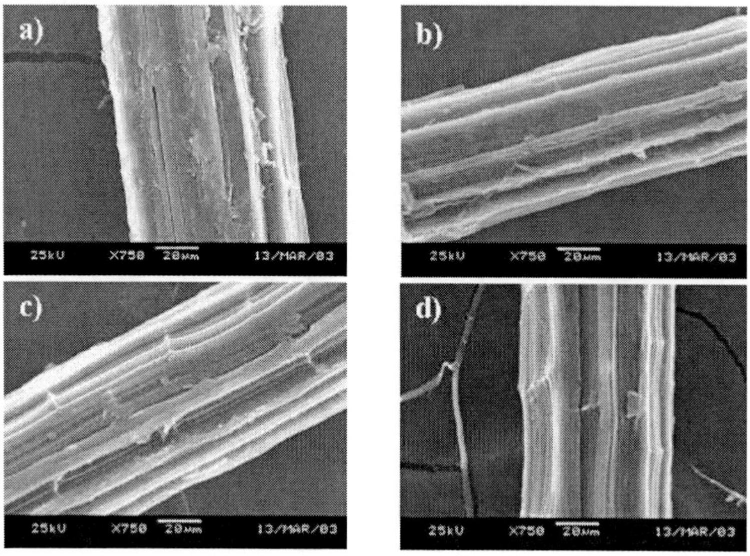

Figure 13. SEM micrographs of curaua fibres surface: (a) curaua in nature, (b) curaua washed with hot water (CW), (c) CH and (d) CP [14].

In order to make fibers more reactive, surface modification has been carried out also with many metal oxides. Mulinari *et al* [65] studied the surface modification of cellulose fibers from sugarcane bagasse bleached and modified by zirconium oxychloride (ZrO_2. NH_2). Results indicate that cellulose fibers from sugar-cane bagasse, observed from SEM photographs, Figure 14a, present flattened forms of fibers of different sizes. The length varies from 100 to 500 lm, and the diameter varies from 10 to 30 lm. In the formation of nanoparticles, it is interesting to observe a large area and roughness on the surface.

Figure 14. (a) Cellulose fibers from sugar cane bagasse characteristics magnification 500x; (b) magnification 2500x; (c) magnification 7500x [65].

4. Thermoplastics and Thermosetting Polymers Reinforced with Natural Fibers

4.1 Thermoplastic Resins

Thermoplastic polymers present increasing applications in several commercial products in combination or in substitution to others material such as metals, wood and glass [66]. Many studies have been performed with the incorporation of fillers or fibers into low cost thermoplastic matrixes to improve mainly mechanical properties of material, allowing substituting thermoplastic engineering matrixes of higher cost. The easy processibility of thermoplastic by extrusion or injection molding allows high production and final products with variable shapes, which are potentially recyclable [17]. However, the incorporation of fibers is not a trivial process. The melt state in thermoplastic polymer is achieved at high temperature (above 150 ^{0}C) and under shear strength that can change the fiber properties. The physical anchorage of fibers in thermoplastic matrix is another frequent difficult that can be minimized with the use of coupling agents [67, 68]. Among the countless thermoplastic, five thermoplastic resins deserve special attention due to their good properties and low cost, representing around 90% of total amount consumed, which are polyethylene (PE), polypropylene (PP), poly(vinyl chloride) (PVC), polystyrene (PS) and poly-(ethyleneterephtalate) (PET). The preparation of composites of natural fibers with thermoplastic matrixes is concentrated mainly with the use of PE, PP and PVC [67, 68]. However, many others thermoplastic matrixes have been used too [17, 67, 68]. Following below is done a brief description about these main thermoplastic matrixes.

4.1.1 Polyethylene (PE)

Polyethylene always motivated interest of researches and industry due to its properties and low cost [69]. Polyethylene is produced with the monomer whose chemical structure is the simplest (ethylene), which is a direct derived from petrol, but can also be produced from removable sources such as ethanol. Figure 15 illustrates the chemical structure of ethylene and polyethylene.

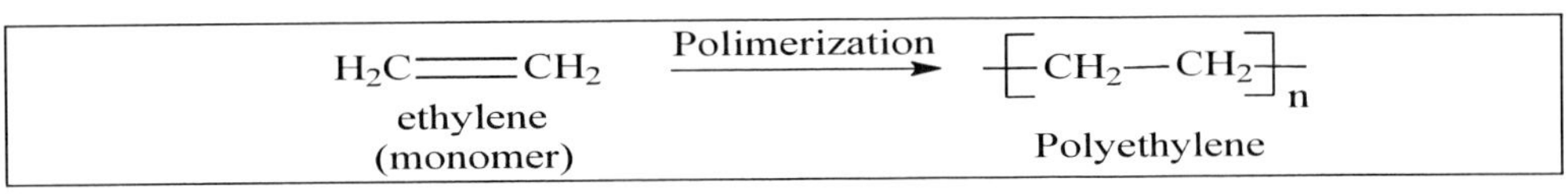

Figure 15. Synthesis Reaction of polyethylene [69].

Polyethylene is a flexible thermoplastic at room temperature, resistant to the impact, with low Young modulus and high plastic deformation. Chemically, it is enough inert, supporting substances such acids or basis without significant structural changes. The control of polymerization can result in several kinds of polyethylenes such as the high density polyethylene (HDPE), low density polyethylene (LDPE) and linear low density polyethylene (LLDPE). The structural distinction between these polymers is related to molecular weight and ramifications present in polymeric chain. HDPE is used for applications that require stiff products and LDPE and LLDPE for flexible products such as films and packings [70]. Due to weak intermolecular interaction between the polymeric chains formed by carbon and hydrogen, typically of Van der Waals kind, the miscibility or compatibility in polyethylene systems such as blends and composites is rare without the use of coupling agents [71].

Natural fibers are based in cellulose structures that have chemical interactions distinct with polyethylene. Thus, the success in compatibility of polyethylene with natural fibers depends of presence coupling agents that have the role of chemically to bond the fiber to the polymeric matrix. The crystallization is other factor that affects the adhesion of natural fibers to the polyethylene matrix. The presence of fiber can induce the crystallization in interface fiber-matrix, hindering the interchange of properties fiber-matrix [70, 71].

The preparation of polyethylene composites with natural fiber is an alternative of low cost for recycling of polyethylene which presents depreciated mechanical properties.

4.1.2 Polypropylene (PP)

Polypropylene has been frequently used as thermoplastic matrix resin in composites with natural fibers. This is due mainly to its cost effectiveness, good mechanical properties, easy processibility, versatile applications, etc. Similarly to the polyethylene, polypropylene is a polyolefin chemically formed only by carbon and hydrogen atoms [67, 71]. Figure 16 presents the chemical structure of monomer propylene and polypropylene.

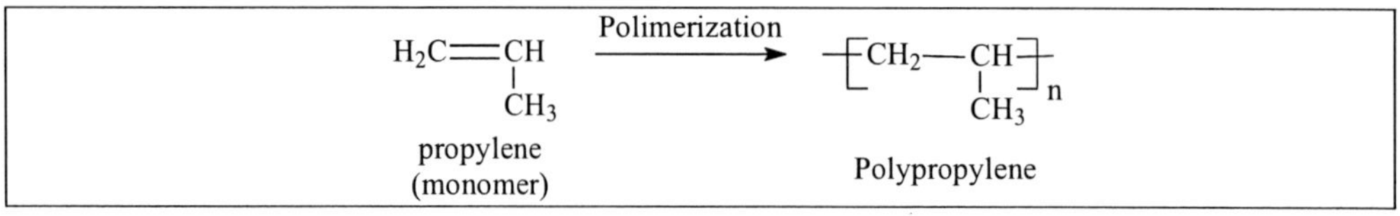

Figure 16. Synthesis Reaction of polypropylene [67].

The presence of lateral group methyl in main polymeric chain of polypropylene allows the syntheses of polymer in three distinct configurations: atatic, syndiotatic and isotatic. Atatic polypropylene (PPa) is fluid at room temperature without important technologic applications. On the other hand, syndiotatic polypropylene (PPs) and isotatic polypropylene (PPi) are semicrystallines with glass transition and fusion around -9 °C and 160 °C, respectively. Thus, both PPs and PPi are flexible and mechanically structured at room temperature, representing the commercial forms of polypropylene [72]. Polypropylene is less flexible than polyethylene, allowing applications as a structural material for confection of appliances, automotive pieces, toy, packing, etc. The incorporation of fillers and filer to the PP with the purpose of to obtain mechanical properties similar to engineering thermoplastic at low cost have been motive of many studies [67, 72, 73].

The chemical interactions between PP and natural fibers as well as in polyethylene are of weak intensity, hindering the fiber-matrix compatibility, which leads to the indispensable use of coupling agents [67, 72].

4.1.3 Poly(Vinyl Chloride) (PVC)

Poly(vinyl chloride) (PVC), as a commodity plastics, has been widely used in industrial fields for many years due to its good chemical and weathering properties, nonflammability, low cost and formulating versatility. PVC is the most versatile polymer among main thermoplastic of wide commercial application and its main applications include pipes, electric wires, window profiles, siding, toys, etc [74]. The chlorine atoms present in each repetitive unit of polymeric chain represent 56 % of total weight of polymer and provide important properties to the material. Figure 17 illustrate the chemical structure of monomer vinyl chloride and PVC.

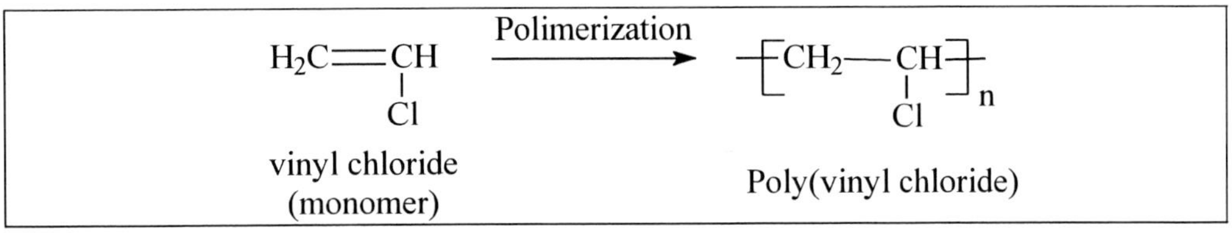

Figure 17. Synthesis Reaction of poly(vinyl chloride) [74].

The chemical polarity of PVC allows intermolecular interactions of higher intensity such as dipole-dipole that facilitate the compatibility with additives, fibers and others polymers. The PVC accept the incorporation of high amount of plasticizers, leading to materials with variable mechanics characteristics that are since rigid material for structural applications until flexible for confection of films [75, 76].

The versatility of PVC to aggregate others material is confirmed in the preparation of composites with natural fibers too. Thus, PVC has been an interesting alternative to mix it with "ecologically friendly" components, such as natural fibers. This could reduce its inconveniences while conserving its advantages. Moreover, applications of polymer composites made from short natural fibers have increased in the last decade because of the advantages they offer, such as low cost, a high relationship resistance/weight and biodegradability. Although the PVC/fiber compatibility is more favorable than Polyolephin/natural fiber, the use of coupling agents is not disrespected, being a viable form to improve the mechanical properties of composites [75, 76].

The automobilist industry has demonstrated great interest in the use of composites PVC/natural fibers. Many studies have also been carried out about PVC/wood composites, substituting the wood in several applications [77].

Thermal degradation is an important limitation of PVC. Thus, special attention should be attributed to the mechanical processing and fiber incorporation to the PVC in the melt state. High temperature and high shear rate during mechanical processing of PVC should be avoid, once it can lead to liberation of chloride acid and formation of conjugated double bond in polymeric chains that depreciate the PVC properties [78].

4.1.4 Polystyrene (PS)

Polystyrene is commercialized in three different forms, which present very distinct properties. Crystal Polystyrene (cPS) is the polymer without the presence of impact modifier or others agents. Despite denominated crystal, cPS has an amorphous structure; it is transparent and has low impact resistance. High impact polystyrene (HIPS) is other form of polystyrene which present around 6 to 11% of polybutadiene as impact modifier. This converts the polymer in a material with high tenacity, high impact resistance, opaque and adequate for applications that demand higher performance, mainly when combined with poly(2,6-dimethyl-1,4-phenylene oxide) (PPO). Both cPS and HIPS can be used to produce composites with natural fibers. The use of polystyrene as matrix to polymeric composites with natural fibers is an alternative to provide mechanical resistance of the material [79, 80]. Chemical structure of polystyrene is presented in Figure 18.

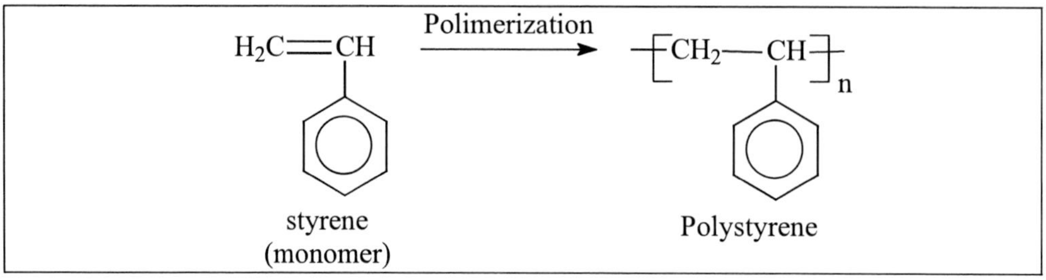

Figure 18. Synthesis Reaction of poly(vinyl chloride) [79].

Polystyrene is also produced as expanded form (ePS) that not present sufficient characteristic to be used as polymeric matrix in composite reinforced with natural fibers.

4.1.5 Others Thermoplastics Matrixes

Others thermoplastic polymer such as poly(ethyleneterephtalate) (PET), polyamides (Nylons), polymethylmetacrylate (PMMA) and polycarbonate (PC) could be used as matrix in composites with natural fibers. However, many properties of these resins such as higher cost, applications that require transparency and intrinsic high mechanical strength lead to a low interest for the preparation of composites with natural fibers.

4.2 Thermosetting Polymers

Thermosetting polymers are historically used as matrix for production of composite with several particulates fillers and fibers. The thermosetting resins before cure reaction are found as a viscous fluid that is be able to involve the fibers or fillers particles, promoting high distribution of reinforcement and excellent homogeneity of the composites [71, 81-83]. After cure reaction of polymeric resin is generated a crosslink chemical structure that physically anchor the reinforcement in the matrix, leading to the excellent combination between matrix and reinforcement properties without to require the utilization of coupling agents [71, 81-83]. On the other hand, the thermosetting composites present some disadvantage such as impossibility of thermo-mechanical reprocessing, which difficult the use of versatile production operations such as extrusion and injection molding. Recycling of thermosetting composites is also a technological challenge. Recently, thermosetting composites have been preferable used for specific applications with lower amount in demand, however with higher cost such as electronic devices and aeronautic industry. The use of natural fibers as reinforcement in thermosetting polymers is an alternative that have been studied to decrease the cost of material without effectively changes theirs mechanical properties [71, 81-83]. Following below is done a brief description about the main thermosetting matrixes used in composites with natural fibers.

4.2.1 Unsaturated Polyester (UPE) Resin

Unsaturated polyester (UPE) resin is the most widely used thermosetting polymer matrix in natural fiber-reinforced polymer composite material fields as well as in conventional glass fiber-reinforced polymer composites industries because of its good mechanical, chemical, and weather resistant properties, especially when reinforced with fibers [71]. Polyester resins is currently used for decorative coatings and for a wide range of commodity composites materials in industrial equipment such as tanks, boat building, truck roofs, etc. UPE resins are a class of thermosetting polymers derived from the polycondensation of a polyol and a polyvalent acid or acid anhydride. Curing reactions are performed through free radical or thermal processes in the presence of unsaturated comonomers such as styrene, leading to a tridimensional network [83]. Figure 19 illustrate the formation of a typical UPE chemical structure.

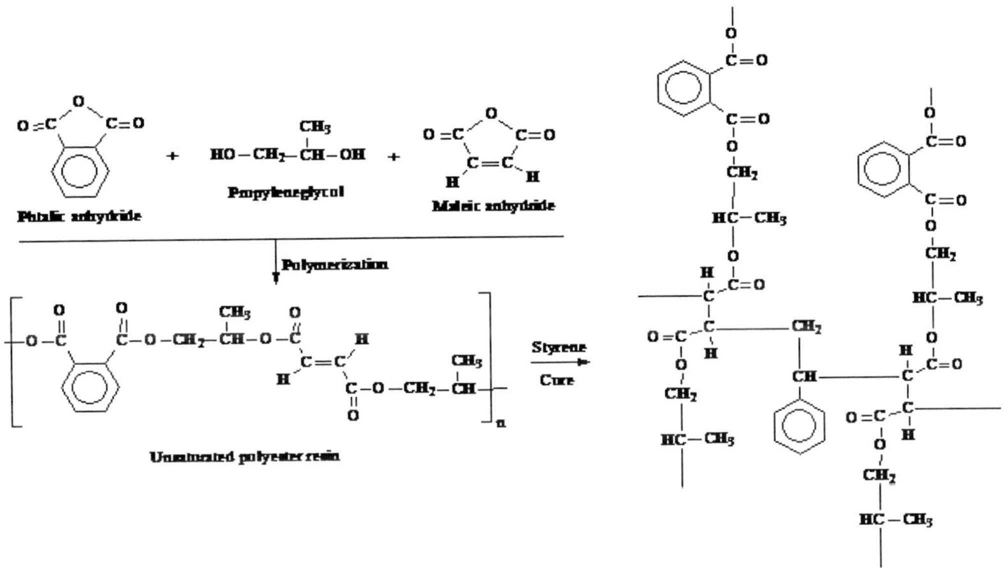

Figure 19. Formation and typical chemical structure of unsaturated polyester resin [71].

Phthalic and maleic anhydrides are the most frequently used as acid monomer, while most used polyols are (di)pentaerythritol, glycerol, ethylene glycol trimethyllolpropane and neopentylglycol. Recently, lignin esters have been proposed for use as unsatured ester thermosettings. Use of lignin esters as an additive in unsaturated thermosetting materials affords several advantages by acting as a toughening agent, improving the connectivity in the network, adding additional stiffening group [83]. Although the wetting and anchorage of natural fibers to be more effective in thermosetting than in thermoplastics matrixes, the presence of coupling agents can still significantly improve the interfacial and mechanical properties. The treatment of Kenaf fibers with silane coupling have showed be more effective in increasing the interfacial shear strength in kenaf/UPE composites than in kenaf/ polypropylene composites [71].

4.2.2. Phenolic Resins

Phenolic resins are the most traditional synthetic thermosetting polymer used as commodity and engineering material in the industry. Their first commercial form was the Bakelite, initially synthesized by Leo Hendrik Baekeland in 1907, which was used in substitution to several materials such as ivory, avoiding the killing of many elephants [84]. Despite the emergence of several new classes of thermosettings, high-performance polymer and new generation materials that are superior in some respects, phenolic resins retain industrial and commercial interest a century after their introduction. Heat resistance, dimensional stability and superior mechanical strength are some of the inherent properties of the phenolic resins [81, 83]. Their synthesis is done from monomers formaldehyde and phenol that can be catalyzed under both acid and alkaline conditions. However, the chemical structures generated are distinct for each condition. Acid catalysis produce the resin called Novalac, whose synthesis is carried out at a formaldehyde/phenol ratio between 0.75 to 0.85, while alkaline catalysis originate the resin called Resol, using a formaldehyde/phenol ratio higher than 1. Others compounds can also be incorporated to the monomers in addition or in substitution to the formaldehyde, resulting in variable chemical structure and properties of material [81, 83]. Figure 20 illustrate the synthesis of phenolic resins.

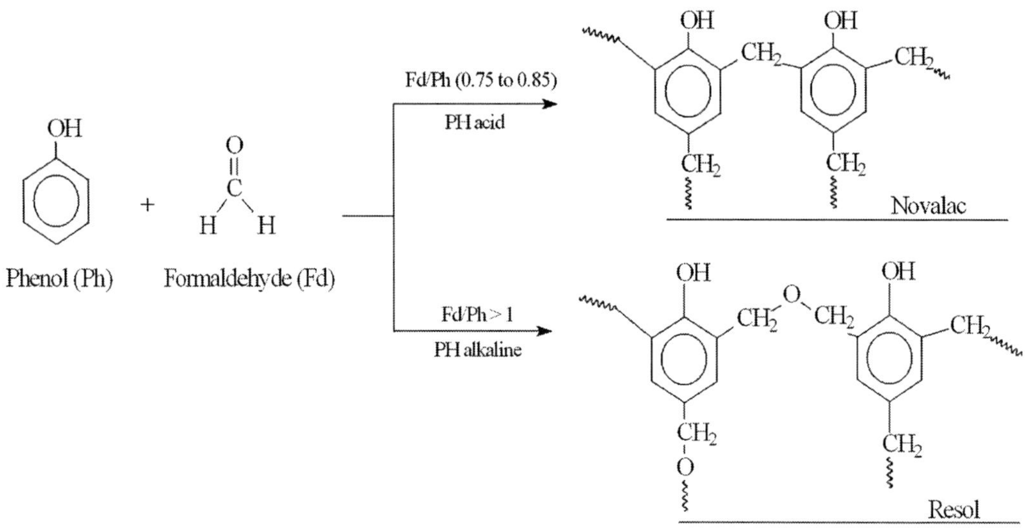

Figure 20. Formation and chemical structure of phenolic resins [84].

Phenolic resins are able to form covalent crosslinks with natural fibers via hydroxyl groups, unlike polypropylene or polyethylene. Due to composite manufacture be achieved using the resin at low viscosity, the fibers incorporation is more effective, being an advantage mainly where long fibers at high content are required. Moreover, these resins cure at room temperature, allowing the use of heat sensible fibers. Although relatively expensive, phenolic resins lead to composites with good properties, justifying their use [81, 83].

4.2.3. Epoxy Resins

Epoxy resins represent a thermosetting polymer class for which the resin precursors contain at least one epoxy function. These epoxy functions are highly reactive with several others chemical groups, being the starting point for crosslink formation or interfacial interaction with others materials [81]. Epoxy resins are widely used for manufacture of circuit board, electronic components encapsulations, adhesives and composites of the highest performance, mainly using carbon fibers as reinforcement. Aeronautic industry, high cost vehicles, frame glasses and competition bikes are some applications of high performance epoxy/carbon fiber composites [85]. The majority of commercial epoxy resin is produced by reaction between 2,2-bis(4-hydroxyphenyl) propane (Bisphenol A) and epichlorohydrin, yielding diglycidyl ether of bisphenol A (DGEBA) [83]. The Figure 21 illustrate this reaction.

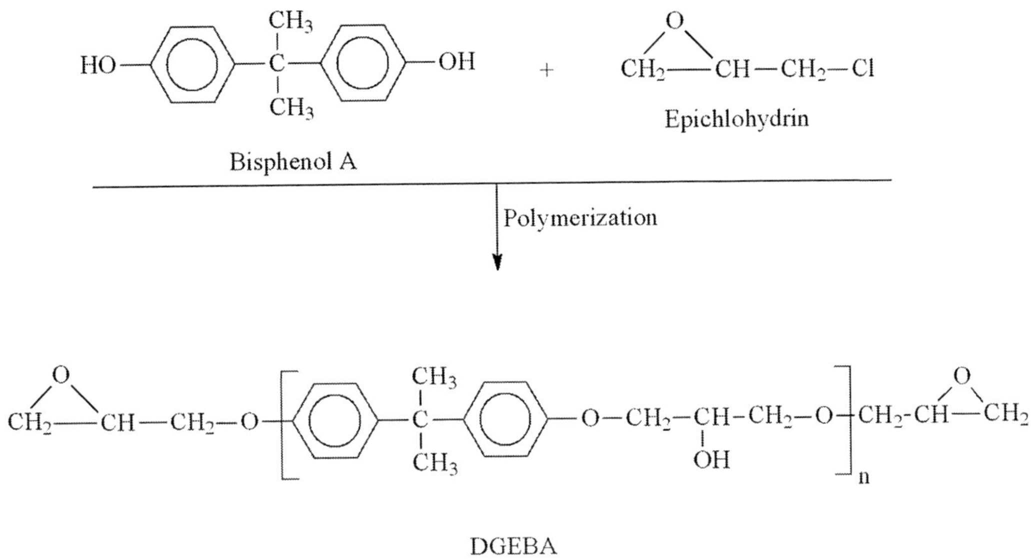

Figure 21. Synthesis Reaction of epoxy resin [81].

The resin cure can be performed with addition of several chemical substances called curing agents such as amines, anhydrides and amides [83]. Epoxy as well as phenolic resins can form covalent crosslinks with natural fibers via reaction of epoxy groups with hydroxyl present in fiber [81]. The considerable improvement of mechanical properties is achieved when epoxy resin is reinforced with natural fibers. However the cost-benefit ratio is a challenger to be wined, mainly when compared to the traditional epoxy composites [86].

4.2.4. Others Thermosetting Resins

Natural fibers have also been incorporated to others thermosetting matrix. Polyurethanes is a versatile class of resin that can be used as a thermoplastic resin as well as a thermosetting resin with applications range from flexible foam to medical devices, adhesives, coatings, etc [83]. This versatility has allowed their used in manufacture of composites with natural fibers. Elastomers thermosetting are others example the polymeric material studied as matrix for natural fibers composites, showing high potential for commercial applications [17].

5. Mechanical Properties of Natural Fibers Composites

In recent years, the effects of natural fibers as reinforcement on the mechanical properties of polymeric composites have been studied. Therefore, the mechanical properties of natural fibers reinforced composites dependent of parameters as volume fraction of the fibers, fibers aspect ratio, fiber–matrix adhesion, stress transfer at the interface, and fiber orientation [16, 63, 87]. Several studies on natural fibers composites involve mechanical properties characterization as a function of fiber content, effect of various treatments of fibers, and the use of external coupling agents 71, 88, 89].

Both the matrix and fiber properties are important to improve mechanical properties of composites. The tensile strength is more sensitive to the matrix properties, whereas the modulus is dependent on the fiber properties. To improve the tensile strength, a strong interface, low stress concentration, fiber orientation is required whereas fiber concentration, fiber wetting in the matrix phase, and high fiber aspect ratio determine tensile modulus. The aspect ratio is very important to determine the fracture properties. In short-fibers-reinforced composites, there exists a critical fiber length that is required to develop its full stress condition in the polymer matrix [38, 90, 91]. Fiber lengths shorter than this critical length lead to failure due to bonding at the interface at lower load. On the other hand, for fiber lengths greater than the critical length, the fiber is stressed under applied load and thus results in a higher strength of the composite.

For impact strength, an optimum bonding level is necessary. The degree of adhesion, fiber pull-out, and a mechanism to absorb energy are some of the parameters that can influence the impact strength of a short-fibers-filled composite. The properties mostly vary with composition as per the rule of mixtures and increase linearly with composition [38, 90, 91].

5.1. Thermosetting Composites

For thermosetting composites, the fibers are combined with phenolic, epoxy, and polyester resins to form composite materials. These thermosetting polymers contain reactive groups, which aid the interface development. The reported work on thermosetting composites covers the effect of process parameters such as curing temperature and various treatments on the properties of composites [90, 92, 93].

De Rosa et at [94] investigated tensile and flexural behavior of untreated New Zealand flax (Phormium tenax) fiber reinforced epoxy composites. Two series of laminates were produced using the same reinforcement content (20 wt.%), arranged either as short fibers or quasi-unidirectional ones. Composites reinforced using quasi-unidirectional fibers showed higher modulus and strength both in tensile and flexural loading, when compared to neat epoxy resin (Figures 22 and 23).

Short fiber composites, although still superior to epoxy resin both for tensile and flexural moduli, proved inferior in strength, especially as concerns tensile strength. These results were supported by scanning electron microscopy (SEM), which allowed characterizing fiber-matrix interface, and by acoustic emission (AE) analysis, which enabled investigating failure mechanisms. In addition, thermal behavior of both untreated phormium fibers and composites were studied by thermogravimetric analysis (TGA), revealing the thermal stability of composites higher than for phormium fibers and epoxy matrix alone [94].

Ramires et al [95] studied glyoxal-phenolic resins for composites using glyoxal, which is a dialdehyde obtained from several natural resources. Resorcinol (10%) was used as an accelerator for curing the glyoxal-phenol resins in order to obtain the thermosets. The impact-strength measurement showed that regardless of the cure cycle used, the reinforcement of thermosets by 30% (w/w) sisal fibers improved the impact strength by one order of magnitude (Table 2). The results of the Izod impact- strength test demonstrated that the presence of fibers greatly improved this property. The fibers were able to efficiently distribute the tension along the matrix, thus improving the impact strength. The composites of glyoxal-phenol matrix reinforced with sisal fibers had an impact strength 10-times higher than that of the phenolic thermoset.

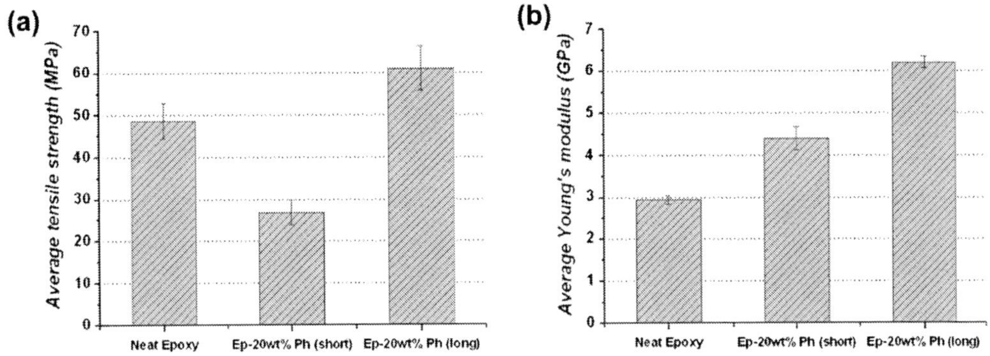

Figure 22. Average tensile strength (a) and Young' s modulus (b) of phormium/epoxy laminates compared to the pure resin [94].

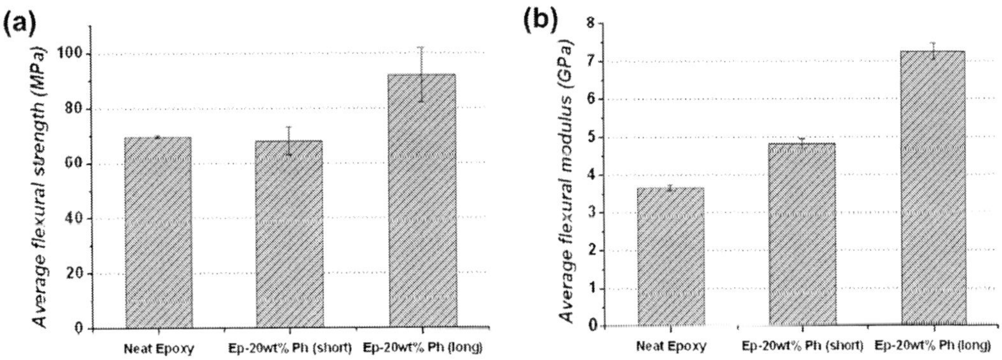

Figure 23. Average flexural strength (a) and modulus (b) of phormium/epoxy laminates compared to the pure resin [94].

Table 2. Izod impact-strenght of glyoxal-phenol-resorcinol composites reinforced with sisal fibers [95].

Glyoxal-phenol composites reinforced with sisal fibers	Izod impact strength (J. m^{-1})
Cure cycle 1	118 ± 13
Cure cycle 2	113 ± 18

Figure 24 shows the SEM images of the composites with cure cycles 1 and 2. The SEM images of the composite with cure cycle 1 showed that the fibers were more adhered to the matrix and also more recovered by the matrix (Fig.24a), when compared to the composite with cure cycle 2 (Figs. 24c and 24d). In addition, it was possible to observe the presence of the matrix inside the fiber (Fig. 24b). In the composite with cure cycle 2, a higher contact among the fibers (Fig. 24c) and a lower fiber/matrix adhesion (Fig. 24d) were observed. Thus, cure cycle 2, which reached higher temperatures, could lead to a larger production of volatile components, generating voids in the matrix. Hence, sisal fibers were less protected by the matrix, which made easier the interaction of the fibers with water molecules, and consequently increased the composite water absorption.

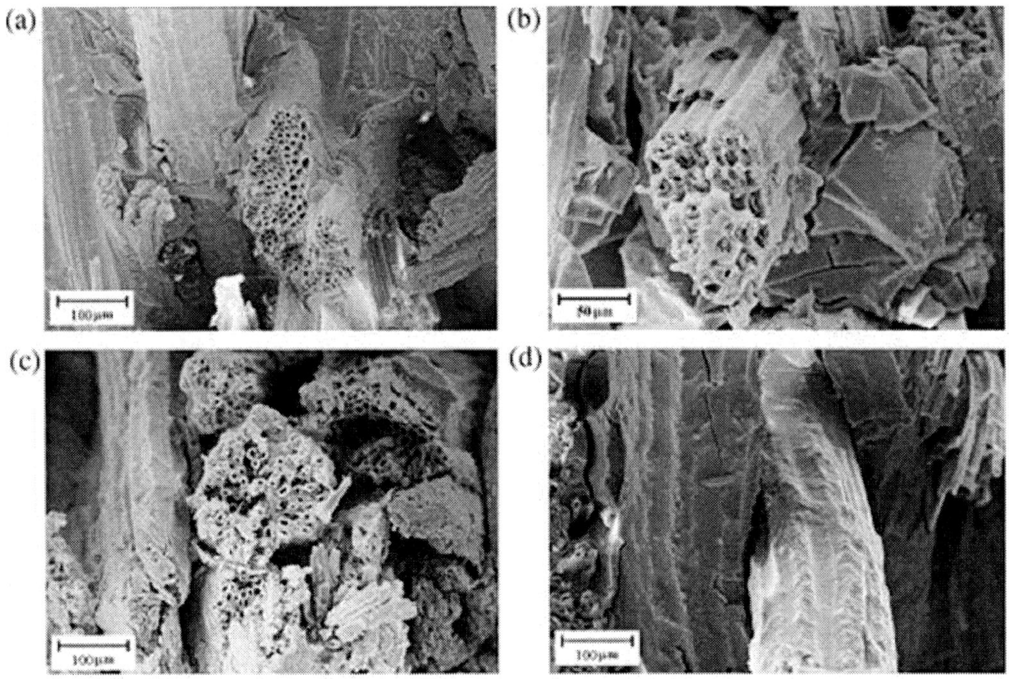

Figure 24. SEM images of the impact fracture surface of glyoxal–phenol–resorcinol composites reinforced by sisal fibers with cure cycle 1 (a) 300X; (b) 1000X and cycle 2 (c) 500 X; (d) 500X [95].

Nirmal et al [96] investigated the use treated betelnut fibers (in mat form) as reinforcement in polyester composites. Different orientation of the fibers mats with respect to the sliding direction of the counterface was considered; anti-parallel (AP), parallel (P) and normal (N). The worn surface morphology was studied using a scanning electron microscope (SEM). An optical microscope was used to observe the wear track surface on the counterface, and the modifications on the counterface roughness were studied. Results revealed that the wear and frictional performance of the composite were enhanced under wet contact conditions by about 54% and 95% compared to the dry.

Youlsif et al [97] also investigated polyester composites based on betelnut fibers for tribological applications using a block on disk machine at different applied loads and sliding distances at 2.8m/s sliding velocity under dry/wet contact conditions. Results revealed that betelnut fibers reinforced polyester composite had better wear and frictional performance under wet contact condition compared to dry (Figures 25 and 26).

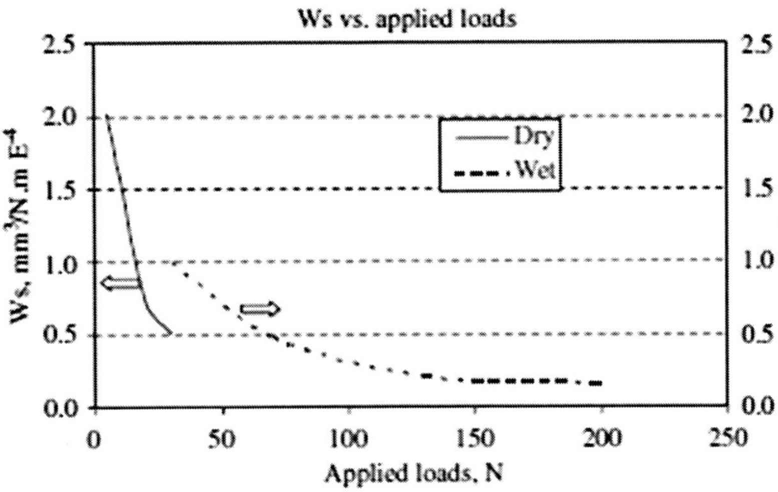

Figure 25. Specific wear rate at 5.0 km vs. applied loads [97].

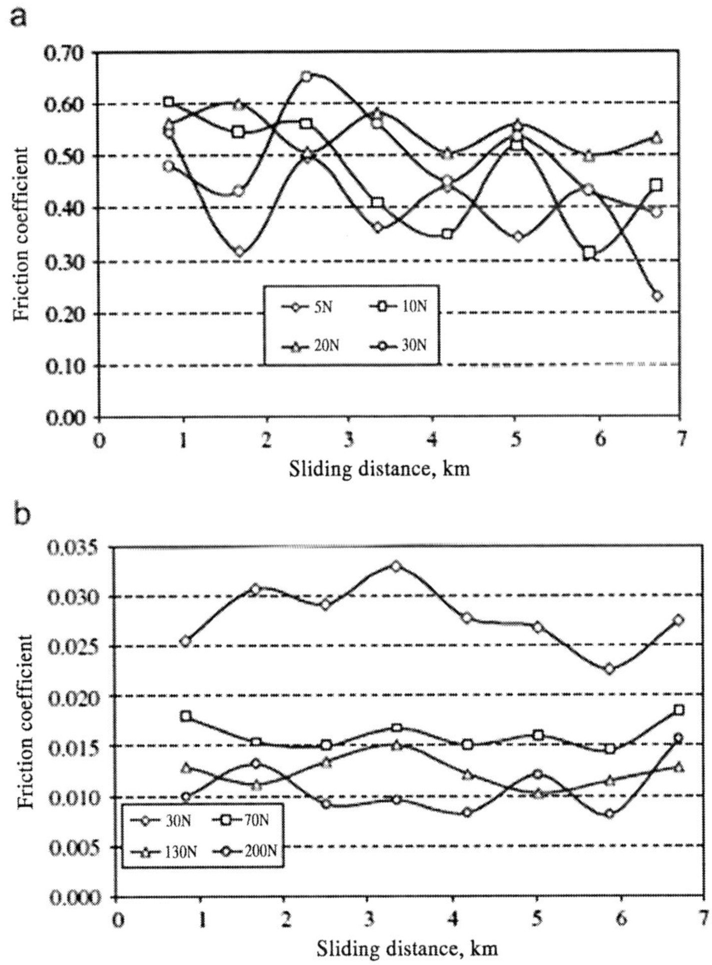

Figure 26. Fricction coefficient vs. sliding distance (a) Fricction coeficcient vs. sliding distance under dry condition and (b) Fricction coefficient Vs. distance under wet condition [97].

Wong et al [98] studied fracture behaviour of short bamboo fibers reinforced polyester composites. The matrix was reinforced with fibers ranging from 10 to 50, 30 to 50 and 30 to 60 vol.% at increments of 10 vol.% for bamboo fibers at 4, 7 and 10 mm lengths respectively. The results reveal that at 4 mm of fiber length, the increment in fiber content deteriorates the fracture toughness. As for 7 and 10 mm fiber lengths, positive effect of fiber reinforcement was observed (Figure 27).

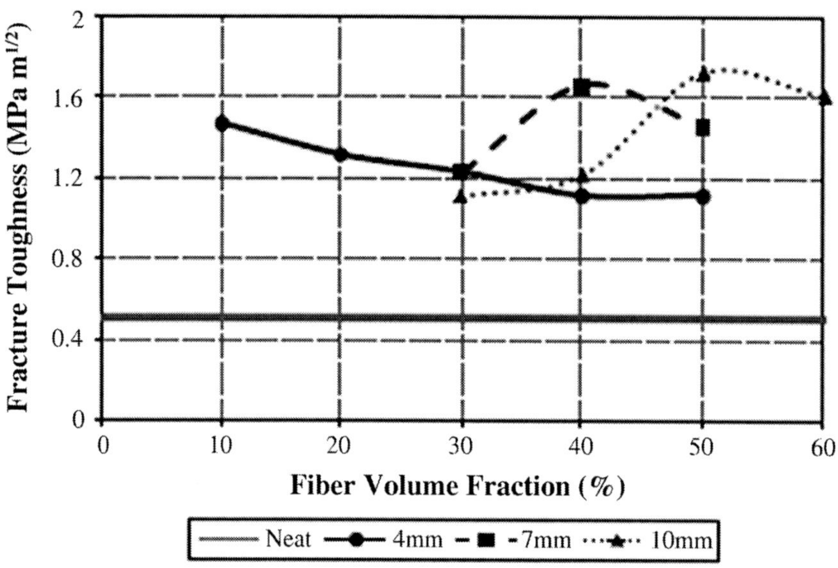

Figure 27. Effects of fibers volume fraction on fracture toughness of the composite [98].

The optimum fiber content was found to be at 40 vol.% for 7 mm fiber and 50 vol.% for 10 mm fiber.

5.2. Thermoplastic Composites

The mechanical properties of thermoplastic composites can be improved by the use of compatibilizers between the fiber and matrix and modification of surface fibers.

Araújo *et al.* [99] studied crystallization phenomenon and the interaction between the degradative processes of the HDPE/curaua fibers composites with and without coupling agent, prepared using an intermeshing corotating extruder. Two different coupling agents (poly(ethylene-co-vinyl acetate), EVA (UE-2866/32 with 28% of vinyl acetate) provided by Polietilenos União (Triunfo, Brazil) and maleic anhydride grafted polyethylene PE-g-MA (MEGHWAX SAW X-01) provided by Megh Ceras & Emulsões (São Paulo, Brazil), with 40–60 mg KOH/g as acidity level were used.

Thermogravimetric results of composites present a decrease in the fiber degradation Temperature in comparison with that isolated fiber, while insignificant differences are associated to the HDPE. It was observed an antagonistic fiber/HDPE interaction, considering that in general an increase in the fiber degradation temperature as reinforce is expected. The composite without coupling agent and that with EVA had part of the fiber weight loss process occurring at higher temperatures, close to the HDPE weight loss, as shown in Table 3. A higher char residue formation was observed when using PE-g-MA as coupling agent.

Table 3. Temperatures when the weight loss reaches 3 wt% (T_i) and when the larger rate of weight loss occurs (T_{max}) [99].

Samples	T_i (value at 3 wt% weight loss)	T_{max} curaua fiber process	T_{max} HDPE process
HDPE	427	------	476
Natural fibre	266	363	------
Calculated	315	363	476
Composite	337 ± 8	346 ± 4	470 ± 6
Composite + PE-g-MA	288 ± 8	350 ± 5	473 ± 6
Composite + EVA	316 ± 8	351 ± 3	471 ± 6

The compatibilized composite shows a higher interfacial interaction due to the reaction between acid groups of the maleic anhydride groups and hydrophilic groups on the fiber surfaces. As a consequence the degradation of one component accelerates the degradation of the other.

Also the increase of the composites crystallinity occurred due to the transcrystallinity effect, as shown in Table 4, which at the same time is decreased by the presence of the coupling agents due to the reactions with the fiber surface OH groups.

Table 4. T_c, ΔH_c, T_m, ΔH_m and X_c for the formulation [99].

Sample	T_c (°C)	ΔH_c (J/mol)	T_m (°C)	ΔH_m (J/mol)			X_c (%)		
				Obtained	Calculated	Desviation	Obtained	Calculated	Desviation
HDPE	113	185	138	190	----	----	65	----	----
Composite	111	179	138	188	152	+ 24%	81	52	+ 56%
Composite + EVA	112	149	135	162	148	+ 10%	62	51	+ 41%
Composite + PE-g-MA	110	129	137	136	148	- 8%	60	51	+ 18%

Carvalho *et al.* [59] studied the modification effect of green coconut fibers for mercerization and bleaching to reinforce HIPS in order to improve mechanical properties.

Green coconut fibers were pre-treated with 1 L alkaline solution containing 10 g sodium hydroxide (10% w/v), for a hour under constant stirring at room temperature. After treatment the solution was filtered in a vaccum filter and fibers were washed with distilled water until neutral pH. Then, fibers were dried in an oven at 50 °C for 24 hours.

The alkali treated fibers (24 g) were bleached with 200 mL solution containing 1 mL acetic acid and 3 g sodium chloride (80%). This solution was stirring for 2 hours at 70 °C, followed by filtration under vacuum and washing with distilled water until neutral pH. Finally, the bleached fibers were dried in an oven at 50 °C for 12 hours.

Figure 28 presents the X-ray diffractogram for untreated, alkali-treated, and bleached green coconut fibers. For fibers untreated and with different treatments the occurrence of two intense peaks was observed, close to values of 2θ = 16 ° and 2θ = 22 °, representing the cellulose crystallographic planes I_{101} and I_{002}, respectively. The X-ray diffraction peaks observed can be attributed to crystalline scattering and the diffuse background associated with disordered regions.

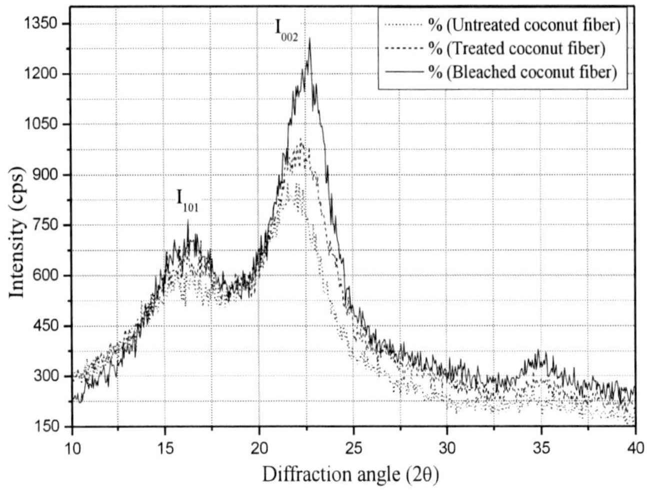

Figure 28. X-Ray diffractograms of cellulose [59].

The spectrum corresponding to untreated fibers shows diffraction peaks at the 2θ angles of 16.15° and 22.01°. The same peaks for alkali-treated fibers were observed at 16.25° and 22.23° and for bleached fibers at 16.25° and 22.77°. The superposition of the X-ray diagrams shows that the signal characteristics of fibers with different treatments were almost similar. However, alkali-treated and bleached fiber peaks were more intense than untreated fibers, peaks, which means that both treatments were able to remove part of the amorphous material covering the fiber, thus exposing the cellulose.

Mechanical properties such as elongation at break, tensile strength, and tensile modulus for pure HIPS and for composites materials containing different green coconut fibers treatment are shown in Table 5. The volume fractions of alkali-treated and bleached green coconut fibers inserted into the polymeric matrix were 10 and 30 wt%.

Table 5. Mechanical Properties of the Composite Materials [59].

Samples (Reinforcement in wt%)	Properties		
	Tensile strength (MPa)	Tensile modulus (MPa)	Elongation at break (%)
HIPS	24.58 ±0.12	3045.68 ±81.42	0.81 ±0.02
Alkali Treated coconut fibers (10%)/HIPS composites	24.51 ±0.65	3146.90 ±242.01	0.78 ±0.04
Alkali Treated coconut fibers (30%)/HIPS composites	24.77 ±0.90	3977.34 ±133.80	0.62 ±0.03
Bleached coconut fibers (10%)/HIPS composites	23.04 ±0.20	3340.49 ±106.75	0.69 ±0.02
Bleached coconut fibers (30%)/HIPS composites	23.32 ±0.37	3986.31 ±244.96	0.51 ±0.03

Figure 29 shows the stress-strain curves obtained for the pure HIPS and composites with different green coconut fibers treated. The curves for the pure HIPS showed a ductile character material with extensive plastic deformation. With the addition of alkali-treated fibers (Fig. 29 (a)) and bleached fibers (Fig. 29(b)) in the polymeric matrix, the curves show that the composites failed after a maximum point, with a small amount of plastic deformation.

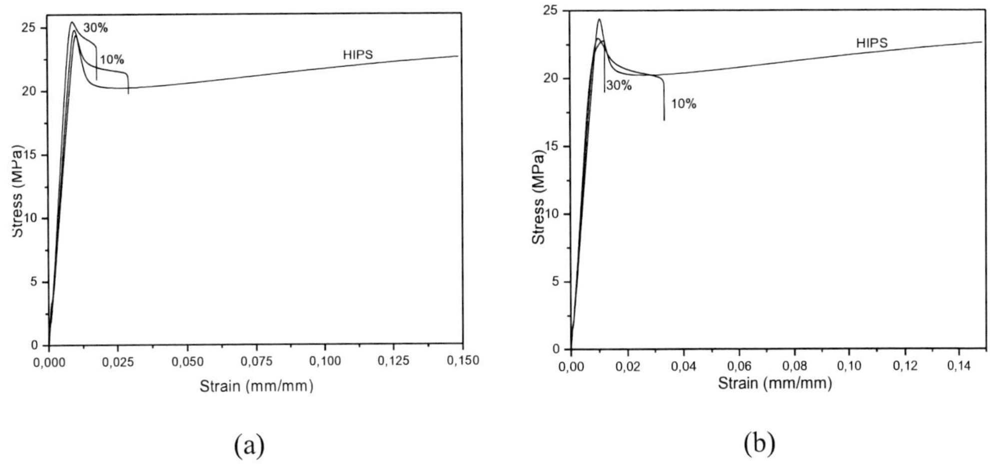

Figure 29. Stress-strain curves for the HIPS matrix and the composites with different (a) alkali-treated fiber contents, and (b) bleached fiber contents.

By analyzing the data in Table 5 and the graphics in Figure 30, it is possible to observe that values of tensile strength increased for composites reinforced with alkali-treated fiber, while they decreased for composites with bleached fiber. Values of tensile modulus increased with the addition of alkali-treated fiber and bleached fibers. Composites with 30% of fibers (alkali-treated and bleached) resulted in an increase of approximately 31% in the tensile modulus values, when compared with pure HIPS. The increase of tensile modulus according to the increase of fiber volume for both chemical treatments can be better viewed in Fig. 30 (b).

Due to more brittle character of reinforced composites, a decrease in the values of elongation at break was observed, especially for those samples reinforced with bleached fiber. The decrease in elongation at break according to the amount of fiber and chemical treatment can also be seen in Fig. 30 (c).

In this study, the reinforcement was more effective in increasing tensile modulus values of composites reinforced with 30% of alkali-treated and bleached fibers.

Mulinari *et al.* [100] studied the modification effect of sugarcane bagasse cellulose with zirconium oxychloride to reinforce HDPE in order to improve mechanical properties.

Cellulose from sugarcane bagasse was obtained by pre-treatment with 10% sulfuric acid solution, followed by centrifugation, deslignification with 1% sodium hydroxide solution and bleaching with sodium chloride. Modified cellulose from sugarcane bagasse was obtained by dissolution of 2 g of zirconium oxychloride in 100 mL of aqueous hydrochloric acid solution (0.5 mol L^{-1}), in which 5 g of cellulose were immersed in this solution. The material was precipitated with ammonium solution (1:3) at pH 10.0, under stirring, filtered under vacuum, exhaustively washed with distilled water for the complete removal of chloride ions (negative silver nitrate test) and dried at 50 °C for 24 h. The resulting material was designated as Cell/$ZrO_2.nH_2O$ [101]. The effect of modification on fibers was evaluated by mechanical properties. Mechanical properties of composites are summarized in Table 6. Modified cellulose fibers reinforced HDPE composites exhibited higher tensile strength in comparison to the non-modified cellulose fibers-reinforced HDPE composites.

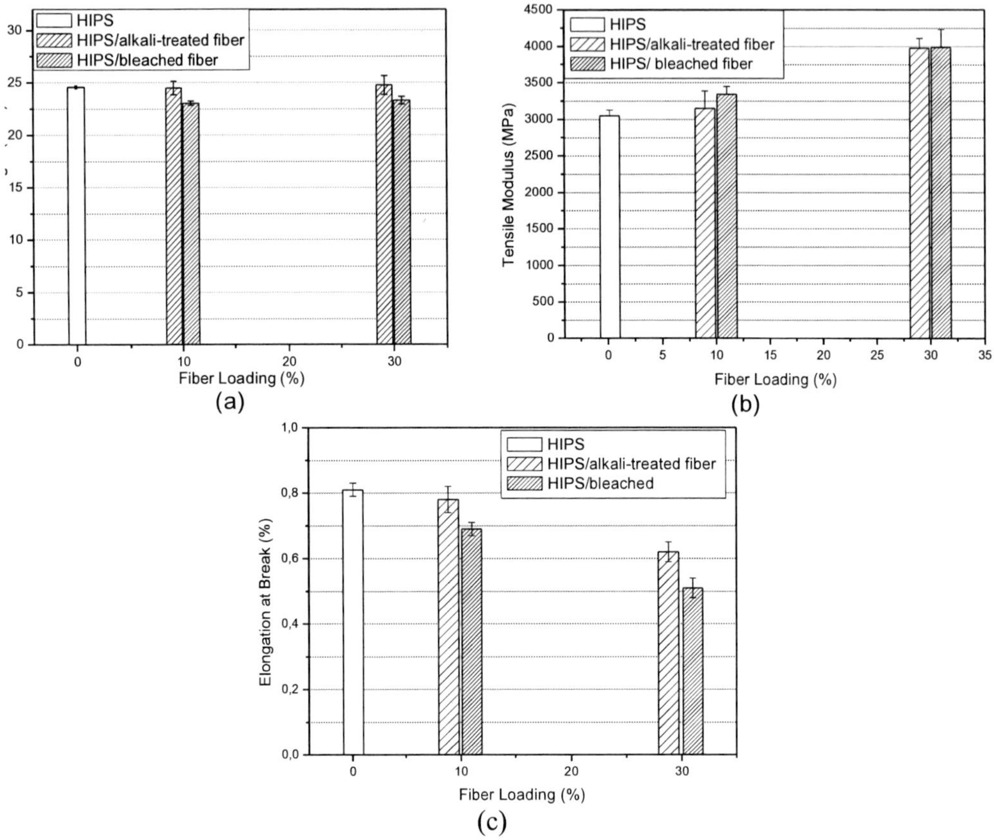

Figure 30. Mechanical properties of pure HIPS and composites with different fiber loadings: (a) tensile strength; (b) tensile modulus; and (c) elongation at break [19].

The amount of added reinforcement contributed to variation of the tensile modulus, as indicated in Table 6, but CM10% composites achieved higher tensile strength. Fibers insertion can contribute to an increase of modulus because the Young's modulus of the fibers is higher compared to the modulus of the thermoplastic matrix. However, to obtain a significant increase, a good interfacial bond between fibers and matrix is necessary.

Experimental results in Table 6 may be explained by the interaction observed between fibers and matrix during the mixing process. The modified cellulose fibers exhibited better tensile strength and adhesion between fibers and matrix compared to the non-modified cellulose, confirming that cellulose modification with zirconium oxychloride improves the adhesion between fibers and matrix.

Ibrahim *et at.* [38] also studied banana fibers and microfibrils as lignocellulosic reinforcements in high density polyethylene. Banana plant waste, as lignocellulosic fiber, was treated with alkaline pulping and steam explosion to produce banana fibers and banana microfibrils. The chemical modification, with maleic anhydride, of the produced particles was further carried out. Composite materials were processed from these natural unmodified and maleated lignocellulosic fibers using polyethylene as the polymeric matrix. The thermal and mechanical properties were studied by differential scanning calorimetry (DSC) and tensile tests, respectively. Better compatibility and enhanced mechanical properties were obtained when using banana microfibrils. The chemical composition of fibers, in terms of lignin and cellulose, as well as their degree of crystallinity, were found to have a strong influence on the mechanical properties of the composites.

Table 6. Mechanical Properties of the Composites [100].

Materials	Properties		
	Elongation at break (%)	Tensile strength (MPa)	Tensile modulus (MPa)
HDPE	8.9 ± 0.8	15.7 ± 1.1	732.45 ± 90.6
CB5%	5.4 ± 0.4	16.2 ± 0.7	942.5 ± 98.6
CB10%	5.5 ± 0.2	15.6 ± 0.3	897.4 ± 27.5
CB20%	5.7 ± 0.3	15.8 ± 0.3	1140.7 ± 62.5
CM5%	7.2 ± 0.1	20.8 ± 0.4	1177.7 ± 25.0
CM10%	7.4 ± 0.4	21.9 ± 0.6	1238.5 ± 41.0
CM20%	6.5 ± 0.2	20.9 ± 0.4	1306.4 ± 26.9

Reinforcement in wt%.

Bledzki *et al.* [102] studied polypropylene composites with enzyme modified abaca fibers. Abaca fibers reinforced PP composites were prepared using a high speed mixer followed by injection moulding with 30 wt.% of fibers load. Prior to composite production, the fibers were modified by fungamix and natural enzyme. The effects of modification of the fibers were assessed on the basis of morphology and thermal resistance and as well as on mechanical, thermal and environmental stress corrosion resistance properties of the resulting composites. Coupling agent (MA-PP) was also used with unmodified abaca fibers to observe the coupling agent effect on resulting composites properties. The moisture absorption of the composites was found to be reduced 20–45% due to modification. Tensile strength found to be 5–45% and flexural strengths found to be 10–35% increased due to modification. Modified fibers composites found to better resistance in acid and base medium.

REFERENCES

[1] Mohanty, A. K.; Misra, M. & Drzal, L. T. (2005). *Natural Fibers, Biopolymers, and Biocomposites.*Taylor &Francis Group, CRC, cap 2, 39-97.

[2] Wallenberger, F. T. & Weston, N. (2004). *Natural Fibers, Plastics and Composites.* Kluwer Academic Publishers, cap 14, 249-274.

[3] Mano, B., Araújo, J. R., Spinacé, M. A. S. & De Paoli, M. A. (2010). *Polyolefin composites with curaua fibres: Effect of the processing conditions on mechanical*

*properties, morphology and fibres dimensions.*Composites Science and Technology,70, 29-35.

[4] Gomes, A., Matsuo, T., Goda, K. & Ohgi, J. (2007). *Development and effect of alkali treatment on tensile properties of curaua fiber green composites.* Composites: Part A, 38, 1811-1820.

[5] Alves, C., Ferrão, P. M. C., Silva, A. J., Reis, L. G., Freitas, M., Rodrigues, L.B. & Alves, D.E. (2010). *Ecodesign of automotive components making use of natural jute fiber composites.* Journal of Cleaner Production, 18, 313-327.

[6] Araujo, J. R., Mano, B., Teixeira, G.M., Spinacé, M.A.S. & De Paoli, M. A. (2010). *Biomicrofibrilar composites of high density polyethylene reinforced with curauá fibers: Mechanical, interfacial and morphological properties.* Composites Science and Technology, In Press.

[7] Bismark, A., Mohanty, A. K., Aranberri-Askargorta, I., Czapla, S., Misra, M., Hinrichsen, G. & Springer, J. (2001).*Surface characterization of natural fibers: surface properties and the water up-take behavior of modified sisal and coir fibers.* Green Chemistry, 3, 100-107.

[8] Mulinari, D. R., Voorwald, H. J. C., Cioffi, M. O. H., Da Silva, M. L. C. P., Cruz, T.G. & Saron, C. (2009). *Sugarcane bagasse cellulose/HDPE composites obtained by extrusion.* Composites Science and Technology, 69, 214-219.

[9] Silva, R., Haraguchi, S. K., Muniz, E.C. & Rubira, A. F. (2009). *Aplicações de fibras lignocelulósicas na química de polímeros e compósitos.* Química Nova, 32, 661-67.

[10] Pietak, A., Korte, S., Tan, E., Downard, A. & Staiger, M. P. (2007). *Atomic force microscopy characterization of the surface wettability of natural fibres.* Applied Surface Science, 253, 3627-3635.

[11] Comeau, G. & Fredericton, N. B. (2006). *A petition concerning options to the long and deeply entrenched practice of flax straw burning on the canadian prairies.* Available: < http://www.oag-bvg.gc.ca/internet/English/pet_186_e_28922.html>. Acessible: 2010 jul 06.

[12] Summerscales, J., Dissanayake, N. P. J.; Virk, A.S. & Hall, W. (2010). *A review of bast fibres and their composites.* Part 1 – Fibres as reinforcements. Composites: Part A, In Press.

[13] Chawla, K. K. (1998). *Fibrous Materials.* New York: Cambridge University Press.

[14] Spinacé, M. A. S., Lambert, C. S., Fermoselli, K. K. G. & De Paoli, M. A. (2009). *Characterization of lignocellulosic curaua fibres.* Carbohydrate Polymers, 77, 47-53.

[15] Fengel, D. & Wegener, G. (1989). *Wood – Chemistry, Ultrastructure, Reactions.* New York: Walter de Gruyter, p.188.

[16] Saha, P., Manna, S., Chowdhury, S. R., Sen, R., Roy, D. & Adhikari, B. (2010). *Enhancement of tensile strength of lignocellulosic jute fibers by alkali-steam treatment.* Bioresource Technology, 101, 3182-3187.

[17] Saheb, D. N. & Jog, J. P. (1999). *Natural fiber polymer composites: A Review.* Advances in Polymer Technology, 18, 351-363.

[18] Textile Class. *Properties & Types of Jute/Jute Fiber Growing Countries List.* Available:< http://articles.textileclass.com/properties-types-of-jute-jute-fiber>. Acessible: 2010 jul. 27.

[19] Acha, B.A., Reboredo, M. M. & Marcovich, N. E. (2007). *Creep and dynamic mechanical behaviour of PP-jute composites: Effect of the interfacial adhesion.* Composites: Part A, 38, 1507-1516.

[20] *Wingham Wool Work.* Available:< http://www.winghamwoolwork.co.uk/fbr_vegetable.php >. Acessible:2010 aug 03.

[21] Wikipedia – *A enciclopedia livre.* Available: <http://pt.wikipedia.org/wiki/Juta>. Acessible: 2010 aug 05.

[22] Andrade, W.; siva, O.R.R>F.; Ribeiro,R.P.; Falcão,V.; Guedes,T.; Bindilat, A.; Souza, E.; Cunha, L.; Gledhill, S.; Borges, N.; Vilela, P. *O Sisal do Brasil- Brazilian Sisal.* Available: http://www.braziliansisal.com/. Acessible: 2010 jul 08.

[23] Martin, A. R., Martins, M. A., Mattoso, L. H. C.& Silva, O. R. R. F. (2009). *Caracterização Química e Estrutural de fibra de sisal da variedade Agave Sisalana.* Polímeros: Ciência e Tecnologia, 19, 40-46.

[24] Kim, J. T. & Netravali, A. N. (2010). *Mercerization of sisal fiers: Effect of tension on mechanical properties of sisal fiber and fiber-reinforced composites.* Composites: Part A, 41, 1245-1252.

[25] LaMMa–*Tecnologie per il Sistema Tessile.*(2010). Available:<http://www.lammatest.rete.toscana.it/lammatest/hbin/fibrenaturali-vegetali.php>. Acessible: 2010 aug 05.

[26] Bakare, I. O., Okieimen, F .E., Pavithran, D., Abdul Khalil, H.P.S. & Brahmakumar, M. (2010). *Mechanical and thermal properties of sisal fiber-reinforced rubber seed oil-based polyurethane composites.* Materials and Design, 31, 4272-4280.

[27] Mothé, C. G. & Araujo, C.R. (2004). *Caracterização Térmica e Mecânica de compósitos de poliuretano com fibras de curauá.* Polímeros: Ciência e Tecnologia, 14, 274-278.

[28] Amarasekera, J.; Santos, P. A.; Giriolli, J. C.; Moraes, G.; Mano, B. & De Paoli, M. A. (2009). *The curaua challenge: optimizing natural fibres.* Plastics Additives & Compounding, 11, 12-15.

[29] Jústiz-Smith, N. G.;Virgo, G.J. & Buchanan,V.E. (2008). *Potential of Jamaican banana, coconut coir and bagasse fibresas composite materials.* Materials Characterization, 59, 1273-1278.

[30] Monteiro, S. N.; Aquino, R. C. M. P.; Lopes, F. P. D.; Carvalho, E. A. & D'Almeida, J. R. M. (2006). *Comportamento Mecânico e Características Estruturais de Compósitos Poliméricos Reforçados com Fibras Contínuas e Alinhadas de Curauá.* Revista Matéria, 11, 197-203.

[31] *Global Pinoy.* (2009). Available:< http://www.globalpinoy.com/pinoyhealth/ph_gamotpinoy/>. Acessible: 2010 aug 05.

[32] Elanthikkal, S.; Gopalakrishnapanicker, U.; Varghese, S. & Guthrie, J. T. (2010). *Cellulose microfibers produced from banana plant waste: isolation and characterization.* Carbohydrate Polymers, 80, 852-859.

[33] Guimarães, J. L., Wypych, F., Saul, C. K., Ramos, L. P. & Satyanarayana, K.G. (2009). *Studies of the processing and characterization of corn starch and its composites with banana and sugarcane fibers from Brazil.* Carbohydrate Polymers, in press.

[34] Liu, H., Wu Q. & Zhang Q. (2009). *Preparation and properties of banana fiber-reinforced composites based on high density polyethylene (HDPE)/Nylon-6 blends.* Bioresorce Technology, 100, 6088–6097.

[35] Pothan, L. A. & George, C. N. (2010). *Dynamic Mechanical and Dielectric Behavior of Banana–Glass Hybrid Fiber Reinforced Polyester Composites.* Journal of Reinforced Plastics and Composites, 29, 1131-1145.

[36] Zainudin, E.S., Sapuan, S. M., Abdan, K. & Mohamad, M. T. M. (2009). *Thermal degradation of banana pseudo-stem filled unplasticized polyvinyl chloride (UPVC) composites.* Materials and Design, 30, 557–562.

[37] Cruz, M. A. R. F., Guerra, A. R. G., Ishikawa, D.N., Alfaya, R. V. S., Alfaya, A. S. (2009). *Farinha da casca da banana: um biossorvente para metais pesados de baixo custo.* In: Sociedade Brasileira de Química, Fortaleza - CE.

[38] Ibrahim, M. M., Dufresne, A., El-Zawawy, W.K. & Agblevor, F.A. (2010). *Banana fibers and microfibrils as lignocellulosic reinforcements in polymer composites.* Carbohydrate Polymers, 81, 811-819.

[39] Paul, S. A., Joseph, K., Mathew, G., Laly, A. & Pothen, S.T. (2010). *Influence of polarity parameters on the mechanical properties of composites from polypropylene fiber and short banana fiber.* Composites: Part A, in press.

[40] Souza, O. C., Melo, R. B., Candido, A. S., Silva, M. G. & Feitosa, T. B. (2009). *Aspectos da comercialização do coco no pólo Petrolina (Pe) e Juazeiro (Ba).* Simpósio de Produtividade em Pesquisa e II Encontro de Iniciação Científica do IFPI.

[41] *Embrapa- Brazilian Agricultural Research Corporation.* 2010. Available: <www.embrapa.br.> Accessible: 2010 july 29.

[42] *Journal da Imprensa Online.* 2010. Available:<www.jornaldaimprensa.com .br>.Accessible: 2010 july. 31.

[43] Monteiro, S. N., Terrones, L. A. H. & D'Almeida, J. R. M. (2008). *Mechanical performance of coir fiber/polyester composites.* Polymer Testing, 27, 591-595.

[44] Ishizaki, M. H, Visconte, L. L. Y., Leite, M. C. A. M., Furtado, C. R. G. & Leblanc, J. L. (2006). *Caracterização mecânica e morfológica de compósitos de polipropileno e fibras de coco verde: Influência do teor de fibras e das condições de mistura.* Polímeros: Ciência e Tecnologia, 16(3), 182-186.

[45] Mulinari, D. R., Voorwald, H. J. C., Cioffi, M O. H., Da Silva, M. L. C. P. & Luz, S.M. (2009). *Preparation and Properties of HDPE/Sugarcane bagasse cellulose composites obtained for thermokinetic mixer.* Carbohydrate Polymers, 75,. 317-321.

[46] Rosa, M. F., Chiou, B., Medeiros, E. S., Wood, D. F., Williams, T. G., Mattoso, L. H. C., Orts, W. J., and Imam, S. H. (2009). *Effect of fiber treatments on tensile and thermal properties of starch/ethylene vinyl alcohol copolymers/coir biocomposites.* Bioresource Technology, 100, 5196-5202.

[47] Harish, S., Michael, D. P., Bensely, A., Mohan Lal, D., and Rajadurai, A. (2009). *"Mechanical property evaluation of natural fiber coir composite".* Materials Characterization,60(1), 44-49.

[48] Esmeraldo, M. A. (2006). *Preparação de novos compósitos suportados em matriz de fibra vegetal/natural.* PhD Thesis, Universidade Federal do Ceará.

[49] Luz, S. M., Gonçalves, A. R. & Del'Arco, A. P. (2007). *Mechanical behavior and microstructural analysis of sugarcane bagasse fibers reinforced polypropylene composites.* Composites: Part A, 38, 1455–1461.

[50] Luz, S. M., Del Tio, J.; Rocha, G. J. R.; Gonçalves, A. R. & Del'Arco, A. P.(2008). *Cellulose and cellulignin from sugarcane bagasse reinforced polypropylene comp-*

osites: Effect of acetylation on mechanical and thermal properties. Composites: Part A, 39, 1362–1369.

[51] Couto, L.O.; Leite, M. C. A. M.; Furtado, C. R. G.; Oliveira, F. L. B. O. & Correia, T. R. (2008). *Avaliação da Biodegradação de Compósitos de Poli(e-caprolactona)/fibra de coco verde*. In CBECimat, Porto de Galinhas – PE.

[52] Vilay, V., Mariatti, M., Taib, R.M. & Todo, M. (2008). *Effect of fiber surface treatment and fiber loading on the properties of bagasse fiber-reinforced unsaturated polyester composites*. Composites Science and Technology, 68, 631-638.

[53] Shenoy, A.V. (1999). *Rheology of filled Polymer Systems*.Dordrecht: Kluwer Academic Publishers, p. 475.

[54] Gurgel, L. V. A., Freitas, R. P. & Gil L. F. (2008).*Adsorption of Cu(II), Cd(II), and Pb(II) from aqueous single metal solutions by sugarcane bagasse and mercerized sugarcane bagasse chemically modified with succinic anhydride*. Carbohydrate Polymers, 74, 922-929.

[55] Mulinari, D. R. & Silva, M. L.C. P. (2008). *Adsorption of sulphate ions by modification of sugarcane bagasse cellulose*. Carbohydrate Polymers, 74,617-620.

[56] Akram, T., Memon, S. A. & Obaid, H. (2008). *Production of low cost self compacting concrete using bagasse ash*. Construction and Building Materials, in press.

[57] Onésippe, C., Passe-Coutrin, N., Toro, F., Delvasto S., Bilba, K. & Arsène, M. A. (2010). *Sugar cane bagasse fibres reinforced cement composites: Thermal considerations*. Composites: Part A, 41,549-556.

[58] *Embrapa – Brazilian Agricultural Research Corporation*. (2008). Available: <http:// www.cnptia.embrapa.br/content/avancam-os-estudos-de-zoneamento-da-cana-de-acucar-290408.htmlSite foto cana de acucar>. Acessible: 2010 aug 04.

[59] Carvalho, K. C. C.; Mulinari, D. R.; Voorwald, H. J. C. & Cioffi, M. O. H.(2010). *Chemical modification effect on the mechanical properties of hips/ coconut fiber composites*. BioResources, 5, 1143-1155.

[60] Mohanty, S., Sushil, K. V. & Sanjay, K. N. (2006). *Dynamic Mechanical and Thermal Properties of Mape Treated Jute/HDPE Composites*. Composites Science and Technology, 66, 538–547.

[61] Troedec, M. L., Sedan, D., Peyratout, C., Bonnet, J.P., Simith, A., Guinebretirre, R., Gloaguen, V. & Krausz, P. (2008). *Influence of various chemical treatments on the composition and structure of hemp fibres*. Composites: Part A, 39,514-522.

[62] Razera, I. A. T. (2006). *Fibras lignocelulósicas como agente de reforço de compósitos de matriz fenólica e lignofenólica*. PhD Thesis, Universidade de São Paulo, São Carlos-SP.

[63] Brígida, A. I. S., Calado, V.M.A., Gonçalves, L.R.B. & Coelho, M. A. Z. (2010). *Effect of chemical treatments on properties of green coconut fiber*. Carbohydrate Polymers, 79, 832-838.

[64] Calado, V., Barreto, D. W. & D'Almeida, J. R. M. (2000). *The effect of a chemical treatment on the structure and morphology of coir fibers*. Journal of materials science letters, 19, 2151-2153.

[65] Mulinari, D.R., Cruz, T.G., Cioffi, M. O. H., Voorwald, H. J. C., Da Silva, M. L. C. P. & Rocha, G. J. M. (2010). *Image analysis of modified cellulose fibers from sugarcane bagasse by zirconium oxychloride*. Carbohydrate Research, in press.

[66] Matabola, K. P., De Vries, A. R., Moolman, F. S. & Luyt, A. S. (2009). *Single polymer composites: A review.* Journal Materials Science, 44, 6213-6222.

[67] Cho, D, Lee, H. S. & Han, S. O. (2009). *Effect of fiber surface modification on the interfacial and mechanical properties of kenaf fiber-reinforced thermoplastic and thermosetting polymer composites.* Composites Interface, 16, 711-729.

[68] Bledzki, A.K., Gassan, J. (1999). *Composites reinforced with cellulose based fibres.* Programs in Polymer Science, 24, 221-274.

[69] Takemura, A. & Tajvidi, M. (2009). *Effect of fiber content and type, compatibilizer, and heating rate on thermogravimetric properties of natural fiber high density polyethylene composites.* Polymer Composites, 30, 1226-1233.

[70] Hussein, I. A. (2004). *Implications of melt compatibility/incompatibility on thermal and mechanical properties of metallocene and Ziegler–Natta linear low density polyethylene (LLDPE) blends with high density polyethylene (HDPE): influence of composition distribution and branch content of LLDPE.* Polymer International, 53, 1327-1335.

[71] Xie, Y., Hill, C. A. S., Xiao, Z., Militz, H. & Mai, C. (2010). *Silane coupling agents used for natural fiber/polymer composites: A review.* Compostes: Part A, 41, 806-819.

[72] Morán, J., Alvarez, V., Petrucci, R., Kenny, J. & Vazquez, A. (2007). *Mechanical properties of polypropylene composites based on natural fibers subjected to multiple extrusion cycles.* Journal of Applied Polymer Science, 103, 228-237.

[73] Biagiotti, J., Fiori, S., Torre, L., López-Machado, M. A. & Kenny, J. (2004). *Mechanical properties of polypropylene matrix composites reinforced with natural fibers: A statistical approach.* Polymer Composites, 25, 26-36.

[74] Bakar, A. A., Hassan, A. & Mohd Yusof, A. F. (2005). *Effect of oil palm empty fruit bunch and acrylic impact modifier on mechanical properties and processability of unplasticized poly(vinyl chloride) composites.* Polymer-Plastics Technology Engineering, 44, 1125-1137.

[75] Crespo, J. E., Sánches, L., Garcia, D. & López, J. (2008). *Study of the mechanical and morphological properties of plasticized PVC composites containing rice husk fillers.* Journal of Reinforced Plastics & Composites, 27, 229-243.

[76] Ayora, M.; Rios, R., Quijano, J. & Márques, A. (1997*). Evaluation by torque-rheometer of suspensions of semi-rigid and flexible natural fibers in a matrix of poly(vinyl choloride).* Polymer Composites, 18, 549-560.

[77] Xu, Y., Wu, E. Q., Lei, Y., Yao, F. & Zhang, Q. (2008). *Natural fiber reinforced poly(vinyl chloride) composites: Effect of fiber type and impact modifier.* Journal of Polymers and the Environment, 16, 250-257.

[78] Starnes Jr, W. H. (2002). *Structural and mechanistic aspects of the thermal degradation of poly(vinyl chloride).* Programs in Polymer Science, 27, 2133-2170.

[79] Haneefa, A., Bindu, P. & Aravind, I. (2008). *Studies on tensile and flexural properties of short banana/glass hybrid fiber reinforced polystyrene composites.* Journal of Materials Science, 42, 1471-1489.

[80] Guyot, A. (1986). *Recent developments in the thermal degradation of polystyrene-A Review.* Polymer Degradation and Stability, 15, 219-235.

[81] Hepworth, D. G., Bruce, D. M., Vincent, J. F. V. & Jeronimids, G. (2000). *The manufacture and mechanical testing of thermosetting natural fibre composites.* Journal of Materials Science, 35, 293-298.

[82] Kaddami, H., Dufresce, A., Khelifi B., Bendahou, A., Taourirte, M., Raihane, M., Issartel, N., Sautereau, H., Gérard, J. & Sami, N. (2006). *Short palm tree fibers – thermoset matrices composites*. Composites: Part A, 37, 1413-1422.

[83] Raquez, J. M., Deléglise, M., Lacarmpe, M. F. & Krawczak, P. (2010). *Thermosetting (bio)materials derived from renewable resources: A critical review*. Progress Polymer Science, 35, 487-509.

[84] Lopes L. *Bakelite, o primeiro plástico sintético*. 1ª Semana de Polímeros do IMA, 2007.

[85] Hodgkin, J. H., Simon, G. P.& Varley, R. J. (1998). *Thermoplastic toughening of epoxy resins: a critical review*. Polymers for Advanced Technologies, 9, 3-10.

[86] Newman, R. H., Clauss, E. C., Carpenter, J. E. P. & Thumm, A. (2007). *Epoxy composites reinforced with deacetylated phormium tenax leaf fibres*. Composites: Part A, 38, 2164-2170.

[87] Yemele, M. C. N., Koubaa, A., Cloutier, A., Soulounganga, P. &Wolcott, M. (2010). *Effect of bark fiber content and size on the mechanical properties of bark/HDPE composites*. Composites: Part A, 41, 131–137.

[88] Krouit, M., Belgacem, M. N. & Bras, J. (2010*). Chemical versus solvent extraction treatment: Comparison and influence on polyester based bio-composite mechanical properties*. Composites: Part A, 41, 703–708.

[89] Bessadok, A., Roudesli, S., Marais, S., Follain, N., & Lebrun, L. (2009). *Alfa fibres for unsaturated polyester composites reinforcement: Effects of chemical treatments on mechanical and permeation properties*. Composites: Part A, 40, 184-195.

[90] Megiatto Jr., J. D., Ramires, E. C. & Frollini, E. (2010). *Phenolic matrices and sisal fibers modified with hydroxy terminated polybutadiene rubber: Impact strength, water absorption, and morphological aspects of thermosets and composites*. Industrial Crops and Products, 31, 178–184.

[91] Bakare, I. O., Okieimen, F. E., Pavithran, C., Abdul Khalil, H. P. S. & Brahmakumar, M. (2010). *Mechanical and thermal properties of sisal fiber-reinforced rubber seed oil-based polyurethane composites*. Materials and Design, 31, 4274–4280.

[92] Zaman, H. U., Khan, A.,Khan, R. A., Huq, T., Khan, M. A., Shahruzzaman, M., Rahman, M. M., Al-Mamun, M. & Poddar, P. (2010). *Preparation and characterization of jute fabrics reinforced urethane based thermoset composites: Effect of UV radiation*. Fibers and Polymers, 11, 258-265.

[93] Kushwaha, P. K. & Kumar, R. (2010). *Bamboo fiber reinforced thermosettimg resin composites: Effect of graft copolymerization of fiber with methacrylamide*. Journal of Applied Polymer Science, 118, 1006-1013.

[94] De Rosa, I. M., Santulli, C. & Sarasini, F. (2010). *Mechanical and thermal characterization of epoxy composites reinforced with random and quasi-unidirectional untreated Phormium tenax leaf fibers*. Materials and Design, 31, 2397–2405.

[95] Ramires, E. C., Megiatto Jr., J. D., Gardrat, C., Castellan, A. &Frollini, E. (2010). *Biobased composites from glyoxal–phenolic resins and sisal fibers*. Bioresource Technology, 101, 1998–2006.

[96] Nirmal, U., Yousif, B.F., Rilling, D. & P.V. Brevern. (2010*). Effect of betelnut fibres treatment and contact conditions on adhesive wear and frictional performance of polyester composites*. Wear, 268, 1354–1370.

[97] Yousif, B. F., Saijod, T. W. & Lau, S. McWilliam (2010). *Polyester composite based on betelnut fibre for tribological applications*. Tribology International, 43, 503-511.

[98] Wong, K. J., Zahi, S., Low, K. O. & Lim, C. C. (2010). *Fracture characterisation of short bamboo fibre reinforced polyester composites.* Materials and Design, 31, 4147–4154.

[99] Araújo, J. R., Waldman, W. R. & De Paoli, M. A. (2008). *Thermal properties of high density polyethylene composites with natural fibres: Coupling agent effect.* Polymer Degradation and Stability, 93, 1170-1175.

[100] Mulinari, D. R., Voorwald, H. J. C., Cioffi, M. O. H., Rocha, G. J. M. & Da Silva, M. L. C. P. (2010). *Surface modification of sugarcane bagasse cellulose and its effect on mechanical and water absorption properties of sugarcane bagasse cellulose/ HDPE composites.* BioResources 5(2), 661-671.

[101] Mulinari, D. R., Da Silva, G. L. J. P. & Da Silva, M. L. C. P. (2006). *Adsorção de íons dicromato nos compósitos celulose/ZrO2.nH2O preparados pelos métodos da precipitação convencional e em solução homogênea.* Química Nova, 29, 496-500.

[102] Bledzki, A. K., Mamun, A. A., Jaszkiewicz, A. & Erdmann, K. (2010). *Polypropylene composites with enzyme modified abaca fibre.* Composites Science and Technology, 70, 854–860.

In: Thermoplastic and Thermosetting Polymers and Composites ISBN: 978-1-61209-264-5
Editors: Linda D. Tsai and Matthew R. Hwang

Chapter 4

PHASE SEPARATION OF PMMA-MODIFIED VINYL-ESTER THERMOSETS: MORPHOLOGY, THERMODYNAMICS AND MECHANICAL PROPERTIES

Walter F. Schroeder*, Julio Borrajo and Mirta I. Aranguren

Institute of Materials Science and Technology (INTEMA), University of Mar del Plata – National Research Council (CONICET), Av. Juan B. Justo 4302, (7600) Mar del Plata – Argentina

ABSTRACT

In this Chapter, the initial miscibility, the developed morphologies, and the final properties of styrene(St)/vinyl-ester(VE) thermosets modified with poly(methyl methacrylate) (PMMA) are discussed. The effect of changing the molecular weight and the polydispersity of the VE oligomer and the PMMA modifier are presented. Firstly, the miscibility of binary and ternary physical mixtures of the components involved in the different studied formulations is analyzed. The experimental liquid-liquid equilibrium curves (e.g. cloud point curves) allow computing the binary interaction parameters, χ, in the framework of the Flory Huggins theory for polydisperse systems. These parameters are used to model the quasiternary phase diagram that represents the initial thermodynamic state of each particular system. This miscibility behavior originates quite different morphologies in the cured materials, generated by polymerization induced phase separation (PIPS) mechanism. For instance, dispersion of thermoplastic-rich particles in a thermoset-rich matrix, cocontinuous structure, dispersion of thermoset-rich particles in a thermoplastic-rich matrix (phase-inverted structure), or typical macrophase morphology characterized by droplets-like domains with secondary phase separation inside the droplets can be observed. These morphological structures are directly related to the thermal and mechanical properties, as well as the volume shrinkage of the final systems. The evaluation of the dynamic mechanical behavior, flexural modulus, compressive yield stress, and fracture toughness shows that the addition of PMMA increases the fracture resistance without significantly compromising the thermal or mechanical properties of the

* To whom the correspondence should be addressed: E-mail: wschroeder@fi.mdp.edu.ar

vinyl-ester networks, which is inevitable when using elastomeric additives. The reason for the existence of an optimum modifier concentration is also discussed.

Keywords: vinyl-ester resins, thermoplastic modifiers, phase separation, low-profile additives, thermoset toughening.

1. INTRODUCTION

Vinyl-ester (VE) precursors have been extensively used in the production of thermosetting polymers for a large variety of applications, in which rigid materials resistant to high temperature and solvents are required. Some specialized applications include the use of the resins in dental implants, bone cements, but also in the construction of ship hulls, or equipment for the chemical industry. In order to obtain crosslinked thermosets, these VE oligomers must be copolymerized with an unsaturated monomer, usually styrene (St). [1].

As it is the case with most thermoset materials, the VE-based networks, are fragile and undergo high volume shrinkage during polymerization. The incorporation of fillers and reinforcements partially alleviates the shrinkage, but the issue is not completely inhibited [2]. On the other hand, the incorporation of additional deformation mechanisms by the generation of a heterogeneous structure is a useful way of improving the toughness of these polymers. Such microstructures can be achieved by the addition of preformed particles, [3, 4, 5] or by the *in situ* formation of particles through the technique called polymerization-induced phase separation (PIPS) [3]. This method consists of preparing a reactive homogeneous mixture of the thermosetting precursors and the modifying additive. As the reaction proceeds, the system phase-separates as a result of the reduction of the entropic contribution to the Gibbs free energy of mixing, which occurs due to the copolymer formation, the change in the relative concentrations of the mixture components, and the increase in the level of crosslinking. The developed morphology, determined by the size, shape, and relative proportions of the present phases, greatly affects the fracture properties of the final material. Moreover, other characteristics such as physical, mechanical, and thermal properties are also affected [6, 7, 8, 9].

During reaction, the system is comprised by four components: unreacted St and VE comonomers, formed (St-*co*-VE) copolymer, and modifier additive. At a given conversion, the phase transformation diagram is represented by a tetrahedron with one component in each vertex. The St-VE-modifier triangular face of the tetrahedron represents the phase behavior for the system at zero conversion and it constitutes the ternary phase diagram for the mixture of the three initial components. Several studies on similar systems formulated with St, unsaturated polyester (UP), and thermoplastic modifier, such as PMMA or poly(vinyl acetate) (PVAc), [10, 11] demonstrated that the morphology developed during polymerization can be correlated with the position of the initial mixture on the corresponding ternary phase diagram. Then, the knowledge of these phase diagrams can help to understand the effect of variables such as temperature, composition, or components molecular weight on the developed morphologies.

VE thermosetting systems have been previously modified with liquid elastomers, such as carboxyl terminated poly(butadiene-co-acrylonitrile) (CTBN) and vinyl terminated poly(butadiene-co-acrylonitrile) (VTBN), to improve their fracture toughness. However, the enhancement in the fracture response was inevitably accompanied by a significant loss in the elastic modulus and yield strength [12, 13]. For this reason, the use of thermoplastic modifiers possessing higher modulus, higher glass-transition temperature, and a higher level of toughness has attracted attention in more recent years.

Another reason for adding thermoplastics to UP resins is the improvement of the surface quality of the molded parts, due to the reduction of the volume shrinkage during curing, [14, 15] with a minimum cost on other properties. PVAc has been reported as an efficient LPA for St-UP systems when its addition leads to co-continuous phase microstructures [16, 17]. Under these conditions, tensile stresses arising from internal thermal and curing-derived contractions in the presence of mechanical constraints (e.g. a closed mold) provoke cavitation in the weak PVAc-rich phase. These microcavities compensate for the shrinkage of the molded part, leading to a better copy of the mold's surface.

In this work, the effect of the addition of PMMA to St-VE networks is investigated. The thermodynamic behavior of the initial systems is studied and the final morphologies are presented and analyzed at the light of the initial miscibility of each system. Finally, the properties of the modified materials are reported and correlated with the observed morphologies.

2. Experimental

2.1. Materials and Sample Preparation

Two vinyl-ester (VE) oligomers were selected for this study. The first one (VE_S) was a low molecular weight oligomer synthesized in our laboratory by reacting a diglycidyl ether of bisphenol A epoxy resin (DER 332, Dow Chemical Co., epoxy equivalent weight 175 g/eq) with methacrylic acid (Norent Plast S.A., laboratory-grade reagent) and triphenylphosphine (Fluka A.G., analytical reagent) as a catalyst. The final conversion reached was higher than 97%, and the final product was stabilized with 500 ppm hydroquinone. The second VE was a commercial resin (VE_C) (Derakane Momentum 411-350, Dow Chemical Co.). The molar masses were measured by size exclusion chromatography (SEC) using a polystyrene calibration. PMMA(41k) supplied by Aldrich Chemical Co., and PMMA(239k) of higher molecular weight and polydispersity index supplied by Subiton Laboratories, were used as thermoplastic modifiers. The characteristics of all the components used in this work are summarized in Table 1.

Crosslinking reactions were carried out with a weight proportion of VE/St = 55:45, an usual commercial formulation. The samples were cured at different temperatures (e.g. 25, 40, and 80 °C), and all of them were post-cured at 150 °C during two hours. The reactions were initiated using 2 wt% benzoyl peroxide (Luzidol 75%, Akzo Chemical S.A.), and 0.4 wt% *N,N*-dimethylaniline (Akzo Chemical S.A.) as a promoter of the curing reactions at 25 and 40 °C. The modified samples were prepared by addition of PMMA in a proportion of 5-20 wt% with respect to the total weight. The PMMA particles were initially dissolved in St and then mixed with the VE and remaining St to reach the final composition.

Table 1. Characteristics of the used chemical components.

	Styrene(0) (St)	VE_S(1) synthesized	VE_C(1) commercial	PMMA(2) (41k)	PMMA(2) (239k)
M_n (g/mol)	104	600[a]	950[a]	41,500[a]	239,000[a]
M_w (g/mol)	-	640[a]	1950[a]	80,000[a]	641,000[a]
M_w/M_n	-	1.06	2.05	1.93	2.68
$Density_{25°C}$ (g/cm^3)	0.89[b]	1.16[b]	1.16[b]	1.19[c]	1.12[c]

[a] Measured by SEC.

[b] Measured using a precision balance for densities.

[c] Taken from the supplier's catalogue.

Compression testing samples were prepared by injecting the mixture of the monomers (with or without PMMA modifier) into glass cylinders 6 mm in diameter that were previously sprayed with a silicone release agent. After removal from the molds, the specimens were carefully machined to reach the final dimensions (length/diameter = 1.5-2) and to obtain perfect parallelism for the upper and lower base surfaces.

Plates for bending tests (3 mm thickness) and fracture measurements (6 mm thickness) were obtained by casting the mixture into a mold consisting of two glass plates coated with a silicone release agent, spaced by a rubber cord of the appropriated thickness, and held together with clamps.

2.2. Cloud-Point Measurements

The cloud-point temperature of binary physical mixtures was determined from solutions with different known compositions, by observing the point of change from transparency to cloudiness, or vice versa, by varying the mixture temperature under gentle agitation. Each measurement was repeated at least five times to account for reproducibility.

For ternary systems, cloud-point compositions were measured by a titration method. A fixed volume of a mixture of known composition was introduced in a glass cylinder with jacket and thermostatized at a fixed temperature by a fluid circulating in the jacket supplied from an external thermostatic bath. The mixture was titrated with a St-VE binary solution of known concentration from a burette with 0.1 cm^3 precision. Because the measurements were carried out at temperatures below room temperature, the system was isolated with a polyethylene wrapping and the water vapor was expulsed by a dry nitrogen current to prevent its condensation inside the system. The amount of solution required to achieve the point of change from transparency to cloudiness, or vice versa, allows calculating the cloud-point concentration in each test.

2.3. Electron Microscopy

Specimens fractured in liquid nitrogen were further sputter-coated with gold-palladium at 5-10 Pa and 45 mA for 30 s. The fracture surfaces were observed with a scanning electron microscope (JSM 6460 Lv, JEOL) at a 15-kV accelerating voltage.

Samples to be explored by transmission electron microscopy (TEM) were sectioned in an LKB ultramicrotome with a diamond knife. The rigidity of the obtained materials was high enough to prepare high quality ultrathin sections at room temperature. The section thickness was 60 nm for all the samples observed. The study was carried out with a JEOL 100 CX transmission electron microscope operated at 80 kV.

2.4. Physical and Mechanical Tests

The densities of the unreacted liquid mixtures were measured by weighing techniques with a density balance (precision 10^{-4} g/cm^3, Becker and Sons) at 25.0 ± 0.5 °C. The densities of the cured materials were obtained from cylindrical specimens carefully machined with parallel bottom and top faces. The density values were obtained from the weight of the specimens and the volume calculated from their measured dimensions (ASTM D 1895-89).

Dynamic mechanical tests were carried out with a PerkinElmer 7e, on rectangular bars 2 ± 0.1 mm thick and 3 ± 0.1 mm wide. A three-point bending geometry was used with a span of 15 mm, at a frequency of 1 Hz, and at a heating rate of 5 °C/min. The applied static stress was 0.5 MPa, and the dynamic stress was 0.35 MPa.

Mechanical tests were performed in an electromechanical INSTRON Universal Testing Machine model 4467. Flexural moduli were measured using three-point bending geometry according to ASTM D 790M-93 specifications. Compression test specimens were deformed between metallic plates lubricated with molybdenum disulfide according to ASTM D 695-91. Fracture strength was measured at room temperature using three-point bending geometry at a crosshead displacement rate of 10 mm/min. Central V-shaped notches were machined in the test specimens, and then a razor blade was positioned in the notch and gently tapped to induce the growth of a natural crack ahead the blade. The critical stress intensity factor at the onset of crack growth (K_{IC}) was calculated according to the ASTM D 5045-93 specifications.

3. Results and Discussion

3.1. Thermodynamic Behavior of the Mixtures

A convenient starting point for the thermodynamic description of polymer blends is the Flory-Huggins expression for the dimensionless mixing Gibbs free energy per mole of lattice sites:[18,19,20]

$$\frac{\Delta G^{mix}}{MRT} = \sum_i \sum_{j=1}^{N_i} \frac{\phi_{ij}}{r_{ij}} \ln \phi_{ij} + \sum_{i<k} \chi_{ik} \phi_i \phi_k \qquad (1)$$

where the symbols i,k = 1,2,3 represent components; j = 1,2,...,N_i the total number of different molecular species corresponding to the i component; ϕ_{ij} is the volume fraction of the j molecular species of the i component, and ϕ_i the total volume fraction of the i component; $\chi_{ik} = \chi_{ki}$ is the binary interaction parameter only temperature-dependent. M represents the total mole number of lattice sites and is given by $M = \sum_i \sum_{j=1}^{N_i} n_{ij} r_{ij}$, where n_{ij} is the number of moles of the j-mer of the i component, and r_{ij} its relative molar volume defined by $r_{ij} = V_{ij}/V_r$, with V_{ij} the molar volume of the j molecular species of the i component, and V_r the reference volume taken in this study as the molar volume of the styrene monomer.

The dimensionless chemical potential difference between the solution and the pure state per mole of lattice sites for the generic ij molecular species is defined by:

$$\frac{\Delta\mu_{ij}}{r_{ij}RT} = \frac{1}{r_{ij}} + \frac{\ln\phi_{ij}}{r_{ij}} - \sum_i \sum_{j=1}^{N_i} \frac{\phi_{ij}}{r_{ij}} + \sum_{k\neq i} \chi_{ik}\phi_k(1-\phi_i) - \chi_{kl}\phi_k\phi_l \tag{2}$$

where $i(k,l) = 1,2,3$ are the components in a quasiternary system, and $i(k) = 1,2$ with $\phi_l = 0$ in a quasibinary one. In equations (1) and (2) the binary interaction parameters were assumed inversely proportional to temperature and composition independent, as shown:

$$\chi_{ik} = d_0 + \frac{d_1}{T} \tag{3}$$

where d_0 and d_1 are characteristic constants for a given ik binary system. However, it has been shown that the interaction parameter may also depend on both temperature and concentration. As proposed by several authors, [21, 22] $\chi(T,\phi)$ can be expressed as the product of two simple functions, one which is a unique function of the temperature $D(T)$, and the other an unique function of the concentration $B(\phi)$,

$$\chi_{ik}(T,\phi_i) = D(T) \times B(\phi_i) = \left(d_0 + \frac{d_1}{T}\right) \times \left(\frac{1}{1-b\phi_i}\right) \tag{4}$$

where d_0, d_1, and b are the characteristic constants of the ik system. Note that equations (1) and (2) require the use of χ-interaction parameters only temperature-dependent (eq 3). Equivalent equations considering $\chi(T,\phi)$ can be found in [20].

Under constant temperature and pressure, thermodynamic requirements for the liquid-liquid phase equilibrium in a mixture with polydisperse components are expressed by the identity of the chemical potentials of each molecular species between the two liquid phases:

$$\Delta\mu_{ij}^{\alpha} = \Delta\mu_{ij}^{\beta} \tag{5}$$

where α and β represent the mother and emergent phases in equilibrium at the cloud-point conditions. In addition, an equation for the mass balance for the β-emergent phase must be included in the analysis. This balance is given by $\sum_i \phi_i^{\beta} = 1$, with $\phi_i^{\beta} = \sum_{j=1}^{N_i} \phi_i^{\alpha} W_{ij} \exp(r_i\sigma_i)$ for a polydisperse polymer component and $\phi_i^{\beta} = \phi_i^{\alpha} \exp(r_i\sigma_i)$ for a monodisperse or single component. W_{ij} is the mass fraction molecular species distribution of the i polymer in the α-

mother phase, and σ_i is the separation factor of the *i* component, whose value determines the extension of the fractionation of each molecular species between the two equilibrium phases.

Numerical solution of the phase equilibrium equations (5) allows computing the values of the binary interaction parameters from the previous knowledge of the quasibinary and quasiternary experimental cloud-points. These values can be fitted with equation (3) or (4) in order to determine the characteristic constants of each particular system. Constants d_0, d_1, and b of the binary interaction parameters are listed in Table 2.

Table 2. Constants d_0, d_1, and b of the binary interaction parameters.

Binary Pair	d_0	d_1(K)	b
St-PMMA(41k)	0.085	131.3	--
St-PMMA(239k)	-1.028	443.4	--
St-VE_S, St-VE_C	-0.330	325.1	--
VE_S-PMMA(41k)	-2.498	705.1	--
VE_C-PMMA(41k), VE_C-PMMA(239k)	-0.252	93.7	-2.815

The obtained expressions for the binary interaction parameters can be used to carry out the inverse calculation and to obtain the theoretical phase diagram of each system studied.

3.1.1. St-PMMA Quasibinary Systems

Figures 1A and 1B show the experimental and calculated cloud-point curves (CPC), as well as calculated shadow and spinodal curves for the St-PMMA(41k) and St-PMMA(239k) quasibinary systems, respectively.

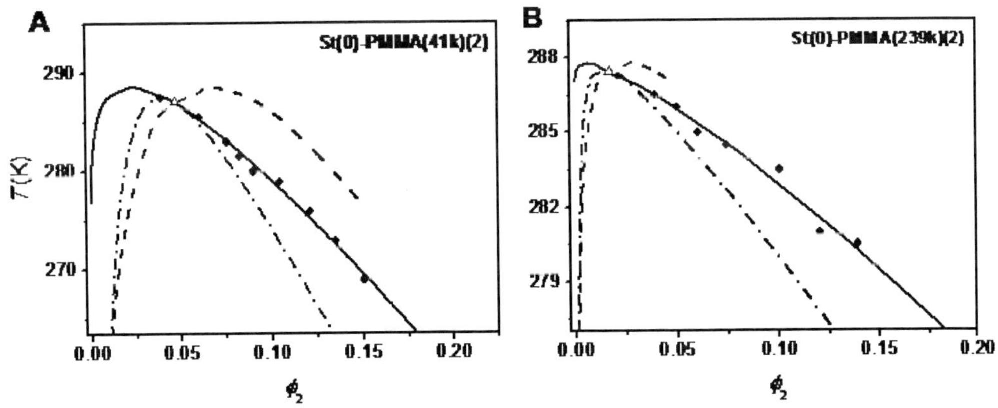

Figure 1. Phase diagrams of the quasibinary systems: A) St(0)-PMMA(41k)(2), and B) St(0)-PMMA(239k)(2). (•) Experimental cloud-points; (—) calculated CPC; (- -) calculated ShC; (- · -) calculated SC; (Δ) calculated CSP. [A) Reprinted from Journal of Applied Polymer Science, 100, Schroeder W.F., Yáñez M.J., Aranguren M.I., Borrajo J., "PMMA-Modified Divinylester/Styrene Resins: Phase Diagrams and Morphologies", 4539-4549, 2006, with permission from John Wiley and Sons. B) Reprinted from Polymer, 46, Schroeder W.F., Auad M.L., Barcia Vico M.A., Borrajo J., Aranguren M.I., "Thermodynamic, morphological, mechanical and fracture properties of poly(methyl methacrylate)(PMMA) modified divinylester(DVE)/styrene(St) thernosets", 2306-2319, 2005, with permission from Elsevier.].

ShC represents the composition of the emergent phase in equilibrium with the mother phase in the cloud-point conditions, whereas SC represents the thermodynamic stability limit of the quasibinary homogeneous solution. The critical solution point (CSP) is the intercept of the CPC, the ShC, and the SC, where the two separated phases become identical in composition and form one phase mixture. The resulting equations used in the calculation of SC and CSP in polydisperse polymer systems have been previously reported [20, 23].

As it can be seen in Figure 1, both St-PMMA systems present an upper critical solution temperature (UCST) behaviour. It is interesting to note that the immiscibility region inside the CPC envelope is very similar in both diagrams. This result suggests that the miscibility of the St-PMMA pair is not strongly affected by the molar mass of PMMA in the studied range.

Calculated CPC, ShC, SC, and CSP for these systems were computed from the inverse calculation using χ-interaction parameters only temperature-dependent (see Table 2). By comparing experimental and calculated CPCs, it can be concluded that the simplest model for the χ-parameter allows describing very well the phase diagrams for these systems.

3.1.2. St-VE Quasibinary Systems

St-VE_C system was analyzed by Auad et al. [24]. This system presents a phase diagram of the UCST type with the cloud-point curve located below room temperature. The experimental cloud-point values were analyzed with the simplest thermodynamic model taking the polydispersity into account. The constants corresponding to the expression of the interaction parameter as a function of temperature are given in Table 2.

Experimental cloud-points corresponding to the St-VE_S system could not be measured during cooling, as the low molar mass of the synthesized VE_S oligomer greatly increases the miscibility of the system. For the wide range of compositions explored, all the solutions remained homogeneous even at temperatures down to 242 K, at which the St crystallization was observed. If it is assumed that the χ-interaction parameter for this system can be considered only temperature dependent and not concentration or molar mass dependent, then it is possible to assign to this system the same χ-parameter expression obtained for the system St-VE_C. The ability of this $\chi(T)$ expression to properly represent the interaction parameter of the system St-VE_S was further confirmed using it in the analysis of the experimental cloud-point values for the St-VE_S-modifier quasiternary system.

3.1.3. VE-PMMA Quasibinary Systems

Cloud-points corresponding to the VE_S-PMMA(41k) solutions could not be measured with a satisfactory reproducibility, because the high viscosity of these solutions at low temperatures, where the phase separation is expected [20], made very difficult the expulsion of the microbubbles generated during stirring, which hindered the cloud-point determination. For this reason, the expression of the χ-parameter of this system was obtained from the analysis of the experimental cloud-point values of the St(0)-VE_S(1)-PMMA(41k)(2) quasiternary system at three different temperatures: 253, 260, and 265 K. For each temperature, a single value of the χ_{12}-parameter was computed averaging the calculated values obtained from each measured quasiternary cloud-point [23]. This procedure was repeated at each temperature. Then, the values were fitted with eq. (3) to determine the constants d_0 and d_1 of the χ-parameter corresponding to the VE_S-PMMA(41k) quasibinary system (Table 2). Figure 2A shows the calculated phase diagram for this system which could not be experimentally measured. Note that the immiscibility region is located at low temperatures as expected.

Figure 2B shows the experimental and calculated CPC for the VE_C-PMMA(41k) system. When the experimental CPC was analyzed in the framework of the simplest thermodynamic treatment, it was found that a simple temperature-dependent χ-interaction parameter is unable to properly describe the system behavior. As a consequence, the use of a χ-parameter depending on both temperature and concentration (eq. 4) was required for this quasibinary system. The calculated constants for the temperature and concentration terms are listed in Table 2. These constants were used to perform the inverse calculation and to compute the CPC, ShC and SC curves, which are shown in Figure 2B. It can be seen that the $\chi(T,\phi)$ parameter describes very well the quasibinary cloud-point curve for this system.

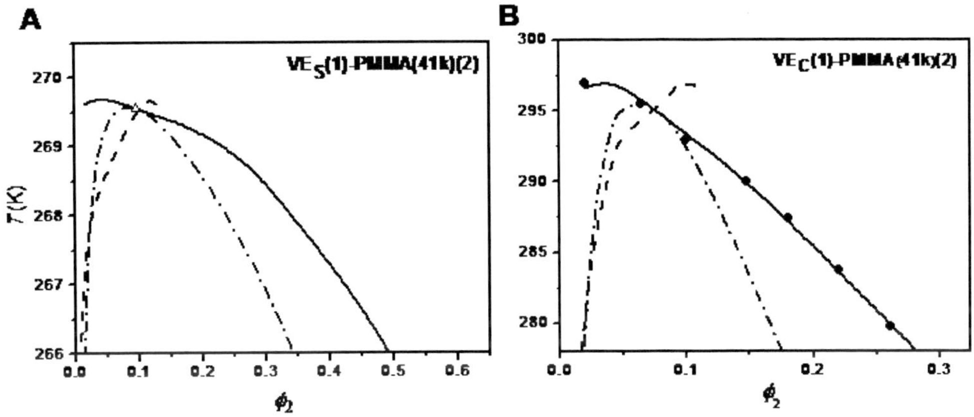

Figure 2. Phase diagrams of the quasibinary systems: A) VE_S(1)-PMMA(41k)(2), and B) VE_C(1)-PMMA(41k)(2). (•) Experimental cloud-points; (—) calculated CPC; (- -) calculated ShC; (- · -) calculated SC; (Δ) calculated CSP. [B) Reprinted from Polymer, 46, Schroeder W.F., Auad M.L., Barcia Vico M.A., Borrajo J., Aranguren M.I., "Thermodynamic, morphological, mechanical and fracture properties of poly(methyl methacrylate)(PMMA) modified divinylester(DVE)/styrene(St) thernosets", 2306-2319, 2005, with permission from Elsevier.].

Comparing Figures 2A and 2B, it can be concluded that the increase of the VE molar mass produces a displacement of the CPC to higher temperatures, increasing in this way the immiscibility range. This effect would be produced by the reduction of the entropic contribution to the mixing Gibbs free energy, caused by the increase of the VE molecular size.

Cloud-points corresponding to the VE_C-PMMA(239k) system could not be experimentally measured due to the elevated viscosity of this system at the cloud-point conditions. To represent the χ-parameter of this system, the expression obtained from the VEc-PMMA(41k) quasibinary system was used, assuming that this parameter is molecular weight independent.

3.1.4. St-VE-PMMA Quasiternary Systems

In general, the phase diagram of the quasiternary physical mixture represents the initial state of the reactive system, when the reactive double bonds conversion equals zero. Figures 3 and 4 correspond to quasiternary phase diagrams of systems with very different miscibility behaviors. Figure 3 shows the calculated cloud-point curves at -20, -13, and -8 °C, for the St-VE_S-PMMA(41k) system. The ternary immiscibility area decreases when temperature increases, as it corresponds to an UCST behavior. Note that the immiscibility region originated in the VE_S-PMMA(41k) binary axis is a stronger function of the temperature than that originated in the St-PMMA(41k) axis. This effect is mainly produced by the higher value of the d_1 constant, which represents a higher endotherm enthalpic contribution to the χ-interaction parameter in the VE_S-PMMA(41k) binary in comparison with that for the St-PMMA(41k). At room temperature and at higher temperatures the ternary system is completely miscible in the whole range of compositions.

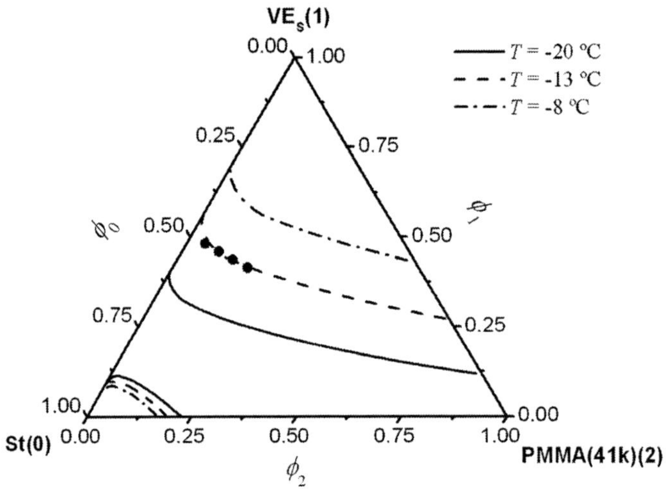

Figure 3. Calculated quasiternary cloud-point curves at different temperatures for the St(0)-VE_S(1)-PMMA(41k)(2) system. (•) Typical industrial formulation with 5, 10, 15, and 20 wt% PMMA(41k). (Reprinted from Journal of Applied Polymer Science, 100, Schroeder W.F., Yáñez M.J., Aranguren M.I., Borrajo J., "PMMA-Modified Divinylester/Styrene Resins: Phase Diagrams and Morphologies", 4539-4549, 2006, with permission from John Wiley and Sons.).

Figure 4 shows the calculated phase diagram for the St-VE_C-PMMA(239k) quasiternary system at 25 °C. It can be seen that when the highest molecular weight components, VE_C and PMMA(239k), are used an increase of the immiscibility is produced. Note that at 25 °C the St-VE_S-PMMA(41k) system is completely miscible in the whole range of compositions whereas the St-VE_C-PMMA(239k) systems presents a immiscibility region originated in the VE_C-PMMA(239k) axis. These results suggest that very different morphologies can be developed in the course of the reaction depending on the molecular weight of the components and the reaction conditions.

3.2. Final Morphologies of the Thermosets

The morphologies of the final cured materials showed differences as observed by SEM and TEM, depending on the molar masses of the PMMA and VE oligomer, content of the modifier, and curing temperature. These differences have a correlation with the thermal, mechanical and fracture properties of the final materials as it will be further discussed.

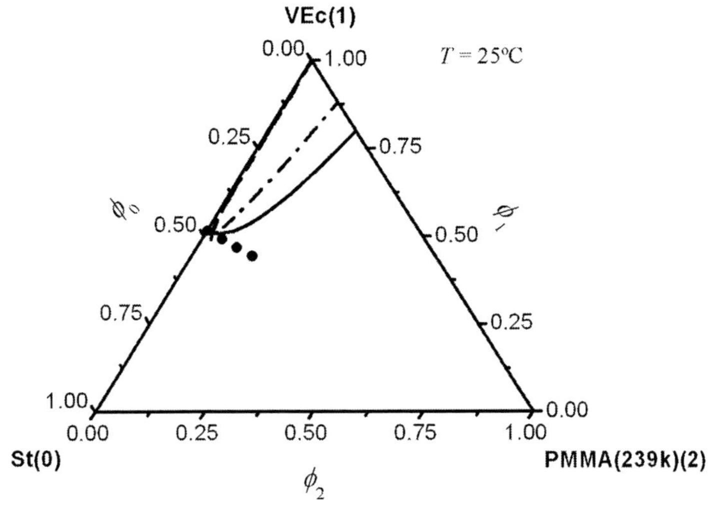

Figure 4. Calculated quasiternary phase diagram for the St(0)-VE_C(1)-PMMA(239k)(2) system at 25 °C: (—) CPC; (- -) ShC; (- · -) SC. (•) Typical industrial formulation with 5, 10, 15, and 20 wt% PMMA(239k).

3.2.1. Neat St-VE Networks

Electron microscopy images showed that the unmodified St-VE_S and St-VE_C systems cured at temperatures in the range between 25 and 80 °C give place to single-phase thermoset materials that fracture in a brittle way without showing any special features in the fracture surface. The homogeneous appearance of the fracture surface is in agreement with the transparent yellow visual aspect of those samples, and the results of the thermodynamic studies.[23]

Analysis of the same samples by Atomic Force Microscopy (AFM), which allows for higher magnification and better definition, has shown that the morphologies of these materials are actually formed by the agglomeration of nodular domains (called microgels) at the nanoscale. The degree of coalescence of the nodular structure depends on the St:VE proportion of the initial mixture, as reported in [25].

3.2.2. PMMA-Modified Networks Cured at 80 °C

Mixtures of St-VE_S and PMMA(41k) are initially transparent at above room temperature. However, after curing, the materials became opaque suggesting that a phase separation process occurs in the course of the polymerization reaction.

The points in Figure 3 represent formulations prepared with 5, 10, 15, and 20 wt% PMMA(41k) in a mixture of St:VE_S = 45:55 (constant weight proportion). As previously mentioned, at 80°C the quasiternary system is completely miscible in the whole range of compositions. As a consequence of the elevated initial miscibility, a significant progress in conversion during the pregel stage of the crosslinking reaction is necessary to widen the immiscibility region. When the growing immiscibility region reaches the points representing the global composition, phase separation begins. Because the mixture was initially homogeneous, and the phase separation starts before the gel point, the whole material suffers this process. This is clearly observed in Figure 5 that shows the SEM and TEM micrographs of the St-VE_S sample modified with 20 wt% PMMA(41k) cured at 80°C.

As it can be seen in Figure 5A, the image shows a heterogeneous material with an irregular fracture surface consisting of aggregates of irregular nodular particles with diameters in the range of 0.1 – 1 μm. The TEM micrograph shown in Figure 5B allows recognizing the main component presents in each phase. Dark gray areas correspond to the copolymer(St-*co*-VE)-rich phase while the light gray areas correspond to the PMMA-rich phase. This gray contrast is a consequence of the different local electronic densities between both phases [26, 27]. Thus, the irregular nodules are formed by St-*co*-VE crosslinked copolymer, while the PMMA-rich phase appears confined to the interstitial spaces surrounding the nodular aggregates. Although the materials prepared with different amounts of PMMA(41k) (between 5 and 20 wt%) presented similar morphological features, the extent of separation of the disperse PMMA phase and the irregularity of the fracture surface increased with the content of PMMA(41k) in the formulation.

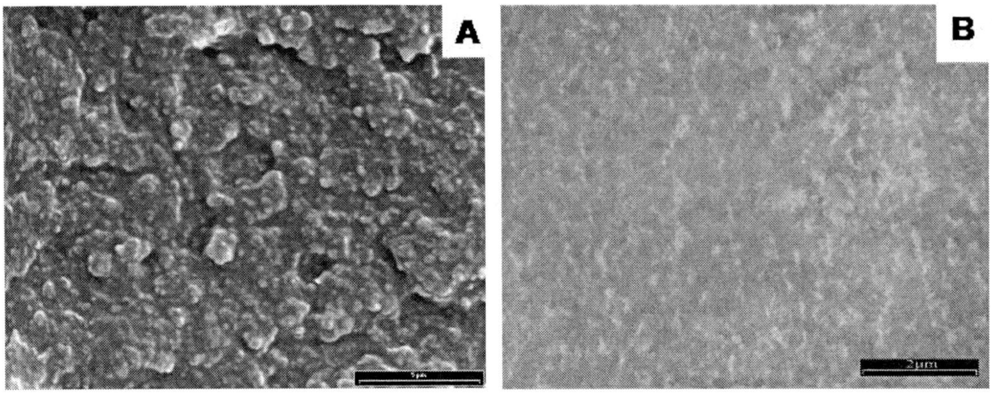

Figure 5. Electron micrographs of the St-VE_S network modified with 20 wt% PMMA(41k) cured at 80 °C: A) SEM, and B) TEM. (Reprinted from Journal of Applied Polymer Science, 100, Schroeder W.F., Yáñez M.J., Aranguren M.I., Borrajo J., "PMMA-Modified Divinylester/Styrene Resins: Phase Diagrams and Morphologies", 4539-4549, 2006, with permission from John Wiley and Sons.).

The developed morphology is a consequence of thermodynamic and diffusional changes in the course of the crosslinking reaction. Initially, the system is a homogeneous solution comprised by the St and VE_S co-monomers and the PMMA(41k) modifier. At low double-bond conversions, nanogels of crosslinked copolymer are formed. These nanogels aggregate and co-react through surface double-bonds giving place to microgels, that further interact and react with each other forming nodules on the order of microns that can be seen by electron microsopy. The PMMA(41k)-modifier that is rejected outside the nodular domains (due to the increasing incompatibility with the formed copolymer) is finally confined to the interstitial spaces left in between the nodular aggregates.

These morphologies were similar to those reported in previous studies on St-VE thermosets modified with elastomers [6, 28].As it has been shown, the use of rubbers leads to materials of low interfacial cohesion and reduced mechanical and thermal properties. In this case, because of the better properties of the thermoplastics at room temperature this problem was somehow alleviated, as it will be further discussed.

The effect of increasing the molecular weight of the VE oligomer on the final morphologies of materials modified with PMMA(41k) cured at 80 °C, was investigated. Samples containing the thermoplastic (5-20 wt%) showed rough fracture surfaces as a consequence of the phase separation, but this process was not as well developed as in the previous materials (Fig. 5). Although thermodynamic considerations predict a bigger extension of phase separation for the St-VE_C-PMMA(41k) materials, [23] the experimental finding suggested that the phase separation process was quickly arrested by diffusion restrictions that prevented species segregation.

3.2.3. PMMA-Modified Networks Cured at 40 °C

The samples obtained with the St-VE_S system modified with 5-15 wt% PMMA(41k) and cured at 40 °C presented morphological features very similar to those of the same formulations cured at 80 °C (Fig. 5). This fact indicates that the final morphology of these materials is not too sensitive to the curing temperature in the range studied.

However, using the high molecular weight VE oligomer (VE_C) resulted in materials with very different morphologies depending on the amount of PMMA(41k) in the initial formulation. Figure 6A shows the SEM micrograph of the sample with 5 wt% PMMA(41k), where a smooth fracture surface containing disperse spherical domains with sizes in the range of 0.7-1.2 μm can be seen. Figure 6B is the TEM image of the same material, which shows that the principal continuous phase is rich in St-*co*-VE_C copolymer, whereas the disperse domains are rich in PMMA(41k). Inside these domains, dark gray zones indicating the presence of copolymer can be distinguished. This suggests that a secondary phase separation inside the PMMA-rich phase took place during polymerization [3].

Figure 6C corresponds to the SEM micrograph of the material formulated with St-VE_C modified with 10 wt% PMMA(41k). A very irregular fracture surface noticeably different from that obtained with 5 wt% PMMA(41k) (Fig. 6A) is observed. The TEM image of the same material (Fig. 6D) shows a cocontinous phase structure with interconnected irregular nodular aggregates rich in copolymer surrounded by the thermoplastic modifier. Figures 6E and 6F are SEM and TEM images, respectively, of the material modified with 15 wt% PMMA(41k). A phase-inverted morphology can be observed, where the irregular aggregates of copolymer are virtually dispersed in a thermoplastic matrix. Obviously, this change in microstructure has a profound effect on the mechanical and fracture properties of these materials, as it will be further discussed.

3.2.4. PMMA-Modified Networks Cured at 25 °C

The St-VE_S thermosets modified with 5 and 10 wt% PMMA(239k) cured at 25 °C present structures consisting of irregular nodular particles of copolymer surrounded by the PMMA-rich phase. These morphologies have similar features to those showed in Figure 5, but with better defined nodules and a greater content of microvoids [20]. As in that case, the ternary cloud-point phase diagram for this system is miscible at all compositions. Thus, a significant progress in conversion in the pregel stage of the crosslinking reaction is required to induce the phase separation process. This thermodynamic behavior gives place to typical nodular structures more or less developed depending on the capacity of segregation of the phases before being arrested by the increase of the viscosity during the pregel stage or by gelation in the postgel stage.

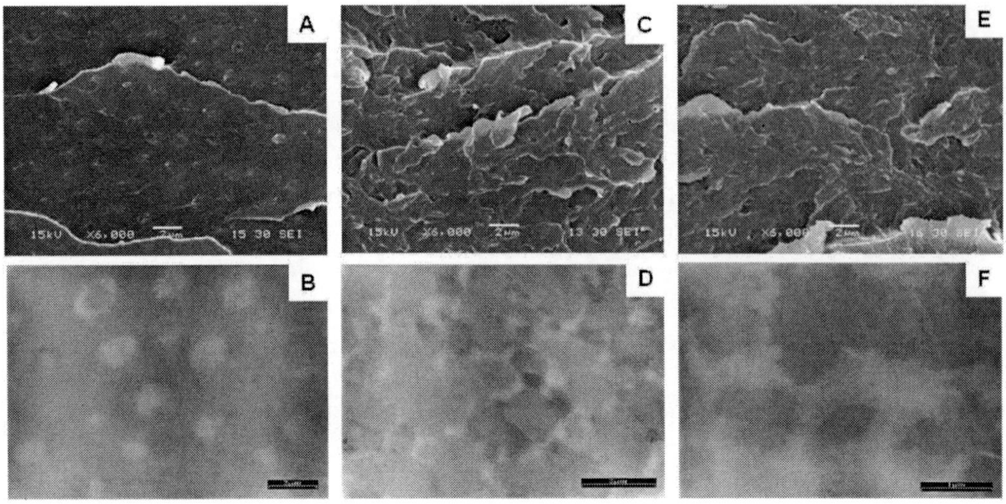

Figure 6. Electron micrographs of the St-VE_C networks modified with different amounts of PMMA(41k) cured at 40 °C. 5 wt% modifier: A) SEM, and B) TEM. 10 wt% modifier: C) SEM, and D) TEM. 15 wt% modifier: E) SEM, and F) TEM. (Reprinted from Journal of Applied Polymer Science, 106, Schroeder W.F., Borrajo J., Aranguren M.I., "Poly(methyl methacrylate)-Modified Vinyl Ester Thermosets: Morphology, Volume Shrinkage, and Mechanical Properties", 4007-4017, 2007, with permission from John Wiley and Sons.).

A very different microstructure is obtained when the higher molecular weight VE is used. Figure 7 shows electron micrographs of the St-VE_C sample with 10 wt% PMMA(239k) cured at 25 °C. Two well different regions are distinguished: a copolymer-rich continuous phase and a discontinuous PMMA-rich phase in the form of drop-like region. The latter shows a very large distribution of sizes, from small inclusions of less than 1 μm to huge droplets close to 100 μm. It can be seen that the PMMA-rich droplets present a very complex internal microstructure, consisting in copolymer nodules surrounded by the thermoplastic. Obviously, this morphology is the result of the initial miscibility of the system and its evolution during curing. As shown in Figure 4, the CPC of the quasiternary system is very close to the overall system composition (points in this figure), and just a small increment in the conversion is enough to initiate phase separation. Since the viscosity is low when the phase separation starts, the emergent phase coalesces. In this way, the reactive system will be phase separated from almost the beginning of the reaction forming two liquid phases. The droplets rich in PMMA(239k) that are formed at the beginning of the curing reaction are not initially neat PMMA but contain also St and VE_C, that react inside the droplets forming microgels and undergoing a secondary phase separation, as observed in Fig. 7.

Similar characteristics were observed in the morphology of the St-VE_C thermoset modified with 10 wt% PMMA(41k) cured at 25 °C, as showed in Figure 8. However, in this case the copolymer-rich phase also contains a distribution of small PMMA-rich domains with sizes in the range from less than 0.1 to 0.2 μm. Evidently, the mixture at the beginning of the reaction has segregated less amount of PMMA(41k) from the main phase than in the previous case (Fig. 7), giving place to secondary phase separation in both phases. This behavior is

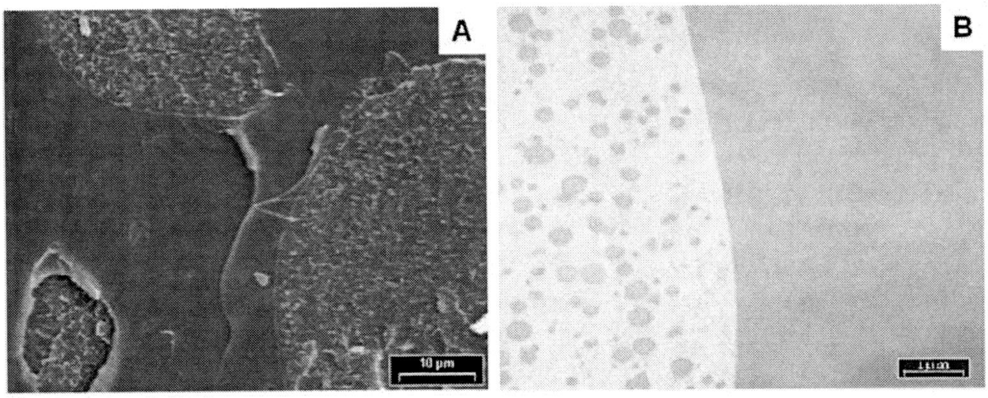

Figure 7. Electron micrographs of the St-VE_C network modified with 10 wt% PMMA(239k) cured at 25 °C: A) SEM, and B) TEM. [A) Reprinted from Polymer, 46, Schroeder W.F., Auad M.L., Barcia Vico M.A., Borrajo J., Aranguren M.I., "Thermodynamic, morphological, mechanical and fracture properties of poly(methyl methacrylate)(PMMA) modified divinylester(DVE)/styrene(St) thernosets", 2306-2319, 2005, with permission from Elsevier. B) Reprinted from Journal of Applied Polymer Science, 100, Schroeder W.F., Yáñez M.J., Aranguren M.I., Borrajo J., "PMMA-Modified Divinylester/Styrene Resins: Phase Diagrams and Morphologies", 4539-4549, 2006, with permission from John Wiley and Sons.].

associated to an increase in the miscibility of the system as the molecular weight of PMMA decreases. However, from the results presented in this section, it can be concluded that the final morphology of St-VE-PMMA thermosets is more dependent on the molecular weight of VE than of PMMA, in the range of molecular weights studied.

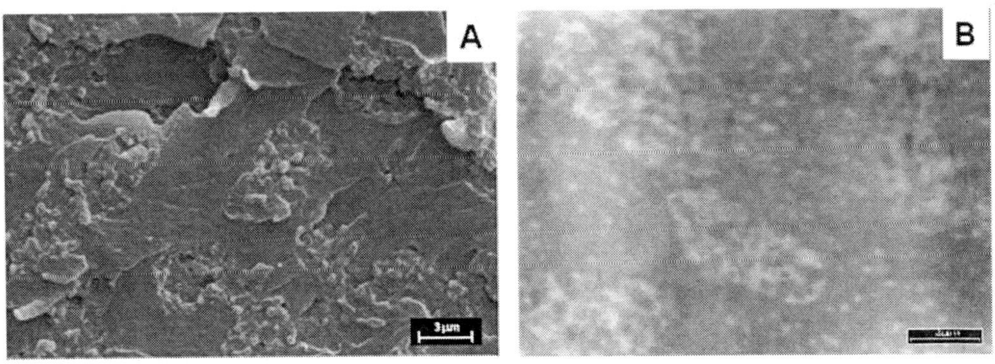

Figure 8. Electron micrographs of the St-VE_C network modified with 10 wt% PMMA(41k) cured at 25 °C: A) SEM, and B) TEM.

3.3. Final Properties of the Thermosets

3.3.1. Volume Shrinkage

Elastomeric or, more recently, thermoplastic modifiers (called Low Profile Additives, LPA) are added to thermosetting precursors, such as VE or unsaturated polyester (UP) resins, to reduce the overall volume shrinkage and to improve the dimensional stability of molded parts. The action of a LPA is based on the formation of a phase-separated structure which promotes the formation of internal microvoids that compensate for the crosslinking shrinkage [14, 16, 17].

The effect of the PMMA concentration on the volume shrinkage control in VE_S and VE_C systems was investigated by measuring the density of the system before and after curing [29]. Figure 9 shows the final volume change against the PMMA(41k) concentration in VE_S and VE_C systems cured at 80 °C. The volume shrinkage was larger during curing of neat St-VE_S (10.22 %) than for neat St-VE_C (8.24 %) because of the higher crosslinking density of the network formed by the lower molecular weight VE. The addition of PMMA(41k) gave place to a reduction of the volume shrinkage of the materials. For instance, with 20 wt% PMMA(41k) the percentage of volume change was reduced to 7.85 % for VE_S and to 6.85 % for VE_C.

The calculated volume changes of both VE systems at different PMMA concentrations, if the last one is considered only as a filler, is shown in Figure 9 with dashed lines [29]. Actually, at low PMMA(41k) concentration the effect of the thermoplastic modifier approaches that of a filler. However, at PMMA concentrations between 15 and 20 wt% for VE_S and between 10 and 20 wt% for VE_C, the effect of PMMA is slightly better than that of a filler.

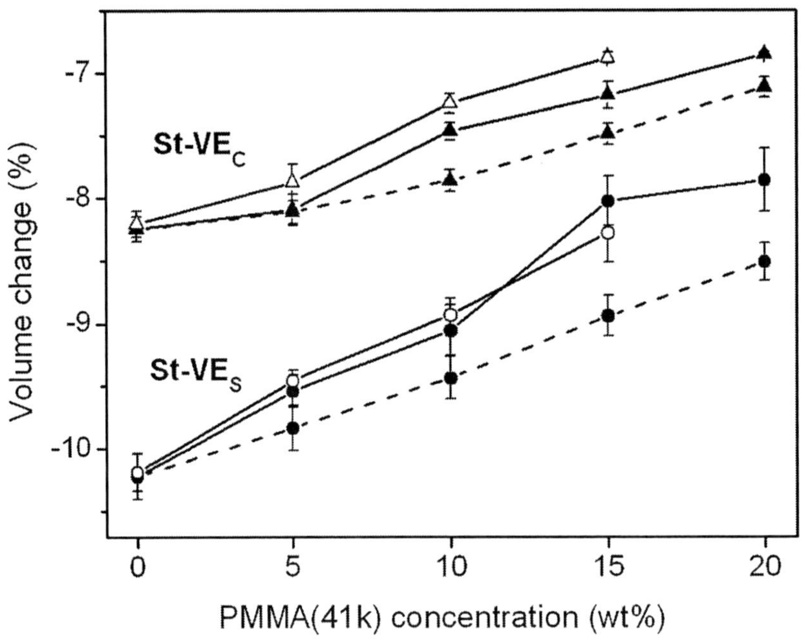

Figure 9. Final volume change against PMMA(41k) concentration for both VE systems: (•) experimental data for VE_S cured at 80 °C; (▲) experimental data for VE_C cured at 80 °C; (○) experimental data for VE_S cured at 40 °C; (Δ) experimental data for VE_C cured at 40 °C; (-•-) calculated behavior considering PMMA as a filler for VE_S; and (-▲-) calculated behavior considering PMMA as a filler for VE_C. (Reprinted from Journal of Applied Polymer Science, 106, Schroeder W.F., Borrajo J., Aranguren M.I., "Poly(methyl methacrylate)-Modified Vinyl Ester Thermosets: Morphology, Volume Shrinkage, and Mechanical Properties", 4007-4017, 2007, with permission from John Wiley and Sons.).

Figure 9 shows that St-VE_S thermosets modified with 5-15 wt% PMMA(41k) cured at 40 °C presented volume shrinkage values very similar to those of the same formulations cured at 80 °C. This behavior responds to the previous observation that the final morphology of these materials is not too sensitive to the curing temperature in the range between 40 and 80 °C. On the other hand, St-VE_C thermosets with 10 and 15 wt% PMMA(41k) cured at 40 °C presented a better volume shrinkage control than the same formulations cured at 80 °C. These materials showed co-continuous (Figs. 6C and 6D) and PMMA-rich matrix (Figs. 6E and 6F) morphologies, respectively, whereas the same formulations cured at 80 °C presented structures with very little phase segregation.

Results reported in the literature for UP resins modified with thermoplastic additives, such as poly(vinyl acetate)(PVAc), have shown that the volume shrinkage of a neat St-UP thermoset (9 %) can be reduced up to 4.5 % with the addition of 4 wt% PVAc [14, 16]. Thus, although co-continous or thermoplastic-rich matrix morphologies have been obtained, which are essential to get suitable volume shrinkage control, they have not been very efficient modifiers of the VE systems studied.

3.3.2. Dynamic Mechanical Tests

The information obtained from dynamic mechanical measurements is directly related to the network structure and the mechanisms involved during its formation. The use of VE of different molecular weights leads to networks with different densities of crosslinking. The neat St-VE_S system gives place to networks with higher crosslinking density than the neat St-VE_C system. This difference is the result of the shorter chains of the VE_S, which produces a tighter network through end groups-crosslinking. Typically, neat St-VE_S networks present higher glass-transition temperature than the neat St-VE_C system, higher modulus in the rubbery state, and lower tan δ peak because of the lower mobility of the chains in the more crosslinked network.

Figure 10 shows representative curves for the series of St-VE_S thermosets modified with PMMA(41k) cured at 80 °C, whose morphology was already shown in Fig. 5. The neat thermoset presents a T_g value of 160 °C, measured as the temperature at the maximum in tan δ, and a modulus in the rubbery state of 36 MPa. The addition of PMMA modifies the dynamic mechanical response of the thermoset. The rubbery modulus decreases from 36 to 17 MPa for 20 wt% PMMA(41k). This behavior suggests that the crosslinking density has decreased with the PMMA addition, due to the diluent effect of the thermoplastic in the crosslinked phase. Tan δ curves in Fig. 10 show that the relaxation of the copolymer-rich phase in the modified networks seems to be comprised by overlapping peaks centered at temperatures closely above and below the T_g of the neat thermoset. These peaks are believed to correspond to the relaxation of copolymers formed at St-VE_S ratios that varied from the initial one. The relatively high viscosity reached at the point when phase separation starts leads to the preferential diffusion of the St to the PMMA-rich phase. This produces a variation in the co-momomer ratios of both phases. The transition of the thermoplastic-rich phase appears as a shoulder at 125 °C.

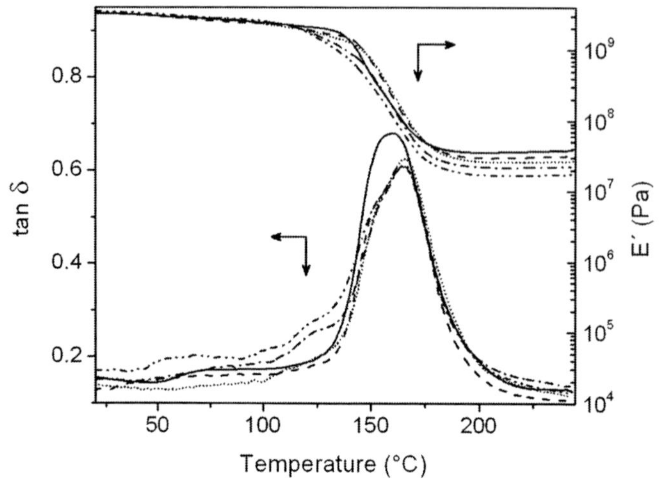

Figure 10. Dynamic mechanical behavior for neat and PMMA(41k)-modified St-VE$_S$ thermosets cured at 80 °C: (—) unmodified system; (– –) 5 wt% PMMA; (···) 10 wt% PMMA; (– · –)15 wt% PMMA; (– · · –) 20 wt% PMMA. (Reprinted from Journal of Applied Polymer Science, 106, Schroeder W.F., Borrajo J., Aranguren M.I., "Poly(methyl methacrylate)-Modified Vinyl Ester Thermosets: Morphology, Volume Shrinkage, and Mechanical Properties", 4007-4017, 2007, with permission from John Wiley and Sons.).

Figure 11 shows the dynamic mechanical responses of St-VE$_C$ thermosets modified with PMMA(41k) cured at 40 °C, whose morphologies were already shown in Figure 6. In this case, the neat thermoset presents a T_g value of 138 °C, and a modulus in the rubbery state of 11 MPa. For the sample modified with 5 wt% PMMA(41k), the tan δ curve shows two resolved transitions corresponding to the two phases formed: the transition of the PMMA-rich phase at the lower temperature and the transition of the copolymer-rich phase at the higher temperature. Moreover, the rubbery modulus for this sample is nearly the same as that of the neat thermoset. These observations, suggest that the two phases are well separated in the final material, and that the crosslinking density of the main phase did not vary significantly with the modifier addition. These findings are supported by the morphology shown in Figures 6A and 6B, where a dispersion of PMMA-rich domains in a virtually neat thermosetting matrix is observed. The sample modified with 15 wt% PMMA(41k) shows also two tan δ transitions resolved in accord with the phase-inverted morphology shown in Figures 6E and 6F. Varying the PMMA(41k) concentration from 5 to 15 wt% produced only a slight increment in the height of the peak corresponding to the PMMA-rich phase, and a small decrease of the height of the peak corresponding to the copolymer-rich phase. These variations are a consequence of the change in the overall composition of the materials. For the thermoset modified with 10 wt% PMMA(41k) the two relaxation peaks were not resolved. This sample presents co-continuous phase morphology as observed in Figs. 6C and 6D.

Although the concentrations of PMMA(41k) are as high as those used in the St-VE$_S$ system (Fig. 10), the height of the peak corresponding to the PMMA-rich phase is much higher, suggesting that the PMMA chains are less restricted during relaxation, because the phase is more segregated than in the previous case.

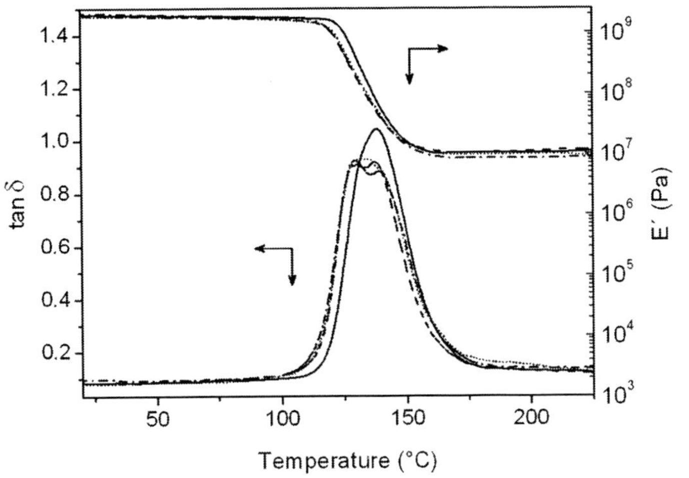

Figure 11. Dynamic mechanical behavior for neat and PMMA(41k)-modified St-VEC thermosets cured at 40 °C: (—) unmodified system; (– –) 5 wt% PMMA; (···) 10 wt% PMMA; (– · –) 15 wt% PMMA. (Reprinted from Journal of Applied Polymer Science, 106, Schroeder W.F., Borrajo J., Aranguren M.I., "Poly(methyl methacrylate)-Modified Vinyl Ester Thermosets: Morphology, Volume Shrinkage, and Mechanical Properties", 4007-4017, 2007, with permission from John Wiley and Sons.).

Figure 12 shows the effect of PMMA(239k) on the St-VE_C system cured at 25 °C. The drop of the modulus in the transition is shifted to lower temperatures, but no major changes appear neither in the value of the glassy nor in the rubbery moduli. Tan δ curves show two transitions clearly separated in the modified materials, corresponding to the two phases

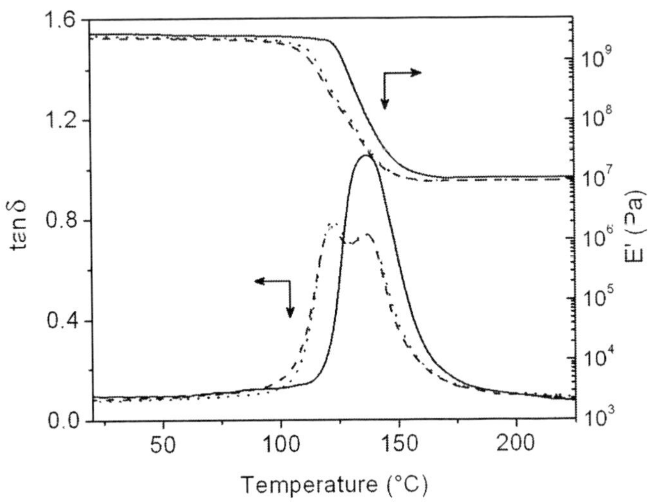

Figure 12. Dynamic mechanical behavior for neat and PMMA(239k)-modified St-VE_C thermosets cured at 25 °C: (—) unmodified system; (– –) 5 wt% PMMA; (···) 10 wt% PMMA. (Reprinted from Polymer, 46, Schroeder W.F., Auad M.L., Barcia Vico M.A., Borrajo J., Aranguren M.I., "Thermodynamic, morphological, mechanical and fracture properties of poly(methyl methacrylate)(PMMA) modified divinylester(DVE)/styrene(St) thernosets", 2306-2319, 2005, with permission from Elsevier.).

formed. The transition corresponding to the copolymer-rich phase appears at the same temperature in all the series, and the same occurs for the transition of the PMMA-rich phase, which does not undergo any shift at different thermoplastic concentrations. Varying the amount of PMMA(239k) from 5 to 10 wt% gave place only to a slight increment in the height of the corresponding transition peak. These observations, as well as the observed invariance of the rubbery modulus of the series suggest that the two phases are well separated in the final material, and that the crosslinking density of the main phase did not vary significantly with the modifier addition. Moreover, the St-VE$_C$ ratio of the material trapped into the separated PMMA-rich phase must be very similar to that of the main phase. This is a consequence of the relative low viscosity of the medium at the beginning of the reaction when phase separation starts, allowing both co-monomers to diffuse between the phases. These results are strongly supported by the morphologies shown in Figure 7.

3.3.3. Mechanical and Fracture Behavior

The obtained thermosetting materials presented the typical behavior of this type of systems; they broke in the initial linear region when tested under flexural forces, and yielded under compression. Tables 3-5 include a summary of the mechanical (flexural and compression) and fracture behavior of the different prepared materials cured at 80, 40, and 25 °C, respectively. Typically, the neat systems prepared with St-VE$_S$ show higher flexural modulus, higher yield stress in compression and lower fracture strength (lower K_{IC} value) than the respective networks prepared from St-VE$_C$ resin. This is expected from a lower molecular weight VE that, as previously discussed, leads to shorter chains between crosslinking points, and thus to a tighter network with less mobile chains.

As it was expected, the flexural and compression behaviors of St-VE thermosets were very little affected by the addition of PMMA, as shown in Tables 3-5. These tables illustrate that the incorporation of the thermoplastic did not significantly reduce either compressive yield stress or flexural modulus, for the different materials prepared. Actually, the mechanical properties were slightly improved in some cases, contrary to the marked reduction observed when elastomeric modifiers are used.

Table 3. Mechanical properties of both VE systems modified with PMMA(41k) cured at 80 °C.†

wt% PMMA (41k)	Flex. Mod., E_b (GPa)		Compres., σ_y (MPa)		K_{IC} (MPa.m$^{1/2}$)	
	St-VE$_S$	St-VE$_C$	St-VE$_S$	St-VE$_C$	St-VE$_S$	St-VE$_C$
0	3.54±0.02	3.51±0.02	120.0±1.0	114.2±1.0	0.63±0.04	0.86±0.06
5	3.58±0.01	3.56±0.04	118.3±1.5	110.6±1.3	0.79±0.05	1.01±0.10
10	3.62±0.02	3.60±0.02	115.4±1.8	111.1±1.8	1.10±0.06	1.13±0.06
15	3.67±0.04	3.57±0.02	116.8±2.2	114.8±0.8	0.95±0.05	1.09±0.04
20	3.68±0.05	3.55±0.03	121.2±2.0	114.1±0.5	1.04±0.16	1.27±0.12

Table 4. Mechanical properties of both VE systems modified with PMMA(41k) cured at 40 °C.†

wt% PMMA(41k)	Flex. Mod., E_b (GPa)		Compres., σ_y (MPa)		K_{IC} (MPa.m$^{1/2}$)	
	St-VE$_S$	St-VE$_C$	St-VE$_S$	St-VE$_C$	St-VE$_S$	St-VE$_C$
0	3.52±0.01	3.48±0.03	117.3±1.8	113.4±0.8	0.65±0.07	0.85±0.05
5	3.54±0.02	3.43±0.04	113.1±1.6	111.5±1.3	1.11±0.13	0.94±0.07
10	3.58±0.02	3.40±0.01	116.2±2.3	109.9±0.7	0.99±0.05	1.36±0.09
15	3.60±0.04	3.38±0.02	120.1±1.5	110.6±1.9	0.83±0.06	1.21±0.14

†Reprinted from Journal of Applied Polymer Science, 106, Schroeder W.F., Borrajo J., Aranguren M.I., "Poly(methyl methacrylate)-Modified Vinyl Ester Thermosets: Morphology, Volume Shrinkage, and Mechanical Properties", 4007-4017, 2007, with permission from John Wiley and Sons.

Table 5. Mechanical properties of both VE systems modified with PMMA(239k) cured at 25 °C.†

wt% PMMA (239k)	Flex. Mod., E_b (GPa)		Compres., σ_y (MPa)		K_{IC} (MPa.m$^{1/2}$)	
	St-VE$_S$	St-VE$_C$	St-VE$_S$	St-VE$_C$	St-VE$_S$	St-VE$_C$
0	3.50±0.03	3.15±0.02	108.7±3.5	104.7±1.6	0.62±0.08	0.83±0.04
5	3.58±0.17	3.49±0.13	119.7±1.1	109.1±3.3	0.50±0.09	1.43±0.05
10	3.63±0.23	3.21±0.10	121.6±1.9	109.8±1.4	--	1.01±0.16

†Reprinted from Polymer, 46, Schroeder W.F., Auad M.L., Barcia Vico M.A., Borrajo J., Aranguren M.I., "Thermodynamic, morphological, mechanical and fracture properties of poly(methyl methacrylate)(PMMA) modified divinylester(DVE)/styrene(St) thernosets", 2306-2319, 2005, with permission from Elsevier.

Stress intensity factors, K_{IC}, measured for both neat St-VE systems cured at 80, 40, and 25 °C presented an inverse correlation with the yield stress values (Tables 3-5). The same trend was previously reported in the literature for neat epoxy networks, and can be interpreted in terms of the crack tip blunting model [30, 31].

Most of the PMMA modified St-VE thermosets showed enhanced fracture strength regarding the respective neat network. Table 3 (materials cured at 80 °C) shows that the modified St-VE$_S$ materials present a maximum in fracture strength for 10 wt% PMMA(41k), with the K_{IC} value increased by 75 % with regard to the neat sample. For the VE$_C$ system cured at 80 °C, the largest enhancement was achieved with 20 wt% PMMA(41k), although this improvement (48 %) was lower than in the previous case. This difference in toughening correlates with the extent of phase separation that each system presents. The PMMA-modified St-VE$_C$ thermosets cured at 80 °C present morphologies with phases much less segregated than the modified-VE$_S$ systems. As a consequence of the more developed heterogeneous morphologies, additional toughening mechanisms are activated in the PMMA-modified St-VE$_S$ materials.

Table 4 lists K_{IC} values for the materials modified with PMMA(41k) cured at 40 °C. The VE$_S$ system presents a maximum in the fracture strength for 5 wt% PMMA, with the K_{IC} parameter increased 71 % regarding the neat system. On the other hand, the largest enhancement in fracture resistance for VE$_C$ systems occurred with 10 wt% PMMA, where K_{IC} increased by 60 %. It is important to remember that as the curing temperature of the PMMA-modified St-VE$_C$ systems was reduced from 80 to 40 °C, an increase in the segregation of species in the course of the phase separation process was observed. The St-VE_C material modified with 10 wt%PMMA (41k) cured at 40 °C (maximum in K_{IC}) presents co-continuous phase morphology, as observed in Figures 6C and 6D. These results show that the development of a co-continuous structure has been more efficient in improving the fracture toughness that other type of structures, such as PMMA particles dispersed in a termoset-rich matrix (Figures 6A and 6B), or copolymer aggregates dispersed in a thermoplastic matrix (Figures 6E and 6F). The existence of a maximum in K_{IC} associated to co-continuous phase morphologies in epoxy networks modified with thermoplastics has been reported.[3,32] It has been shown that good interfacial adhesion is a requirement for this response.

Table 5 shows K_{IC} values for the materials modified with PMMA(239k) cured at 25 °C. It is especially important to see the improvement obtained in the fracture strength of the St-VE_C materials by addition of PMMA(239k). The morphology of these materials was already discussed and shown in Figure 7. An optimum concentration is observed at 5 wt% thermoplastic, where K_{IC} was increased by 72 % regarding the neat network. Further addition produces a reduction of K_{IC}, because of the increment of the amount of droplets containing microvoids, which facilitates the crack advance. Conversely, the St-VE_S systems do not show this improvement. These materials present morphologies with similar features to those showed in Fig. 5, but with more defined nodules and a greater content of microvoids. Because microvoiding is generalized through the sample, any crack finds an easy way to propagate through the voids plane.

Finally, it can be concluded that although the improvement in the fracture strength of St-VE thermosets modified with PMMA was not as important as that produced by elastomeric modifiers [6], the mechanical (flexural and compression) properties were little affected by the addition of the thermoplastic modifier. Actually, the mechanical properties were slightly improved in most cases, instead of the typical reduction produced when elastomers are added to the networks.

CONCLUSIONS

The addition of PMMA leads to phase separated morphologies developed during the St crosslinking of two vinyl-ester oligomers of different molecular weight. Clearly, the resultant morphologies, were dependent mainly on the molecular weight of the VE oligomer, but also on the molecular weight of the PMMA thermoplastic utilized as modifier.

The thermodynamic analysis of the miscibility of the binary and ternary systems gave an adequate framework for the understanding of the final morphologies observed. The modeling made use of the Flory-Huggins theory, but considering the polydispersity of the components. The increasing conversion during reaction and the corresponding increasing viscosity also had an effect on the process, since the last one can strongly reduce chain mobilities and freeze a given morphology before the end of the reaction. Thus, polymer induced phase separation produced an overall nodular structure of crosslinked VE surrounded by PMMA and general microvoiding, when the low molecular weight VE and PMMA components are used. On the other hand, if the high molecular weight VE and PMMA components are considered the system phase separates almost at the beginning of the reaction and the morphology corresponds to large PMMA drops, with secondary phase separation, immersed in a crosslinked resin.

Experimental and theoretical modeling could be correlated with the final morphologies observed by electronic microscopies (scanning and transmission) and further on with the thermo-mechanical, mechanical and fracture properties of the thermoplastic-modified thermosets. It was confirmed that the use of thermoplastic modifiers produced materials with similar tensile and compression properties as those of the corresponding matrices. However, the fracture properties showed improvement by addition of PMMA, although less than that obtained with elastomers (which also reduced the mechanical properties). The exception to this trend was the system with generalized microvoiding. Finally, the co-continuous structure appeared as the most efficient morphology to improve fracture strength.

Acknowledgments

The financial supports of the National Research Council (CONICET), the University of Mar del Plata, and the National Agency for the Promotion of Science and Technology (ANPCyT) are gratefully acknowledged.

References

[1] Selly J.L., *"Encyclopedia of Polymer Science & Engineering"*, 2nd Edition, Mark_Bikales-Overberger-Menges, Jhon Wiley & Sons, Inc., 12 (1998).
[2] Huang Y.J., Su C.C. *J Appl Polym Sci*, 55:4, 305-318 (1995).
[3] Pascault J.P., Sautereau H., Verdu J., Williams R.J.J. *"Thermosetting Polymers"*, Marcel Dekker: New York, (2002).
[4] Karger-Kocsis J., Fröhlich J., Gryshchuk O., Kautz H., Frey H., Mülhaupt R. *Polymer* 45:4, 1185-1195 (2004).
[5] Gryshchuk O., Jost N., Karger-Kocsis J. *Polymer* 43:17, 4763-4768 (2002).
[6] Auad M.L., Frontini P.M., Borrajo J., Aranguren M.I. *Polymer* 42:8, 3723-3730 (2001).
[7] Dong J.P., Huang J.G., Lee F.H., Roan J.W., Huang Y.J. *J Appl Polym Sci* 91:5, 3369-3387 (2004).
[8] Dong J.P., Huang J.G., Lee F.H., Roan J.W., Huang Y.J. *J Appl Polym Sci* 91:5, 3388-3397 (2004).
[9] Wang S., Wang J., Ji Q., Schultz A.R., Ward T.C., McGratz J.E. *J Polym Sci Part B: Polym Phys* 38:18, 2409-2421 (2000).
[10] Suspene L., Fourquier D., Yang Y.S. *Polymer* 32:9, 1593-1604 (1991).
[11] Huang Y.J., Jiang W.C. *Polymer* 39:25, 6631-6641 (1998).
[12] Auad M.L., Frontini P.M., Borrajo J., Aranguren M.I. *Polymer* 42:7, 2723-2730 (2001).
[13] Yee A.F., Du J., Thouless M.D. In *"Polymer Blends"*, Paul D.R., Bucknall C.B., Eds.; Wiley: New York (2000); Vol. 2, Chapter 26.
[14] Li W., Lee L.J. *Polymer* 41:2, 697-710 (2000).
[15] Dong J.P., Chiu S.G., Hsu M.W., Huang Y,J. *J Appl Polym Sci* 100:2, 967-979 (2006).
[16] Bucknall C.B., Partridge I.K., Phillips M.J. *Polymer* 32:4, 636-640 (1991).

[17] Lucas J.C., Borrajo J., Williams R.J.J. *Polymer* 34:9, 1886-1890 (1993).
[18] Kamide K. In *"Thermodynamics of Polymer Solutions: Phase Equilibria and Critical Phenomena"*; Jenkis A.D., Ed.; PolymerScience Library 9; Elsevier: Amsterdam (1990); Chapters 4,5.
[19] Kamide K., Matsuda S., Shirataki H. *Eur Polym J* 26:4, 379-391 (1990).
[20] Schroeder W.F., Auad M.L., Barcia Vico M.A., Borrajo J., Aranguren M.I. *Polymer* 46:7, 2306-2319 (2005).
[21] Bae Y.C., Shim J.J., Soane D.S., Prausnitz J.M. *J Appl Polym Sci* 47:7, 1193-206 (1993).
[22] Choi J.J., Bae Y.C. *Eur Polym J* 35:9, 1703-1711 (1999).
[23] Schroeder, W.F., Yáñez M.J., Aranguren M.I., Borrajo J. *J Appl Polym Sci* 100:6, 4539-4549 (2006).
[24] Auad M.L., Aranguren M.I., Borrajo J. *Polymer* 42:15, 6503-6513 (2001).
[25] Mosiewicki M.A., Schroeder W.F., Leite F.L., Hermann P.S.P., Curvelo A.A.S., Aranguren M.I., Borrajo J. *J Mater Sci* 41:18 ,6154-6158 (2006).
[26] Sawyer L.C., Grubb D.T. In *"Polymer Microscopy"*; Chapman & Hall: London (1996); Chapter 3.
[27] Williams D.B., Barry Carter C. In "*Transmission Electron Microscopy*"; Plenum: New York (1996); Vol. 3, Chapter 22.
[28] Auad M.L., Proia M., Borrajo J., Aranguren M.I. *J Mater Sci* 37:19, 4117-4126 (2002).
[29] Schroeder W.F., Borrajo J., Aranguren M.I. *J Appl Polym Sci* 106:6, 4007-4017 (2007).
[30] Kinloch A.J., Williams J.G. *J Mater Sci* 15:4, 987-996 (1980).
[31] Huang Y., Kinloch A.J. *Polymer* 33:24, 5338-5340 (1992).
[32] Pascault J.P., Williams R.J.J. In *"Polymer Blends"*; Paul D.R., Bucknall C.B., Eds.; Wiley: New York (2000); Vol. 1, Chapter 13.

In: Thermoplastic and Thermosetting Polymers and Composites ISBN: 978-1-61209-264-5
Editors: Linda D. Tsai and Matthew R. Hwang

Chapter 5

THERMOPLASTIC AND THERMOSETTING POLYMERS FROM VEGETABLE OILS

Gerard Lligadas, Juan C Ronda, Marina Galià and Virginia Cádiz
Departament de Química Analítica i Química Orgànica, Universitat Rovira i Virgili, Marcel.lí Domingo s/n, 43007 Tarragona, Spain

ABSTRACT

Vegetable oils are excellent renewable raw materials for polymers. This chapter discusses the application of environmentally friendly and high efficient processes to plant oils and several approaches to new biobased materials and is organized as a function of the thermoplastic or thermosetting nature of the final polymers. Starting from undecylenic or oleic acid, we obtained difunctional telechelic diols by thiol-ene "click" coupling, that can were further polymerized to linear polyurethanes. Moreover, we synthesized thermoplastic flame retardant phosphorus-containing polyesters by ADMET polymerization. To obtain thermosetting materials, we explored the cationic polymerization of triglyceride double bonds of the fatty acid chain. We carried out the copolymerization of soybean oil with styrenic monomers containing silicon, boron or phosphorus producing materials with improved mechanical and flame retardant properties. Moreover, we obtained organic-inorganic hybrid materials with promising properties for optical applications by the hydrosilylation of alkenyl-terminated fatty acid derivatives. The presence of double bonds in triglycerides makes possible to attach some functional groups through chemical modification and we described various chemical pathways for functionalising triglycerides and fatty acids. Epoxidation is one of the most interesting chemical modifications that leads to epoxidized vegetable oils. We reported the preparation of biobased polyhedral oligomeric silsesquioxanes-nanocomposites from epoxidized linseed oil. Moreover, we obtained new fatty-acid derived compounds that could find applications as flame retardant materials in biobased epoxy resins.

We described the preparation of a new family of epoxidized methyl oleate-based polyether polyols which were used in the synthesis of polyurethanes with specific applications: silicon-containing polyurethanes with enhanced flame-retardant properties and polyurethane networks with potential applications in biomedicine. An enone-containing triglyceride derivative was obtained by an environmentally friendly chemical procedure from high oleic sunflower oil, that could be crosslinked with diamines. In a similar way, triglycerides containing secondary allylic alcohols can be obtained, that can be further functionalised with acrylate or phosphorus-containing derivatives to obtain flame retardant thermosets.

INTRODUCTION

Recent years have witnessed an increasing demand on natural products in industrial applications for environmental issues, waste disposal and depletion of non-renewable resources.[1] Renewable resources can provide an interesting sustainable platform to substitute petroleum-based polymers through the design of bio-based polymers that can compete or even surpass the existing materials on a cost-performance basis with high eco-friendliness values.[2,3] Plant oils are considered as the most important renewable raw materials for the production of bio-based polymers. In recent years, a variety of polymeric materials have been developed and tested with plant oils as feedstock.[4,5,6,7]

Vegetable oils are one of the cheapest and most abundant biological sources available in large quantities. Their main components are triglycerides, made up of three fatty acids joined at a glycerol junction. Most of the common oils contain fatty acids that vary from 12 to 22 carbons in length, with 0 to 3 double bonds per fatty acid. Triglycerides contain several reactive positions that are amenable to chemical reactions: ester groups, C=C double bonds, allylic positions and the α-position of ester groups. In general, there are three main routes for the preparation of polymers from plant oils (Figure 1). The chemical transformation of plant oils produces platform chemicals which can be used to obtain monomers for the polymer synthesis. Alternatively, to obtain polymers from vegetable oils the direct polymerization through the double bonds, or other reactive functional groups present in the fatty acid chain, or the chemical modification of the double bonds, introducing functional groups which are easier to polymerize, can be carried out.

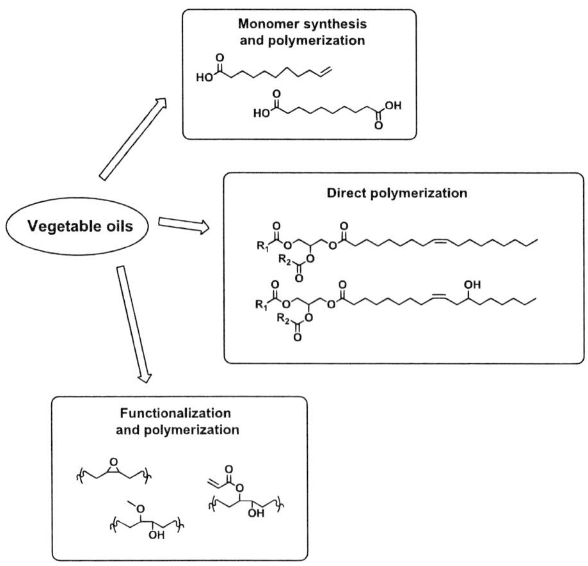

Figure 1.

Like other organic polymeric materials, the flammability of vegetable oil based materials is a shortcoming in some applications. Most of the flame retardant resins used in industry come from petroleum-based chemicals. The synthesis and characterization of flame-retardant polymers from bromoacrylated plant oil triglycerides has been reported, [8] however, the flame-retardant resins that contain bromine release hydrogen bromide during combustion, which causes corrosion and toxicity. The concept of sustainable development requires fire retardant technologies to be developed which have minimum impact on health and the environment throughout the life cycle of the fire-retardant material. These considerations mean that the search is now on new environmentally friendly flame-retardant polymeric materials. Phosphorus- boron- and silicon-containing polymers are well recognized for their flame retardant properties [9], and they are increasingly becoming more popular than their halogen counterparts, as they generally give off nontoxic combustion products [10]. Phosphorous compounds influence the reaction via a solid-state fire-protection mechanism [11] and can also act in the gas state as a catalyst radical scavenger, when degradation produces volatile P-containing moieties. Silicon compounds, when present in a polymer, have a fire retardant effect arising partly from vapour phase action, dilution of combustible organic gases in the flame zone, and partly from the formation of a barrier to heat and mass transfer that silicaceous residues can form behind the flame front.[12,13] Boron compounds thermally decompose producing boron oxide in the condensed phase and alter the decomposition process of the polymer in favour of carbonaceous residues rather than CO or CO_2.[14,15]

To further extend the application of renewable resources as well as of the environmentally friendly and high efficient processes to obtain sustainable polymers, the purpose of our research is to develop new biobased polymers from vegetable oils. This chapter discusses the approaches above mentioned and deals with the preparation and structure-properties relationships in thermoplastic and thermosetting polymers from different vegetable oils.

1. Thermoplastic Polymers

The chemical transformation of plant oils can produce platform chemicals with well defined structure and functionality which can be used to produce polymers. In this context, our research is focused in the preparation of difunctional monomers from undecylenic acid and oleic acid. Undecylenic acid is a C11 fatty acid with a terminal carbon-carbon double bond. It can be easily obtained from castor oil and is commercially available in high purity [16]. Oleic acid is a C18 fatty acid containing a carbon-carbon double bond at position C 9. It can be found in several natural oils such as olive oil (71%), canola oil (61%), sunflower oil (42%) and palm oil (39%). Modern genetic engineering techniques have been able to develop natural oils with much higher content of an individual fatty acid. For example, "high oleic" sunflowers have been developed with an oleic acid content up to 90%. We used 10-undecenoic acid and oleic acid to synthesize monomers capable of homo and copolymerizing in a controlled way by using two emerging and powerful methodologies; the acyclic diene metathesis ADMET polymerization and the thiol-ene coupling.

1.1 Acyclic Diene Metathesis (ADMET) Polymerization

ADMET polymerization of α,ω-dienes has been shown to be an efficient and powerful tool for the synthesis of a wide variety of linear polymers and polymer architectures that are not available using other polymerization methods [17]. The development of new highly active and robust metathesis catalysts allows polymerizing dienic monomers in presence of several functional groups. It has been demonstrated that ADMET polymerization can proceed in the presence of heteroatoms, as long as the terminal olefins are far enough apart from them [18]. There are many examples of ADMET of heteroatom-containing α,ω-dienes in the literature, since the introduction of functionalities in the backbone of the polymers provides different properties and permits further modifications [19,20 ,21].

We synthesized the undecenoyl ester of the 10-undecenol 1 by transesterification of the methyl 10-undecenoate with 10-undecenol (Figure 2). In order to obtain thermoplastic polyesters with improved flame retardance, we designed a new α,ω-dienic monomer containing a phosphorus moiety [22,23]. First, a phosphorus containing diphenol was obtained by addition of 9,10-dihydro-9-oxa-10-phosphaphenanthrene-10-oxide (DOPO) to p-benzoquinone. The novel biobased P-containing diene 2 was prepared by esterification of this diphenol and 10-undecenoyl chloride [24]. The DOPO-containining derivatives have been used to make many synthetic resins flame retardant: for example epoxy, [25] polyurethanes, [26] polyesters [27] and novolac resins.[28] This rigid and bulky group contain an unusually high thermally stable P-O-C bond which can be attributed to the O=P-O group being protected by phenylene.

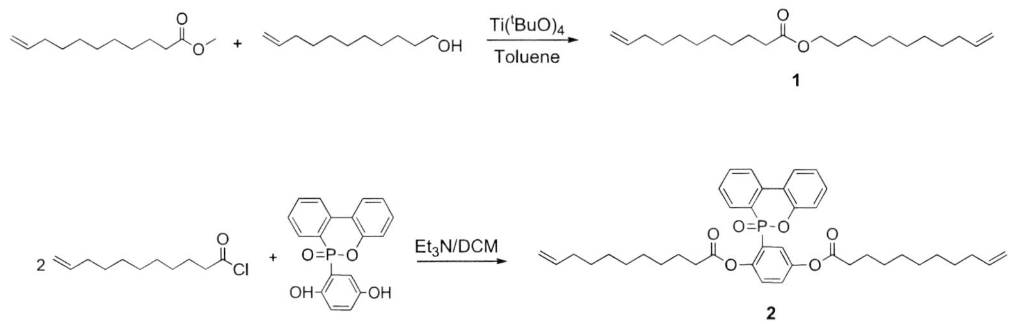

Figure 2.

Dienes 1 and 2 were copolymerized in absence and in presence of different amounts of methyl undecenoate as chain stopper using Grubbs 1st generation catalyst, Grubbs 2nd generation catalyst and Hoveyda-Grubbs 2nd generation catalyst working in bulk at 80ºC and 100ºC (Figure 3). The best results were obtained at 80ºC using the Grubbs second generation catalyst. In this way, varying the comonomer molar ratio we obtained a set of linear polyesters with molecular weights ranging from 9x10^3 to 40x10^3 Da and with phosphorus contents ranging from 0 to 4.7%[22]. These polyesters were semicrystalline or amorphous depending on the percentage of the bulky P-comonomer which impedes the crystallization of the polyester chains. So the copolymers containing more than 50% of 2 were amorphous whereas the rest were semicrystalline. The flame retardant behavior of these linear polyesters was evaluated using the Limiting Oxygen Index test (LOI) which measures the minimum concentration of oxygen required to support the combustion of a sample under specific conditions. LOI values were found to increase from 19 for the phosphorus-free polymers to 23 for polymers having 3% of P and remain almost constant for higher P content which is a quite common behavior for P-based systems. The significance of this result is that polyesters whose LOI values are higher than 21 no longer burn in ambient air if there is no supplementary oxygen and are very interesting materials for applications that require fire resistance.

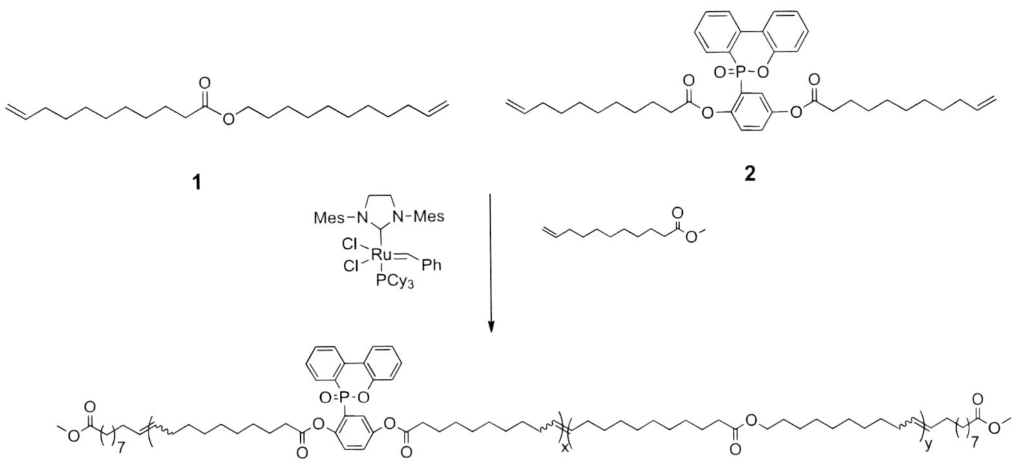

Figure 3.

1.2 Thiol-Ene Coupling

Click chemistry concept, introduced by Sharpless and collegues in 2001, describes chemistry tailored to generate substances quickly and reliably by joining small units together.[29] Recently, the addition of thiols to double bonds, defined as thiol-ene click coupling, has emerged as a powerful tool for synthetic purposes.[30,31] In particular, thiol-ene click chemistry of fatty acids obtained from plant oils is a promising tool that can be used to easily prepare biobased diols.

We first prepared two aliphatic diols, 3 and 4 (Figure 4), applying photoinitiated thiol-ene coupling with 2-mercaptoethanol to methyl esters of undecylenic and oleic acids respectively, and the subsequent reduction. [32] As expected, the addition of 2-mercaptoethanol to undecylenic acid methyl ester reached 100% in few minutes, whereas the addition to oleic acid methyl ester required longer reaction times (90 min to reach 99%). This difference in reactivity is due to ene susceptibility to thiyl attack and subsequent hydrogen abstraction. [33] Similar procedure was applied to their allyl ester derivatives leading to 5 and 6. Thus, four suitable monomers 3, 4, 5 and 6 for linear polyurethane synthesis were prepared from renewable resources. It is noteworthy to mention that fatty acid allyl ester derivatives are 100% renewable monomers as they can be easily prepared by esterification of the corresponding fatty acid with allyl alcohol, which can be obtained from glycerol, the main byproduct in triglyceride transesterification process for biodiesel manufacture. [34]

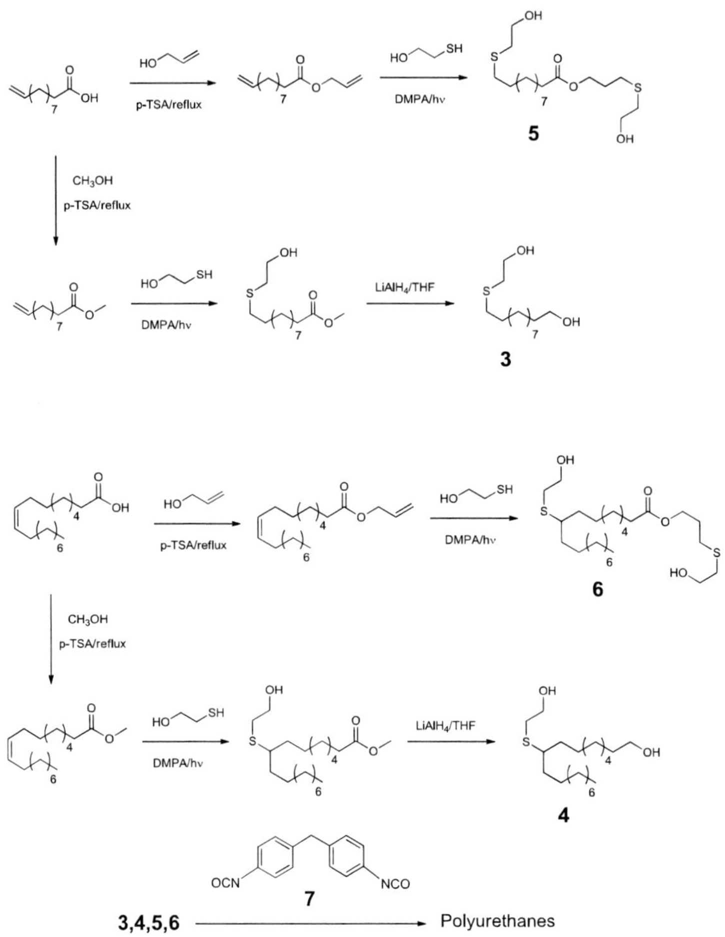

Figure 4.

High molecular weight linear polyurethanes were obtained in high yields by polymerization of the above mentioned diols with 4,4'-methylenebis(phenylisocyanate) 7. High intensity ultrasound irradiation was also applied to the preparation of polyurethane derived from 5. In this case, the sonochemical reaction proceeded faster in the early stages and led to higher molecular weight polyurethanes. These polyurethanes showed an excellent hydrolytic and chemical resistance (no weight loss and decrease in the molecular weight were observed after 6 months in a sodium phosphate buffer solution at pH 7.4 and 60ºC) due to its high hydrophobicity. This hydrophobic character can be exploited in moisture-sensitive environments and long-term applications.

To further explore the potential of thiol-ene coupling in the functionalization of plant oil derivatives, we developed an efficient and versatile "onepot" method for the accelerated preparation of well-defined telechelics diols from undecylenic acid via two sequential thiol-ene click processes: step-growth photopolymerization and end group post-polymerization modification (Figure 5).[35] A major problem for the preparation of telechelic polymers and particularly for the transformation of end-groups is the incompleteness of the reactions. Thus, it is essential to develop synthetic methodologies involving high efficient reactions such as thiol-ene coupling.

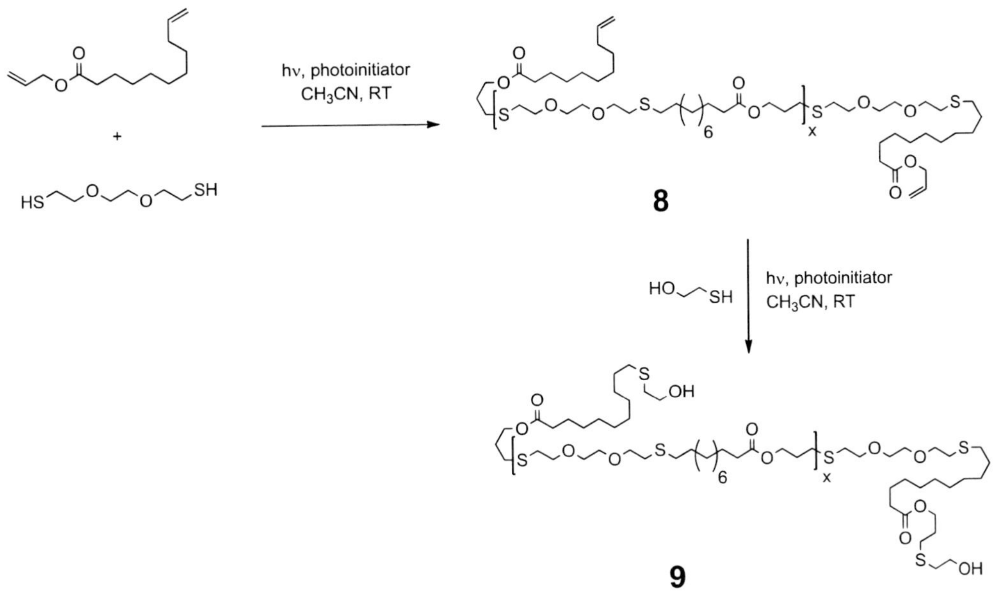

Figure 5.

Allyl ester of undecylenic acid and 3,6-dioxa-1,8-octanedithiol were "clicked" to prepare alkene-functionalized linear polymers 8 with variable molecular weight by thiol-ene click stepgrowth polymerization. Thereafter, the modification at the polymer terminus of 8 has been done using 2-mercaptoethanol to prepare new biobased telechelic diols 9 with targeted molecular weight ranging from 1000-3000 g/mol. An exhaustive ^{1}H NMR and MALDI-TOF MS analyses demonstrated the highly end-group fidelity of this methodology being an interesting procedure for the accelerated preparation of telechelics derived from divinyl monomers. Thelechelic diols prepared using this methodology were reacted with diisocyanate 7 and 1,3-propanediol as the chain extender to obtain multiblock poly(ester urethane) with phase separated morphology.

2. Thermosetting Polymers

2.1 Direct Polymerization of C=C

2.1.1 Cationic Polymerization of C=C

Vegetable oils contain internal C=C double bonds capable of being polymerised. As is well known, these double bonds can be polymerised through a free radical or a cationic mechanism.[36,37] The free-radical polymerization of triglyceride double bonds has received little attention due to the presence of chain-transfer processes to the many allylic positions in the molecule, however, vegetable oils such as linseed and tung oils have been classically exploited as drying oils and used in paints and coatings. The cationic polymerization of the C=C double bonds has been studied and the preparation of thermosetting polymers ranging from rubbers to hard plastics has been reported.[7] The copolymerization of a variety of oils with petroleum-based comonomers such as styrene, divinylbenzene and dicyclopentadiene in the presence of boron trifluoride diethyl etherate as the initiator has been described and due to the poor miscibility between soybean oil and the initiator and the large difference in the reactivity of the soybean oil and the comonomer, heterogeneous reactions occur at the early stages of the copolymerization, leading to a phase separated materials. Homogeneous copolymerization of soybean oils with styrene and divinylbenzene can be achieved using BF_3OEt_2 modified with methyl oleate [38] or Norwayfish oil ethyl ester [39] as initiator, as both are miscible with vegetable oils and styrenic monomers.

However, in order to achieve the incorporation of soybean oil into the polymeric network the polymerizations had to be carried out at high temperatures, up to 140ºC and for long periods of time up to 60h because of the low reactivity of the internal double bonds. So for practical reasons it would be advisable to enhance somehow the reaction rates in the production of these materials. Microwave irradiation has been shown to be a promising alternative heat source for organic transformations and polymerization reactions [40,41]. Microwave technology mainly owes its popularity to the enhanced reaction rates, higher yields, and greater purity of the products. The enhanced reaction rates can be explained by the high reaction temperature that stems from a list of advantages over conventional heating such as noncontact heating, circumventing the decomposition of molecules close to the walls of reaction vessel; instantaneous and rapid heating, resulting in a uniform heating of the reaction mixture; and highly specific heating, with the material selectivity emerging from the wavelength of microwave irradiation that intrinsically excites dipolar oscillation and induces ionic conduction. The use of microwave irradiation in polymer chemistry is an emerging field of research and a number of examples of step-growth, ring-opening, and radical polymerizations can be found in the literature [42,43,44].

In this way, we explored the cationic polymerization of soybean oil 10 and its copolymerization with styrene 11 and divinylbenzene 12 under microwave assistance to investigate how microwave irradiation influences reaction rate, monomer reactivity ratios, and polymer properties. [46] (Figure 6). We tested different heating and microwave irradiation cycles and compared both processes measuring the percentage and composition of soluble fraction and the thermal and mechanical properties of the final crosslinked materials. In the case of control soybean oil the cationic homopolymerization reach its maximum reaction degree after only one hour whereas it needs at least 24h when cured at 140ºC. In the case of the copolymers with 11 and 12, all the analysis performed indicate that microwave irradiation produces thermosets with comparable composition and crosslinking density but 25 times faster than the conventional heating demonstrating that in small scale, microwaves are a more efficient energy source of energy for this cationic polymerization.

To further extend the application of renewable resources to the field of flame retardant polymers, our approach consisted of the cationic copolymerization of 10 with 11, 12 and different amounts of 4-trimethylsilylstyrene 13 [38] (Figure 6). The amount of oil and 12 as the crosslinking agent were kept constant and the molar ratio of 11 and the silylated styrene 13 were modified to obtain polymers with increasing silicon content (1, 3, 5 and 7%). In this way, tough ductile materials with Tg values between 54 and 62ºC were obtained. Solvent extraction was used to determinate the amount of crosslinked material that was found to decrease (from 92% to 68%) with the increase of the amount of sililated monomer which can be attributed to its lower reactivity. The soluble fraction was found to consist of polystyrene oligomers and unreacted soybean oil.

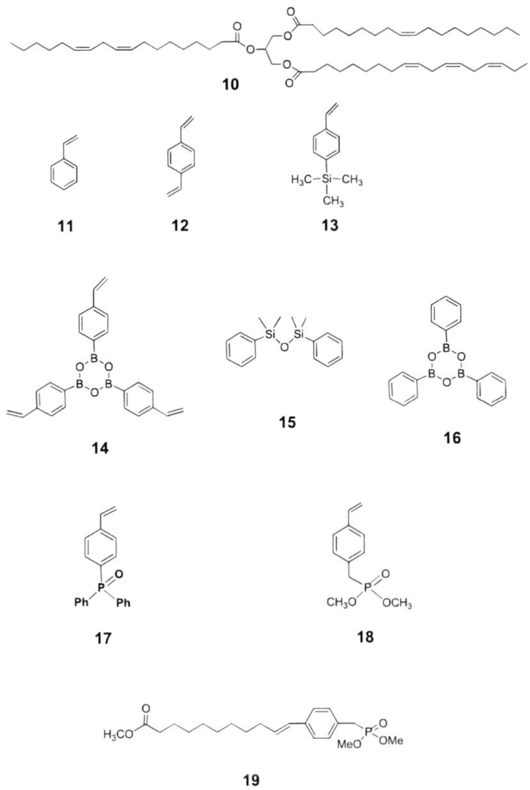

Figure 6.

In a similar way, we studied the preparation and properties of boron containing soybean oil copolymers by cationic copolymerization of 10, 11 and 12 with different amounts of tris(4-vinylphenyl)boroxine 14 (Figure 6). Materials with a 1 and 3% of boron content, with Tg ranging from 43 to 60 °C which are thermally stable below 50°C, were obtained. [47]

Flame retardancy of the silicon and boron containing soybean oil copolymers was evaluated using the Limiting Oxygen Index test. So, whereas the LOI for the heteratom-free thermoset is 19, the LOI values for all the Si containing thermosets are higher than 21 and increase with the Si content reaching a value of 29.7 for the copolymer containing 7% of silicon. For the B containing polymers, a LOI value of 25.6 for the copolymer containing 3% of boron.

Whilst measurement of LOI is a useful, small-scale test that correlates to the ignitability in polymers, it is not a reliable indicator of how a material will perform once ignited in a real fire. The most widely used method for this is the cone calorimeter, which gives information on both the ignitability, and the burning behaviour a polymer. The fire retardance and thermal stability of the above mentioned soybean-based copolymers reactively modified by copolymerization with 13 and 14 have been compared with those prepared with equivalent amounts of the additive 1,3-diphenyl-1,1,3,3-tetramethyldisiloxane 15 and tris-(phenylboroxine) 16 and with the heteroatom free soybean based copolymers. [48] The cone calorimetry experiments gave much clearer evidence than the LOI measurements that incorporation of Si or B into the soybean oil copolymer resulted in fire retardation. The boron-containing copolymers were found to be more efficient flame retardants for this system than were the silicon-containing copolymers. Moreover, the reactive or additive approach is a significant factor in terms of the level of fire retardance achieved and significant improvements are obtained when reactive flame retardants are used.

To obtain phosphorus-containing soybean oil copolymers we investigated two different approaches.[49] First, two phosphorus-containing styrene derivatives diphenyl styryl phosphine oxide 17 and dimethyl-p-vinylbenzylphosphonate 18 (Figure 6) where tested as comonomers in the cationic copolymerization of 10, 11 and 12 obtaining heterogeneous systems in all cases. To overcome this drawback, the cross-metathesis reaction of methyl 10-undecenoate and 18 was carried out to link the phosphorus moiety to the vegetable oil derivative. This second approach permitted the synthesis of a new reactive phosphorus-containing plant oil derivative 19, which was incorporated to the soybean oil system. The cationic copolymerization was investigated and the structure, thermal stability and mechanical and flame retardant properties of the resulting copolymers were studied. Thermosets with moderate glass transition temperatures were obtained showing that cross-metathesis reaction is a convenient way to produce oil compatible monomers able to undergo homogeneous polymerization reactions. The resulting thermosets with 1% of phosphorus had LOI values about 24.0 indicating an improvement on fire retardant properties on the soybean oil-based copolymers.

2.1.2 Hydrosilylation of C=C

In an attempt to find out some new applications and uses of triglycerides, we investigated the synthesis of new organic-inorganic hybrid materials via hydrosilylation reaction [50]. Although the hydrosilylation of olefins has been widely studied, in the last decade only a few studies on fatty acid esters and oils have been reported [51]. They all involved adding a monofunctional silane compound to the double bond, and introducing a certain silicon reagent to the ester or the oil. The reaction known as hydrosilylation proceeds when, after certain hydrosilanes have been activated, they undergo addition across the carbon-carbon multiple bonds. This reaction usually requires a catalyst, the most commonly used of which are the transition metal complexes (Co(I), Rh(I), Pd(0) or Pt(0)). It is well known that the reactivity of terminal C=C is higher than that of internal ones. Terminal C=C-containing fatty acid derivatives are available from unsaturated fatty acids by methathesis with ethylene or by pyrolisis.

We developed hybrid organic-inorganic materials from 10-undencenoyl triglyceride 20, and methyl 3,4,5-tris(10-undecenoyloxy)benzoate 21, a new aromatic fatty acid-derivative compound (Figure 7) by hydrosilylation with several polyfunctional hydrosilylating agents; 1,4-bis(dimethylsilyl)benzene 22, tetrakis (dimethylsilyloxy) silane 23 and 2,4,6,8-tetramethy-lcyclotetrasiloxane 24 catalyzed by Pt(0)-divinyltetramethyldisiloxane complex (Karstedt catalyst). Elastomeric materials with good thermal stabilities were obtained in all cases. The resulting cured hybrid networks showed good transparency according to the good miscibility of the organic and inorganic components.

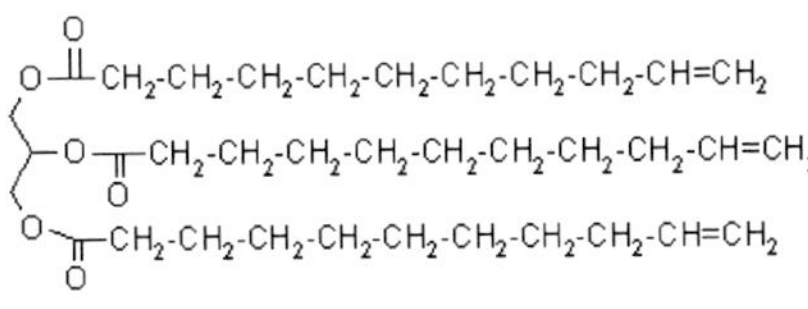

20

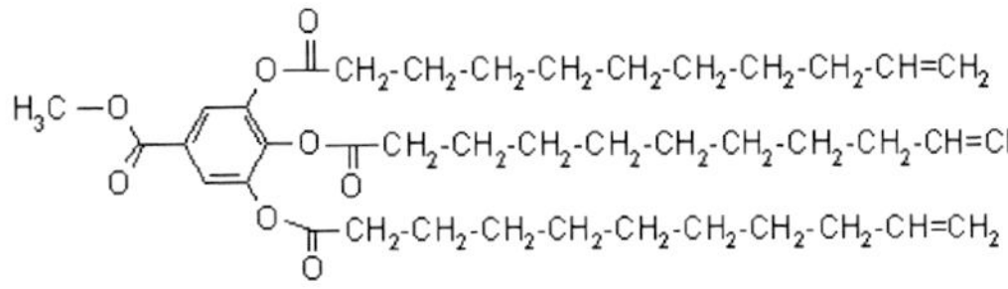

21

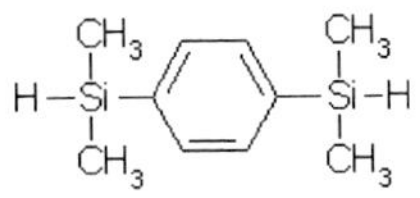

22

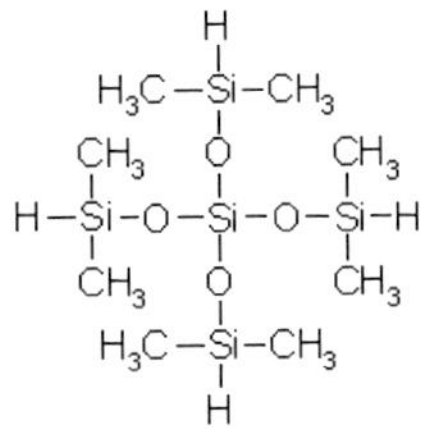

23

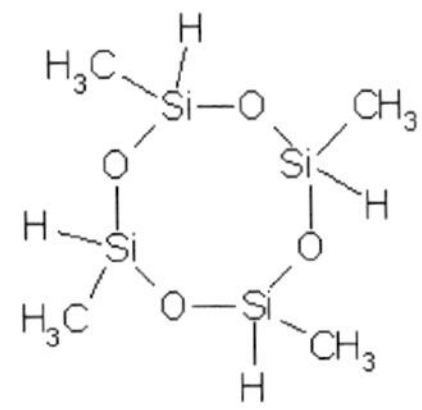

24

Figure 7.

2.2 Functionalization of C=C and Further Crosslinking

As commented above, the functionalization of the triglyceride double bonds to introduce readily polymerizable functional groups is another common strategy for obtaining high performance polymeric materials. Since the internal double bonds in the triglyceride structure are not sufficiently reactive for any viable polymerisation process, except for cationic polymerisation, considerable efforts have been devoted to modification of C=C into more reactive functional groups that could facilitate the polymerisation of the triglycerides. Various chemical pathways for functionalising triglycerides and fatty acids have been described [4].

2.2.1 Biobased Epoxy Resins

Epoxidation is one of the most important functionalization reaction involving C=C double bonds and epoxidized vegetable oils show excellent promise as inexpensive, renewable materials for industrial applications. In recent years, extensive work has been done to develop polymers from epoxidized triglycerides or fatty acids [2]. The polymerization of epoxidized vegetable oils has been investigated using either diamine hardeners [52] or thermal latent cationic catalysts [53] to produce biobased epoxy resins with promising properties.

Polymers reinforced with well-defined nanosized inorganic clusters have attracted a tremendous amount of interest because of their versatility; among these systems polyhedral oligomeric silsesquioxane (POSS) compounds, which possess a unique cage-like structure and nanoscale dimensions, are of particular interest. POSS compounds are 1-3 nm in diameter and their inorganic cage framework is made up of a fixed proportion of silicon and oxygen: $(SiO_{1.5})_n$, where n=8, 10, or 12. A different polymerizable POSS macromers have been employed to copolymerize with organic monomers, via the formation of covalent bonds, to afford a variety of polymer/POSS nanocomposites [54]. Epoxy resins/POSS nanocomposites are the most studied of these nanocomposites, and various mono- or polyepoxide POSS monomers have been used to modify epoxy networks [55]. The incorporation of POSS derivatives into polymeric materials can lead to substantial improvements in polymer properties including increases in use temperature and mechanical properties, as well as reductions in flammability, heat evolution and viscosity during processing.

We described the first example of the preparation of biobased POSS-nanocomposites from plant oil derivatives [56]. In this study, epoxidized linseed oil 25 and 3-glycidylpropylheptaisobutyl-T_8-polyhedral oligomeric silsesquioxane 26 were used as organic monomers and inorganic nanoparticules, respectively (Figure 8). The bionanocomposites were synthesized by curing both epoxy monomers, using a thermally-latent cationic catalyst. All the cured nanocomposites were homogeneous and transparent, implying that no phase separation occurred. Most of the resulting POSS-containing networks displayed slightly enhanced glass transition temperatures. The storage moduli of the networks at the glassy and the rubbery plateau were observed to be somewhat higher than that of POSS-free network. These results can be adscribed to the nanoscale reinforcement effect of POSS cages on the crosslinked matrix.

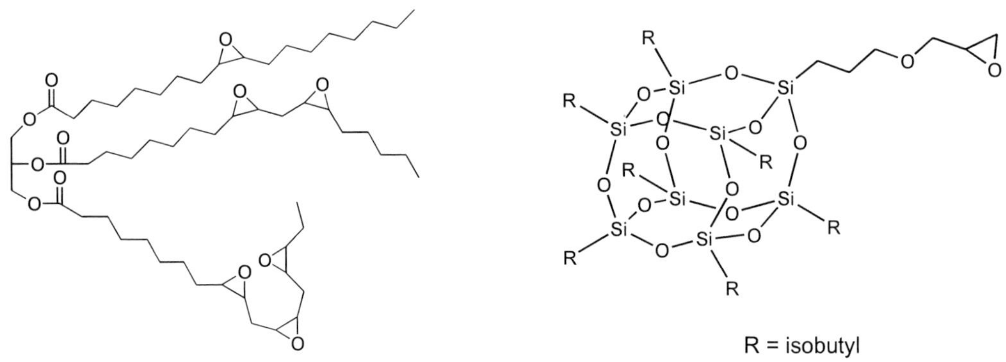

25 **26**

Figure 8.

In an attempt to obtain new fatty acid-derived compounds that could find applications in flame retardant materials, we synthesized a new diepoxy compound 27 from the DOPO derivative 2 (Figure 9).[24] Novel epoxy resins were prepared from 27 and two fatty acid derivatives; epoxidized 10-undecenoyl triglyceride 28 and epoxidized methyl 3,4,5-tris(10-undecenoyloxy)benzoate 29 by curing with 4,4'-diaminodiphenylmethane 30 and bis(m-aminophenyl)methylphosphine oxide 31 as crosslinking agents [57]. In this way, a series of biobased epoxy resins with phosphorus contents ranging from 1.8 to 5.7% were obtained. The materials showed Tg values from 56 to 98°C and exhibit good thermal stabilities, with initial temperatures of weight loss above 300°C. To further investigate the thermal degradation mechanism, samples were heated in an oven at 350°C for 3 h with nitrogen or air as the purge gas. Volatile products were trapped at the liquid nitrogen temperature and subsequently analysed by GC/MS and the solid residues of pyrolyzed polymers were characterized by FTIR and ^{31}P MAS NMR spectroscopies. The most noticeable feature of these analyses was that phosphorus-containing species were detected only in the solid residues.

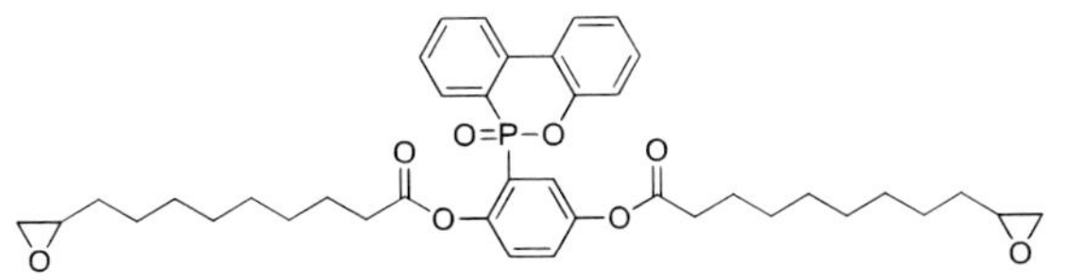

27

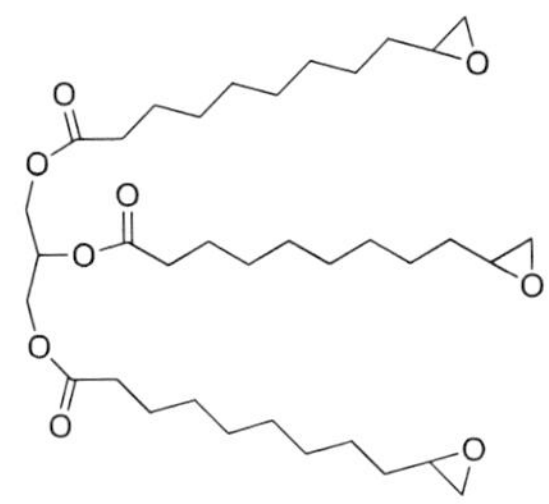

28

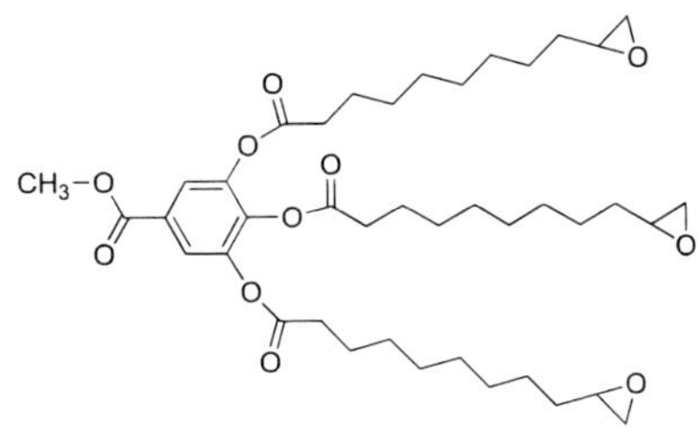

29

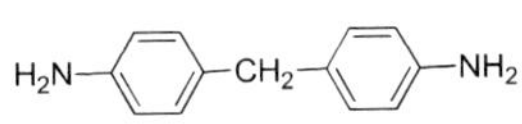

30

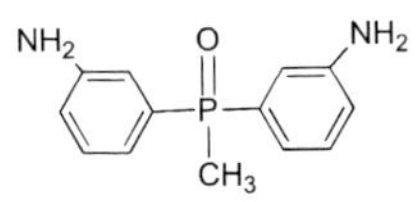

31

Figure 9.

The presence of phosphorus increased the LOI values from 21.9 to 32.0, leading to epoxy resins with improved flame retardancy. To better understand the role of phosphorous in the flame retardant properties of the polymer, element mapping was performed with energy-dispersive X-ray spectroscopy (EDX) on the surfaces of the initial samples and of the samples after the LOI test. The P distribution shows that the phosphorous density increased towards the top burned surface and that a phosphorous-rich layer formed. When they are heated, phosphorous compounds can form glass-like polyphosphoric acid, which protects the burning surface, or they can form inflammable phosphorous-carbon char by reacting with organic components. This protective layer is resistant to even higher temperatures and shields the underlying polymer from attack by oxygen and radiant heat as well as prevents the combustible gases from transferring to the surface of the materials and feed the flame, thus improving the fire resistance.

2.2.2 Biobased Polyols and Polyurethanes

The preparation of polyols from fatty acids and oils for general polyurethane use has been the subject of many studies [58], but limited attention has been paid to the preparation of polyether polyols from this kind of compounds. Polyether polyols with molecular weights of 200 to 10000 g/mol are important building blocks for polyurethane applications. Polyols with molecular weights of about 3000 or more are used to produce flexible polyurethanes, and polyols of about 200 to 1200 g/mol are used for rigid polyurethanes.

Polyether polyols are usually produced by the anionic ring-opening polymerization of alkylene oxides such as ethylene oxide or propylene oxide. Longer chain oxiranes or functionalized epoxides show lower reactivity due to a higher sterical hindrance and side reactions, and high molecular weight polymers cannot be obtained by using anionic or cationic catalysts. Low molecular weight polyethers can be synthesized by cationic ring-opening polymerization which requires an acid catalyst. Several classes of cationic initiators have been developed such as photosensitive onium salt initiators which are inactive under ambient conditions and can release the acid initiator by UV-radiation or heat and were applied to epoxidized fatty compounds.[59] High molecular weight polyethers can be obtained by coordinative ring-opening polymerization that has been described for ω-epoxy alkanoates by using the Vandenberg catalyst. [60]

We reported the synthesis and characterization of polyether polyols from epoxidized methyl oleate 32. Polyols 33 were prepared through combining its polymerization and the controlled reduction of the carboxylate groups to hydroxyl moieties (Figura 10). Two different polymerization initiators were used in order to modulate molecular weight of the final products. Low molecular weight polyether polyols were synthesized by cationic ring-opening polymerization of 32 [61] whereas coordinative ring-opening polymerization using ionic-coordinative initiators led to higher molecular weight polyether polyols. [62]

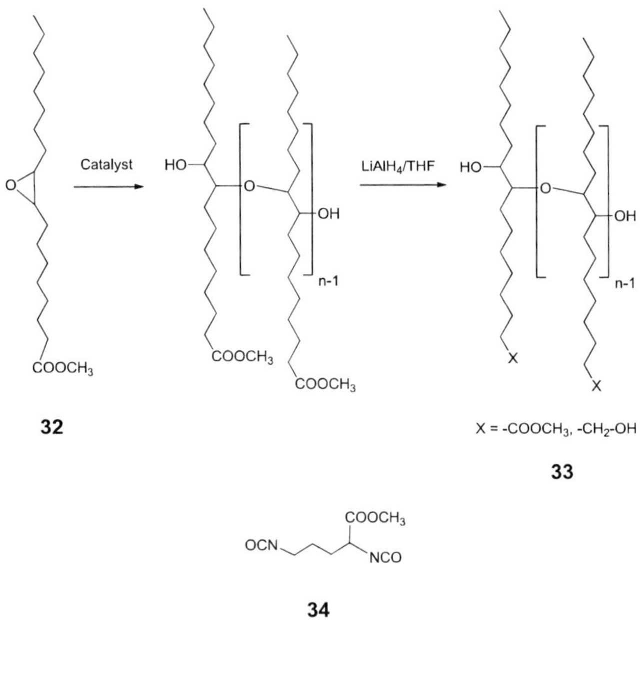

Figure 10.

Low molecular weight polyols were prepared through the cationic polymerization in the presence of 0.5wt.-% of $HSbF_6$ at room temperature and the further partial reduction of the carboxylate groups to hydroxyl moieties using lithium aluminum hydride as reducing agent. The catalyst was completely soluble in the epoxidized methyl oleate at room temperature, and the oligomerization was performed homogeneously in absence of solvent, thus being an advantageous process from an environmental viewpoint. In this way, after reduction with the appropriate amount of lithium aluminum hydride, polyols with a broad range of functionalities were obtained, which are from clear liquids to white waxy solids at room temperature.

When ionic-coordinative catalysts such as tetraisobutyldialuminoxane and Vandenberg catalyst $(C_2H_5)_3Al/H_2O$ were used, polyols with higher molecular weight than above mentioned cationic catalyst were obtained. These materials were found to consist of a complex mixture of cyclic and linear chains with different chain ends which were related to the catalyst nature and the occurrence of two main polymerization mechanisms, the cationic and the ionic-coordinative.

A series of segmented and non-segmented crosslinked polyurethanes were synthesized from the above mentioned polyols using 4,4'-methylenebis(phenylisocyanate) 7 or L-lysine diisocyanate 34 as coupling agents and 1,3-propanediol as a chain extender [61,63]. Segmented polyurethanes are elastomeric block copolymers that generally exhibit a phase-segregated morphology made up of soft rubbery segments and hard glassy or semicrystalline segments. As expected, mechanical properties of segmented polyurethanes were improved as hard segment concentration was higher, whereas non-segmented polyurethanes behave as soft rubbers showing low Tg values that increase with polyol functionality.

Starting from low molecular weight polyols 33 we synthesized different polyurethane systems: aromatic fatty acid-based polyurethanes with improved mechanical properties, silicon-containing polyurethanes with flame retardant properties and polyurethanes entirely from renewable resources with potential applications in the biomedical field.

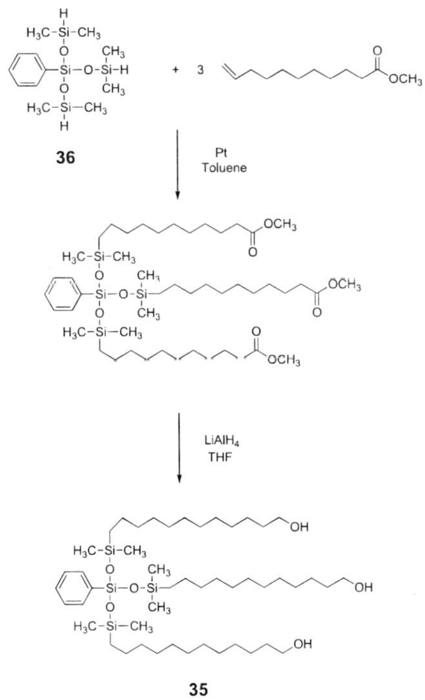

Figure 11.

Novel silicon-containing biobased polyurethanes were obtained from polyol 33, the silicon-containing triol 35 and 7 as a crosslinking agent [64]. 35 was obtained by hydrosilylation of methyl 10-undecenoate with a trifunctional hydrosilylating agent, phenyltris(dimethylsiloxy)silane 36 and the subsequent reduction of carboxylate groups to obtain primary hydroxyl groups (Figure 11). A series of five different polyurethanes with a silicon content between 1.7 and 9.0% was prepared by varying the polyols molar ratio and the LOI increased with the silicon content, so, therefore they are very interesting materials for applications that require fire resistance.

The introduction of aromatic comonomers into the polymer structure would appear to be suitable in the search of new viable polymeric materials. Aromatic compounds derived from fats can be obtained by transition-metal catalyzed cyclotrimerization of the respective alkyne fatty derivatives. We reported the synthesis and characterization of the biobased triols 37 and 38 (mixtures of 1,3,5 and 1,2,4 isomers) prepared by the trimerization of methyl 10-undecynoate or methyl 9-octadecynoate, respectively and subsequent reduction of carboxylate groups to give primary hydroxyl groups (Figure 12) [65]. Polyurethanes were obtained from these triols, 1,4-butanediol as a chain extender, and diisocyanate 7 as coupling agent.

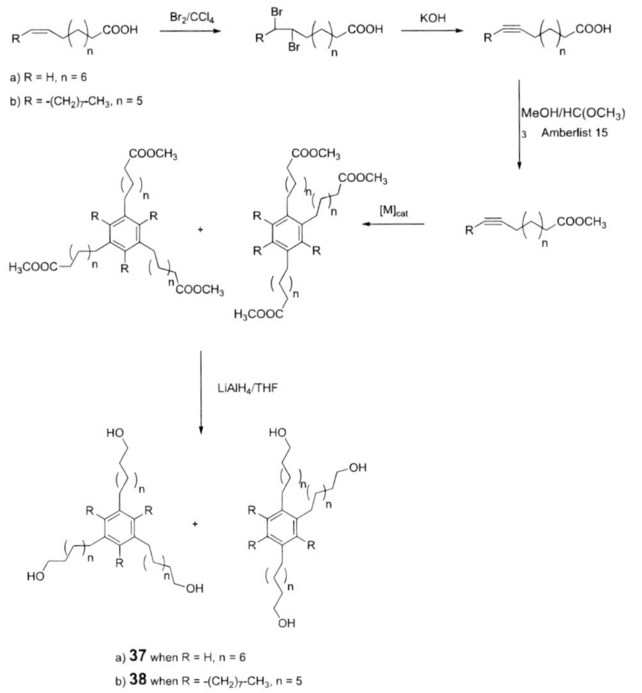

Figure 12.

Segmented polyurethane elastomers have been used as biomaterials for several decades in the fabrication of medical implants such as cardiac pace makers and vascular grafts because their physical properties and relatively good biocompatibility [66]. We prepared novel segmented biobased polyurethanes using the one-shot technique from 33, amino acid derived diisocyanate 34, and 1,3-propanediol as a chain extender [67]. In view of the future application of the synthesized polyurethanes in the biomedical field, water uptake and in vitro degradation studies were carried out and the morphology of the degraded polyurethanes were observed by scanning electronic microscopy (SEM). Water absorption was measured to determine the polyurethane bulk hydrophilicity because this parameter was expected to have a substantial impact on hydrolytic degradation. The hard segment content is the main factor that controls the amount of absorbed water. As the hard segment content increases, a more hydrophilic character in the final network due to the presence of higher amount of urethane groups can be expected, thus increasing the water uptake. The in vitro degradation experiments of the synthesized polyurethanes were carried out by immersion of the samples in PBS (pH 7.4, 0.1 M) at 37 °C, and the degradation rate was evaluated by the weight loss of the polymers over predetermined time intervals. After 72 days of degradation, the weight losses of polyurethanes are all below 12 %. The hard segment content has some influence on the hydrolytic degradation rate. With the increase of the hard segment content, the degradation rate increased, which is in agreement with the hydrophilicity of the polymers. The reason may be that the resulting polyurethanes containing higher hard segment content have more hydrophilicity and water diffusion is relatively easy.

The visual examination of the surface of the degraded polyurethanes was carried out using SEM. For all samples, surface appeared spotted with round pits where material had been removed and showed more extensive cracks and numerous pores in progressive weeks, indicating a larger extent of degradation with time. Moreover, with the increase of the hard segment content, the erosion was more serious. The spotted surface is due to the presence of areas with marked differences in hydrolytic stability. The hydrophobic character of soft-segments prevented the entry of water molecules, resulting in greater hydrolytic stability of the polyurethane. Hard-segments increase the hydrophilicity and promote the susceptibility of urethane bonds to hydrolysis, leading to the formation of significant amounts of water soluble products leaching into the solution, and to a higher degradation rate.

2.2.3 Enone-Containing Triglyceride Derivatives

The challenge to progressively replace fossil feedstocks by materials arising from plant-derived renewable sources implies not only the development of new original reactions and catalysts but also the application of well established reactions to the production of new tailor made compounds capable to produce competitive performance materials. The singlet oxygen "ene" reaction is one of the highest investigated processes in organic chemistry to functionalize the allylic C-H bonds of unsaturated compounds. This reaction was discovered in 1948 by Schenck [68], who demonstrated that allylic hydroperoxides are handily prepared by reaction of alkenes with photochemically generated singlet oxygen. The mechanism of this reaction has been widely studied and it is actually well established [69]. For synthetic applications, the unsaturated substrate can be photoxygenated "in situ" with singlet oxygen generated by means of a high pressure sodium-vapor lamp and tetraphenylphorphyrin as sensitizer in an oxygen saturated medium, to give a mixture of isomeric allylic hydroperoxides. This reaction has been used to oxidize the allylic position of fatty acids and their derivatives [70,71]. The mild conditions utilized and the use of oxygen, as the only reagent, makes this process particularly favorable from both economical and ecological viewpoint.

The allylic hydroperoxides resulting from photoperoxidation of the allylic positions can undergo a number of different transformations [72,73]. One of the most interesting reaction is the conversion of these hydroperoxides into a regioisomeric mixture of enones, that can be carried out in the presence of acetic anhydride and pyridine or tertiary amines [75,76]. This reaction has been scarcely used with fatty acids and their derivatives.

We have applied for first time this environmentally friendly chemical procedure to obtain enone-containing triglycerides 39 from high oleic sunflower oil 40 (Figure 13) [77]. The resulting material contains an average of 2.6 α,β-unsaturated ketone groups per triglyceride molecule with 80% overall yield.

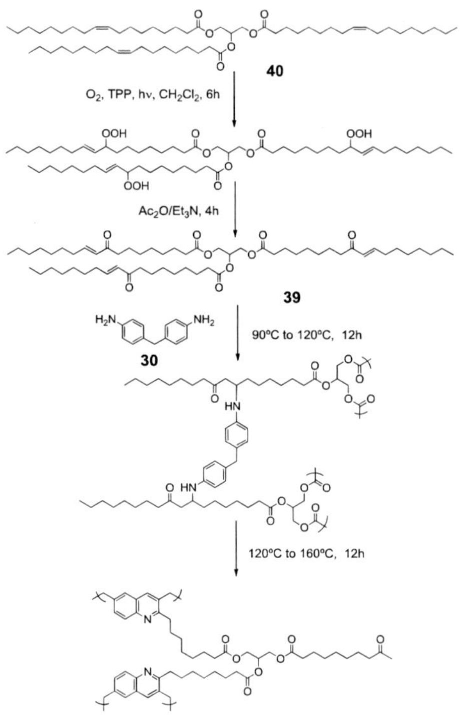

Figure 13.

This enone-containing triglyceride could be an interesting alternative to epoxidized vegetable oils to produce thermosets by crosslinking with conventional aromatic diamines. Naturally occurring epoxi oil, as vernonia oil, or epoxidized vegetable oils from soybean, linseed or castor oils have been cured cationically with conventional hardeners as diamines or dianhydrides. Amine-cured ω-epoxy fatty acid triglycerides have been shown to yield robust networks with good adhesive characteristics similar to those of conventional thermosets based on diglicidylether of bisphenol A. However, epoxidized linseed and soybean oils, which contain oxirane groups that are hindered at both carbons, react sluggishly with nucleophilic curing agents [78] On the other hand, the Michael addition reaction is a valuable tool in the synthesis of polymeric networks [79]. The aza-Michael reaction, a variation in which an amine acts as the nucleophile, has been used in the synthesis of improved bismaleimide networks [80], but this reaction has not been applied to the synthesis of crosslinked polymers from vegetable oils.

The reactivity of these enone groups towards aromatic amines was evaluated performing kinetic experiments with p-toluidine and methyl enone-oleate as model compound and compared with the reactivity of the methyl epoxy-oleate. The results indicated that the aza-Michael addition to the unsaturated ketone proceeds much faster than the corresponding nucleophilic attack over the epoxide ring even when a cationic ring-opening catalyst was used to activate the ring opening. In this way, enone-containing triglycerides could be considered as interesting alternatives to epoxidized oils to produce thermosets by croslinking with amines, providing a promising route to obtain polymeric networks under mild conditions and without the aid of a catalyst.

The efectivity of aza-Michael addition as curing reaction could be proved by following the crosslinking reaction of the enone-containing triglyceride 39 and diamine 30 by FTIR spectroscopy which allows confirming the complete reaction of the enone groups and the formation of the expected β-aminoketone product. So, aza-Michael reaction has been used to produce thermosets under quite mild conditions (90-120ºC for 12h) (Figure 13). The resulting materials were dark yellow soft rubbers with Tg values below room temperature indicating a light crosslinking density. Unexpectedly, when these materials were heated at higher temperatures (up to 160ºC) or when the curing process was carried out in presence of a Lewis acid like boron trifluoride, much tougher materials were obtained. (Tg by DMTA = 64ºC).The analysis of these thermosets by FTIR spectroscopy showed structural changes that evidenced the existence of secondary reactions. We carried out a detailed study of the reactions involved throughout this high temperature curing process using model compounds, isolating the intermediate species by column chromatography and characterizing their structure by NMR spectroscopy.

This study let us to conclude that, after the initial aza-Michael addition, a set of cascade reactions occur [81] (Figure 14). First, the imine of the Michael adduct 41 experiment a retro-Manich fragmentation leading a ketimine 42 and a highly reactive aldimine 43.

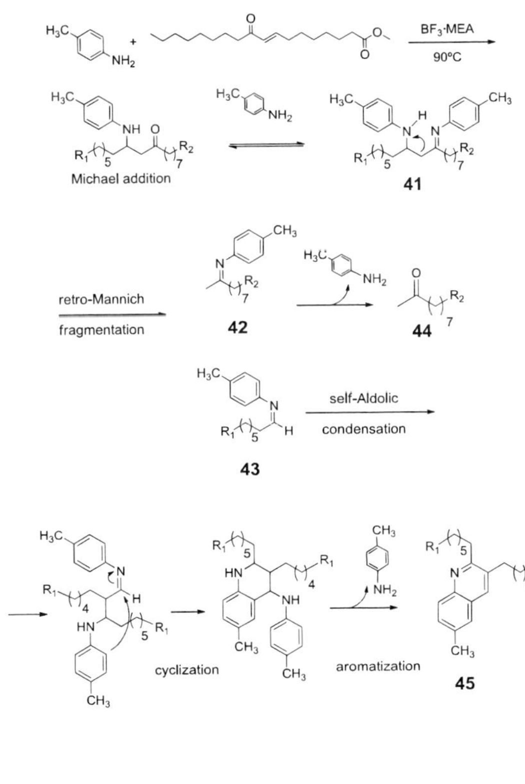

Figure 14.

The less reactive ketimine moiety is finally hydrolyzed leading to methylketone groups 44 which can be detected in the final products. But the highly reactive aldimine groups experiment a fast self-aldol condensation that gives an intermediate product that undergoes cyclization and aromatization reactions that finally leads to substituted quinoline ring 45. Thus, this new high temperature crosslinking process involves two enone groups per every aromatic amine and produces a trisubstituted quinoline that acts as crosslinking site and explains the improvement of the thermal and mechanical properties of these thermosets (Figure 14).

This aromatization process could be followed by UV spectroscopy when the low temperature aza-Michael croslinked product is progressively heated at higher temperatures. Thus a batocromic displacement of the UV absorption peaks can be observed and this effect is more important as higher is the postcuring temperature which can be related with the formation of an increasing percentage of the quinoline groups.

According to the proposed mechanism, the quinoline moieties are produced from an aldimine which we demonstrated to be formed by fragmentation of the aza-Michael addition product. Obviously the necessary aldimine can be also produced in a more direct way by reaction between an aldehyde and the aromatic amine. So an aldehyde-containing triglyceride should be the ideal starting product to produce thermosets with high quinoline content using this cyclization-aromatization process.

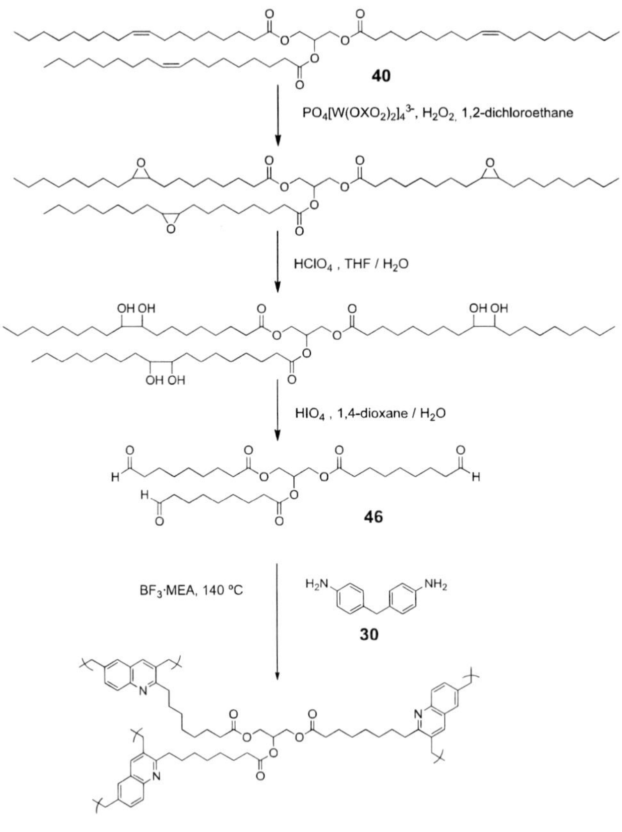

Figure 15.

In this way, we synthesized an aldehyde-containing triglyceride 46 starting from high oleic sunflower oil 40 [81] (Figure 15). The reaction sequence consists of the epoxidation of the double bonds with hydrogen peroxide, followed by the acidic ring opening with water and further periodate cleavage of the resulting diol to produce the desired product plus nonanal which is eliminated by distillation under reducted pressure. A triglyceride 46 containing 2.5 aldehyde groups per molecule was obtained with 90% overall yield. This product was cured with diamine 30 at 140°C yielding a material with better properties due to its higher quinoline content. The improvement in the mechanical properties could be followed by the evolution of the Tan δ plots of the DMTA analysis. The enone-containing triglyceride croslinked with diamine 30 at low temperature (90°C) shoved a Tg of 16°C. The same material postcured at 140°C showed a Tg of 64°C due to the formation of the quinoline groups. Finally, the aldehyde-containing triglyceride cured with 30 at 140°C showed a Tg of 92°C due to its higher quinoline content.

2.2.4 Biobased Acrylate Oils

Acrylate triglycerides are usually low viscosity monomeric liquids that can be free radically polymerised and can be copolymerised easily with other commercial comonomers because of the high reactivity of the acrylate group. The special properties of these chemically modified triglycerides offer a widespread field of applications and so they are excellent candidates for use as thermosetting liquid molding resins in techniques such as vacuum assisted resin transfer molding, composite fabrication processes or pressure sensitive adhesives. The traditional method used to obtain acrylate oils is to convert the triglyceride double bonds first to an epoxide and then open the epoxy groups with acrylic acid to yield oil with vic-hydroxiacrylate ester groups [82]. The direct acrylation by esterification of hydroxyl groups in the fatty oil is a scarcely described pathway as there are not many hydroxy-containing natural oils.

We reported a new general and environmentally friendly route to introduce acrylate groups in unsaturated vegetable oils. A common transformation of allylic hydroperoxides is their reduction to the corresponding hydroxylic compounds. Thus, we developed hydroxyl-containing triglycerides 47 by a three step synthetic pathway that first uses the singlet oxygen photooxygenation of high oleic sunflower oil 40 to led to a mixture of allylic hydroperoxides, which second can be reduced to a mixture of secondary allylic alcohols 47 (Figure 16) [83]. The new hydroxyl containing triglyceride was esterified with acryloil chloride to obtain 48 which was radically crosslinked in presence of different amounts of pentaerythritol tetraacrylate. Alternatively, the unsaturated alcohols can be further reduced to saturated alcohols 49 which can be also functionalised with acrylate groups to 50 and further crosslinked. All the cured materials showed high toughness and good transparency. In this way, new renewable based acrylates were obtained, showing similar properties and reactivity than other reported acrylate triglyceride-based materials and consequently are candidates to produce similar good performance thermosets.

Among the desirable properties thermosets and elastomers should have, flame retardancy is one of the most important concerning the security of the final users of these materials. An approach to flame retardancy is the design of new intrinsically flame retardant polymers based on phosphorus containing vegetable oils. To obtain these compounds we choose the simple and straightforward [2,3]-sigmatropic rearrangement of allylic phosphinites [84]. So, using the secondary alcohol containing-triglyceride derivative 47 as starting material and by reaction with chlorodiphenylphosphine 51, intermediate allylic phosphinites 52 were obtained (Figure 17) [85].

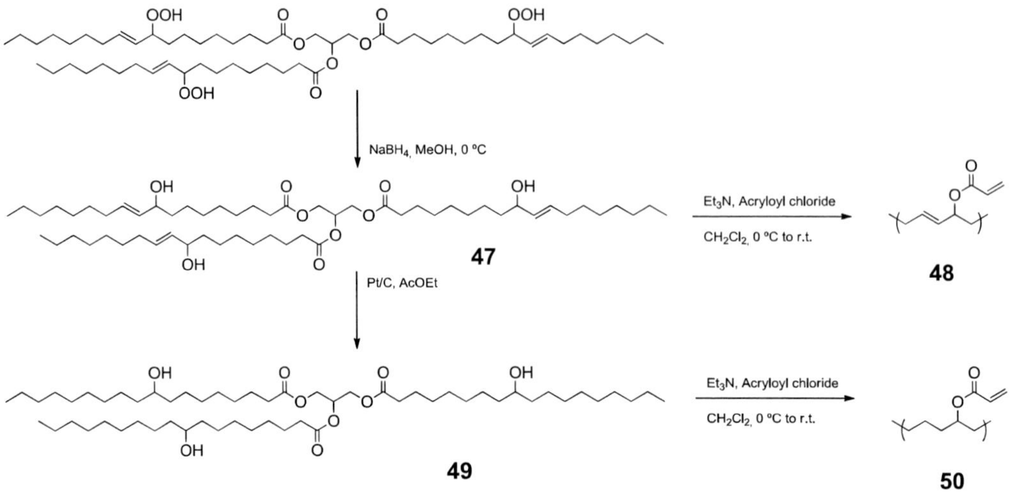

Figure 16.

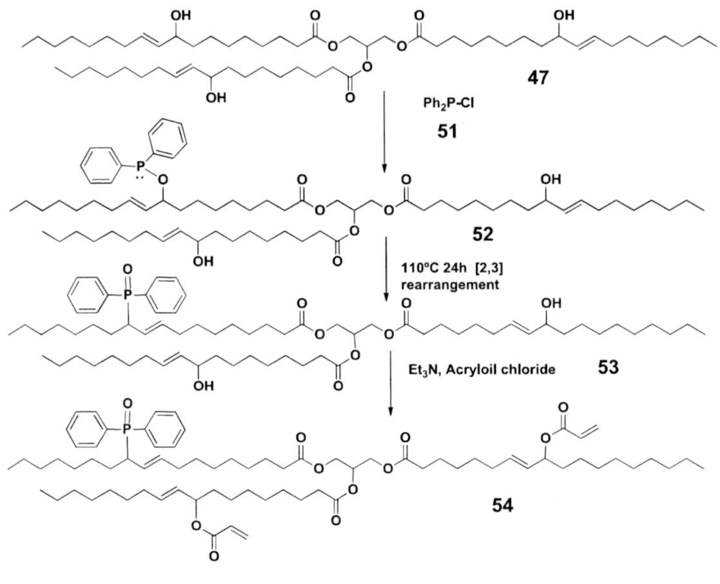

Figure 17.

When heated to 120°C, these phosphinites undergo a [2,3]-sigmatropic rearrangement to give a tertiary phosphine oxide 53 with new P-C bonds that link the phosphorus directly to the fatty acids chain. To the best of our knowledge, this thermal rearrangement has not been described for a triglyceride derivative, as a very efficient procedure to obtain allylic phosphine oxides. The remaining hydroxyl groups were esterified with acryloyl chloride in presence of triethylamine yielding 54, thus introducing the necessary reactive sites for further crosslinking. Pentaerythritol tetraacrylate was used as a crosslinking agent and in this way, four crosslinked materials were synthesized with 0.0, 1.4, 2.0 and 2.8 % w/w of phosphorus content. The flame retardancy of these materials was evaluated using LOI values. The addition of 1.4% of P gave a LOI of 21.2 and the maximum content of P (2.8%) gave a LOI of 22.4. These values show that slight improvement on the flame retardant properties is related to the increase in the phosphorus content

To prepare flame retardant polyester thermosets [23], we also used the bis-10-undecenoyl ester of glycerol 55 which is a mixture of glycerol esters containing primary and secondary hydroxyl groups in a 4 to 6 ratio. This monomer was homopolymerized and copolymerized with the phosphorus-containing monomer 2 and the methyl 10-undecenoate as chain stopper (Figure 18). For the homopolymerization the best results were achieved with the Hoveyda-Grubbs second generation catalyst but for the copolymerization the best results were obtained with the Grubbs second generation catalyst. Following this procedure and varying the molar ratios between monomers and chain stopper we were able to prepare a set of polyester-polyols 55 with molecular weights ranging from $3.7x10^3$ to $7.3x10^3$ Da and containing differents amounts of P (0, 1.6, 2.9, and 3.9%).

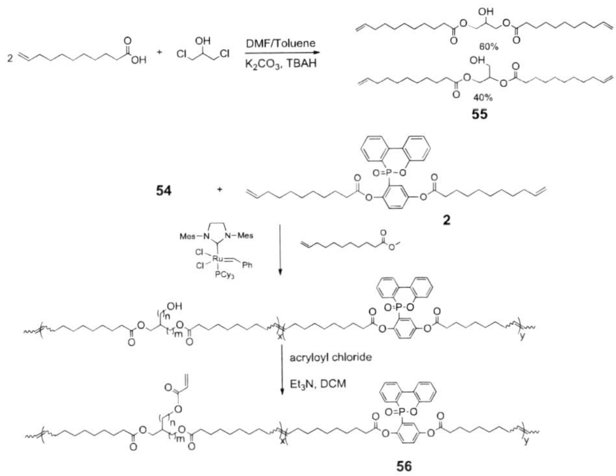

Figure 18.

In this case, with the exception of the 55 homopolymer, these materials are amorphous and due to its low molecular weight presented a good solubility in solvents like dichloromethane what facilitated the next step of chemical modification to introduce radicallary croslinkable functional groups. So, we esterified the hydroxyl groups of these polymers using acryloil chloride and triethylamine to produce acrylate phosphorus-containing polyesters 56. The curing of these reactive materials with dicumyl peroxide at 150ºC led to a set of colorless materials with Tg values that increase as the P content does. The flame retardancy of these thermosets was evaluated using the LOI test which showed that in spite of its high aliphatic content, the presence of about 3% of phosphorus infers a notable fire resistance reaching LOI values up to 25.6.

CONCLUSION

The concepts and examples summarized in this chapter show that different strategies can be used to produce valuable thermoplastic and thermosetting polymers starting either from the vegetable-oil triglycerides or their derivatives. The variety of the synthesized materials demonstrates the versatility and value of plant oils as platform chemicals for polymer synthesis. Moreover, the use environmentally friendly procedures and new efficient and green synthetic methodologies together with the switch to renewable resources is an important initiative in getting the world onto a sustainable trajectory.

REFERENCES

[1] Raston C. *Green Chem* 2005, *7*, 57-57; Jenck J.F., Agterberg F., Droescher M.J. *Green Chem* 2004, 6, 544-556

[2] Kaplan D.L. *Biopolymers from renewable resources*. Springer: New York; 1998.

[3] Belgacem, M.N.; Gandini, A. *Monomers, polymers and composites from renewable resources,* Elsevier: Oxford, 2008.

[4] Meier M.A.R.; Metzger, J.O.; Schubert, U.S. *Chem Soc Rev* 2007, 36, 1788-1802

[5] Sharma, V.; Kundu, P.P. *Prog Polym Sci* 2008, *33*, 1199-1215. Sharma, V.; Kundu, P.P. *Prog Polym Sci* 2006, 31, 983-1008

[6] Güner, F. S.; Yagci, Y.; Erciyes, A. T. *Prog Polym Sci* 2006, 31, 633-670.

[7] Lu, Y.; Larock, R.C. *ChemSusChem* 2009, 2, 136-147

[8] Eren, T.; Kusefoglu, S.H. *J Appl Polym Sci* 2004, 91, 2700-2710.

[9] Jain, P.; Choudhary, V.; Varma, I. K. *J Macromol Sci Polym Rev* 2002, 42, 139-183.

[10] Lu, S. Y.; Hamerton, I. *Prog. Polym. Sci.* 2002, 27, 1661-1712.

[11] Hörold, S. *Polym Deg Stab* 1999, 64, 427-431.

[12] Kashiwagi, T.; Gilman, J.W. In: *Fire retardancy of polymeric materials*. Grand, A.F.; Wilkie, C.A. Eds. Marcel Dekker Inc:New York, 2000.

[13] Ebdon, J.R.; Hunt, B.J.; Jones, M.S.; Thorpe, F.G. *Polym Deg Stab* 1996, 54, 395-400.

[14] Flammability Handbook for Plastics. Hilado, C.J. Ed. *Technomic Publications:* Lancaster, PA 1974.

[15] Armitage, P.; Ebdon, J.R.; Hunt, B.J.; Jones, M.S.; Thorpe, F.G. *Polym Degrad Stab* 1996, 54, 387-393.

[16] Van der Steen, M.; Stevens, C. V. *ChemSusChem* 2009, 2, 692-713.

[17] Schwendeman, J. E.; Church, A. C.; Wagener, K. B. *Adv Synth Catal* 2002, 344, 597-613.

[18] Wagener, K. B.; Brzezinska, K.; Anderson, J. D.; Younkin, T. R.; Steppe, K.; DeBoer, W. *Macromolecules* 1997, *30, 7363-7369.*

[19] Baughman, T. W.; Wagener, K. B. *Adv Polym Sci* 2005, 176, 1-42

[20] Shultz, G. V.; Zakharov, L. N.; Tyler, D. R. *Macromolecules* 2008, 4, 5555-5558.

[21] Terada, K.; Berda, E.B.;Wagener, K.B.; Sanda, F.; Masuda, T. *Macromolecules* 2008, 41, 6041-6046.

[22] Montero de Espinosa, L.; Ronda, J.C.; Galià, M.; Cádiz, V.; Meier, M.A.R. *J Polym Sci Part A Polym Chem* 2009, 47, 5760-5771.

[23] Montero de Espinosa, L.; Meier, M.A.R.; Ronda, J.C.; Galià, M.; Cádiz, V. *J Polym Sci Part A Polym Chem* 2010, 48, 1649–1660.

[24] Lligadas, G.; Ronda, J. C.; Galià, M.; Cádiz, V. *J Polym Sci Part A: Polym Chem* 2006, 44, 5630-5644.

[25] Levchik, S. V.; Weil, E. D. *Polym Int* 2004, 53, 1901-1929

[26] Wu, C. S.; Liu, Y. L.; Chiu, Y. S. *J Appl Polym Sci* 2002, 5, 2254-2259.
[27] Shan, W. C.; Hsuan, L. C. *Polymer* 1999, 40, 4387-4398.
[28] Espinosa, M. A.; Cádiz, V.; Galià, M. *J Polym Sci Part A: Polym Chem* 2004, 42, 279-289
[29] Kolb, H. C.; Finn, M. G.; Sharpless, K. B. *Angew Chem Int Ed* 2001, 40, 2004-2021.
[30] Hoyle, C.E.; Bowman, C.N. *Angew Chem Int Ed* 2010, 49, 1540-1573.
[31] Hoyle, C.E.; Lowe, A.B.; Bowman, C.N. *Chem Soc Rev* 2010, 39, 1355-1387.
[32] González, R.J.; Lligadas, G.; Ronda, J.C.; Cádiz, V. *Polymer* Submitted.
[33] Roper, T. M.; Guymon, C. A.; Jönsson, E. S.; Hoyle, C. E. *J Polym Sci Part A: Polym Chem* 2004, 42, 6283-6298.
[34] Arceo, E.; Marsden, P.; Bergman, R.G.; Ellman, J.A. *Chem Comm* 2009, 23, 3357-3359.
[35] Lluch, C.; Ronda, J.C.; Galià, M.; Lligadas, G.; Cádiz, V. *Biomacromolecules,* 2010 In press.
[36] Bozell, J.J.; Patel M.K. In *Feedstocks for the future: Renewables for the production of chemicals and materials.* ACS Symposium series 921, American Chemical Society: Washington D.C, 2006.
[37] Wool, R.P.; Sun, X.S. *Bio-based polymers and composites*, Elsevier,:Amsterdam, 2005.
[38] Sacristan, M.; Ronda, J. C.; Galià, M.; Cádiz, V. *Biomacromolecules* 2009, 10, 2678-2685.
[39] Li, F.; Hanson, M.V.; Larock, R.C. *Polymer* 2001, 42, 1567-1579.
[40] Loupy, A. *Microwaves in Organic Synthesis.* Wiley VCH: Weinheim, 2002
[41] Bogdal, D.; Prociak, A. *Microwave enhanced polymer chemistry and technology.* Blackwell Publishing: Ames, Iowa, 2007.
[42] Sinnwell, S.; Ritter, H. *Aust J Chem* 2007, 60, 729-743.
[43] Wiesbrock, F.; Hoogenboom, R.; Schubert, U.S. *Macromol Rapid Commun* 2004, 25, 1739-1764.
[44] Hoogenboom, R.; Schubert, U.S. *Macromol Rapid Commun* 2007, 28, 368-386.
[45] Bardts, M.; Gonsior, N.; Ritter, H. *Macromol Chem Phys* 2008, 209, 25-31.
[46] Sacristan, M.; Ronda, J. C.; Galià, M.; Cádiz, V. *Macromol Chem Phys* 2010, 211, 801–808.
[47] Sacristan, M.; Ronda, J. C.; Galià, M.; Cádiz, *Polymer* Submitted.
[48] Sacristan, M.; Hull, T.R.; Stec, A.A.; Ronda, J. C.; Galià, M.; Cádiz, *Polym Deg Stab* 2010, 95, 1269-1274.
[49] Sacristan, M.; Ronda, J. C.; Galià, M.; Cádiz *J Appl Polym Sci* Submitted
[50] Lligadas, G.; Callau, L.; Ronda, J.C.; Galià, M.; Cádiz, V. *J Polym Sci Part A: Polym Chem* 2005, 43, 6259-6307.
[51] Berh, A.; Naendrup, F.; Obst, D. *Eur J Lipid Sci Technol* 2002, 104, 161-166.
[52] Earls, J.D.; White, J.E.; López, L.C.; Lysenko, Z.; Dettloff, M.L.; Null, M.J. *Polymer* 2007, 48, 712-719.
[53] Park, S.J.; Jin, F.L.; Lee, J.R. *Macromol Rapid Commun* 2004, 25, 724-727.
[54] Kannan, R.Y.; Salacinsqi, H.J.; Butler, P.E.; Seifalian, A.M. *Acc Chem Res* 2005, 38, 879-884
[55] Abad, M.J.; Barral, L.; Fasce, D.P.; Williams, R.J. *Macromolecules* 2003, 36, 3128-3135.

[56] Lligadas, G.; Ronda, J.C.; Galià, M.; Cádiz, V. *Biomacromolecules*. 2006, 7, 3521-3526.
[57] Lligadas, G.; Ronda, J.C.; Galià, M.; Cádiz, V. *J. Polym. Sci. Part A. Polym. Chem* 2006, 44, 6717-6727
[58] Petrovic Z.S. *Polymer Reviews* 2008, 48, 109-155.
[59] Warwel, S.; Fehling, E. Kunz. M. *Eur J Lipid Sci Technol* 2001, 103, 133-184
[60] Muggee, J. J *Polym Sci Polym Chem Ed* 1986, 24, 2327-2349
[61] Lligadas, G.; Ronda, J.C.; Galià, M.; Biermann, U.; Metzger, J.O. *J Polym Sci Part A: Polym Chem* 2006, 44, 634-645.
[62] del Rio, E.; Galià, M.; Cádiz, V.; Lligadas, G.; Ronda, J.C. *J Polym Sci Part A: Polym Chem* Submitted.
[63] del Rio, E.; Lligadas, G.; Ronda, J.C.; Galià, M.; Cádiz, V. *J Polym Sci Part A: Polym Chem* Submitted.
[64] Lligadas, G.; Ronda, J.C.; Galià, M.; Cádiz, V. *Biomacromolecules* 2006, 7, 2420-2426.
[65] Lligadas, G.; Ronda, J.C.; Galià, M.; Cádiz, V. *Biomacromolecules* 2007, 8, 1858-1864
[66] Lamba, N.M.K.; Woodhouse, K.A.; Cooper, S.L. *Polyurethanes in biomedical applications*. CRC: Boca raton, FL, 1998.
[67] Lligadas, G.; Ronda, J.C.; Galià, M.; Cádiz, V. *Biomacromolecules* 2007, 8, 686-692
[68] Schenck, G. O. *Naturwissenschaften* 1948, 35, 28–29.
[69] Greer, A. *Acc Chem Res*. 2006, 39, 797–804.
[70] Hui, S.P. ; Yoshimura, T. ; Murai, T.; Chiba, H.; Kurosawa, T. *Anal Sci* 2000, 16, 1023-1028.
[71] Samadi, A.; Martinez, L.A.; Miranda, M.A., Morera, I.M. *Photochem Photobiol* 2001, 73, 359-365.
[72] *Frimer, A.A.* Chem Rev 1979, 79*,* 359-387.
[73] Montero de Espinosa, L.; Ronda, J.C.; Galià, M.; Cádiz, V. *J Polym Sci Part A Polym Chem* 2009, 47, 1159–1167.
[74] Montero de Espinosa, L.; Ronda, J.C.; Galià, M.; Cádiz, V. *J Polym Sci Part A Polym Chem* 2009, 47, 4051–4063.
[75] *Kornblum, N.; DeLaMare, H.E.* J Am Chem Soc *1951, 73, 880-881.*
[76] Ohloff. G. *Pure Appl Chem* 1975, 43, 481-502.
[77] Montero de Espinosa, L.; Ronda, J.C.; Galià, M.; Cádiz, V. *J Polym Sci Part A Polym Chem* 2008, 46, 6843–6850.
[78] L.S. Corley: U.S. Patent 4,962,179 (1990).
[79] Mather, B.D.; Viswanathan, K.; Miller, K.M.; Long, T.E. *Prog Polym Sci* 2006, 31, 487-531.
[80] Wu, C.S.; Liu, Y.L.; Chiu, Y.S. *Polymer* 2002, 43, 1773-1779.
[81] Montero de Espinosa, L.; Ronda, J.C.; Galià, M.; Cádiz, V. *J Polym Sci Part A Polym Chem* 2010, 48, 869–878.
[82] La Scala, J.; Wool, R.P. *Polymer* 2005, 46, 61-69.
[83] Montero de Espinosa, L.; Ronda, J.C.; Galià, M.; Cádiz, V. *J Polym Sci Part A Polym Chem* 2009, 47, 1159-1167.
[84] Liron, F. ; Knochel, P. *Chem Commun* 2004, 304-305.
[85] Montero de Espinosa, L.; Ronda, J.C.; Galià, M.; Cádiz, V. *J Polym Sci Part A Polym Chem* 2009, 47, 4051-4063.

In: Thermoplastic and Thermosetting Polymers and Composites ISBN: 978-1-61209-264-5
Editors: Linda D. Tsai and Matthew R. Hwang

Chapter 6

Mechanisms of Impregnation in Compression Molding of Thermoplastic Matrix Composites

Antonio Greco*, Riccardo Gennaro and Alfonso Maffezzoli
Department of Engineering for Innovation, University of Salento, Via per Monteroni, 73100, Lecce, Italy

Abstract

Continuous fiber reinforced thermoplastic matrix composites, based on commodity polymers, are attracting a growing interest in many industrial application due to distinct advantages over thermosetting matrix composites. Commingled yarn semi-pregs, or stackings of thermoplastic films and dry reinforcements, lead to fast processing by compression molding. In the forming process the material is placed between matched die, to promote the flow of the matrix into the interstices of the fibers, due to the applied pressure and temperature.

The aim of this paper is to study the different phenomena involved in the consolidation of thermoplastic matrix composites, and to model the impregnation process in both commingled yarn semipregs and in composites produced by film stacking technique.-Optical microscopy showed that for both systems there are two mechanisms of impregnation: macro-scale impregnation, associated with the formation of a homogeneous molten pool around each fiber bundle, and micro-scale impregnation, associated to the flow of the matrix inside each bundle. Since the two mechanisms occur at different temperatures, their contribution to the reduction of void fraction can be considered separately.

Macro-scale and micro-scale impregnation during film stacking were simulated by two different finite element models, taking into account the non Newtonian rheological behavior of the matrix. The micro-scale impregnation of fibers was simulated by using a randomly spaced and non-overlapping unidirectional filaments. The results obtained showed that at low molding pressures the polymer melt exhibits a Newtonian behavior during micro-scale impregnation, which makes it possible to predict the tow permeability

* antonio.greco@unisalento.it

by the Darcy law. At high molding pressures and during macro-scale impregnation, the high shear rates and non-Newtonian behavior of the melt required the introduction of a permeability coefficient that is also dependent on the rheological properties of the melt.

The combination of on-line consolidation measurement, thermomechanical analysis and numerical analysis showed that the macro-scale impregnation during molding of commingled yarn fabrics, taking place at lower temperatures, is governed by matrix fiber deformation and sintering. At a temperature higher than the onset of the flow region of the matrix, micro-scale impregnation occurs, due to matrix flow inside each bundle.

Keywords: thermoplastic matrix composite, consolidation, matrix flow, sintering

INTRODUCTION

Fiber reinforced polymer composites are candidates as structural materials for lightweight and fuel efficient cars of the future. Potential applications of these materials include body side, interior and floor panels, and other light load bearing parts. Long fiber reinforced thermoplastic composites, in the form of the so called GMT (Glass Mat Thermoplastic) already find many applications in the automotive industry. Continuous fiber reinforced thermoplastic matrix composites, based on commodity polymers, are attracting a growing interest essentially thanks to their processability, high impact and delamination strength, abrasion and chemical resistance, low moisture absorption, unlimited shelf life of raw materials and low cost. In addition, the ability of thermoplastic matrix composites to be recycled is one of the main advantages over the thermosetting matrix composites in the automotive industry [1, 2, 3,4]. Among others, a key drawback is the high viscosity of the molten thermoplastic resin compared to thermosetting, which can prevent a thorough impregnation of fibers. The use of commingled fabrics can overcome this problem, thanks to the intimate physical blending of reinforcing and matrix fibers directly into the bundles [4, 5,6].

The use of commingled yarn semi-pregs, or stackings of thermoplastic films and dry reinforcements, can lead to fast processing of composites by compression molding [5,7]. In the forming process, the material is placed in a matched die, where the flow of the matrix wets out the fibers, as a results of the applied pressure and temperature. Many experimental studies have dealt with the correlation between molding parameters and the final properties of composites [2, 8,9], often neglecting the mechanism of fiber impregnation. In contrast to this, many theoretical and experimental studies have been devoted to the basics of flow mechanism of thermosetting resins in RTM like processes, mainly using the Darcy's approach [10,11].

In the film stacking process for thermoplastic matrix composites, as well as in compression molding of commingled fabrics, impregnation occur by two different mechanisms: macro-scale impregnation, which leads to the formation of a homogeneous polymer pool around each bundle of fibers, and micro-scale impregnation, associated to the reduction of the void fraction inside each bundle [12,13]. Although the two phenomena occur at different time scales and at different shear rates, most of the works reported in the literature assume that the flow of molten matrix is governed by Darcy's law, regardless the scale of impregnation considered [14, 15, 16, 17, 18, 19,20]. Moreover, the determination of a permeability coefficient, for non-Newtonian matrices, requires special considerations, since in these cases the permeability coefficient is not only dependent on the nature of the reinforcement, but also on the rheological properties of the molten matrix [21,22]. The flow conditions during impregnation of fibers in thermoplastic matrix composite molding are substantially different from those prevailing in RTM processes with thermosetting resins with respect to the following:

1) The thermoplastic melts have a viscosity at least 1000 times higher than that of thermosetting resin.
2) The rheological behavior of thermoplastic melts is non-Newtonian.
3) The impregnation process is dominated by an out of plane or through thickness flow.

Whereas it is usually assumed that macro-scale impregnation is related to the molten polymer flow around fibers bundle, it has also been recently recognized that, during processing of commingled fabrics, the macro-scale impregnation occurs through sintering, which is strictly related to the surface tension of molten polymer [23,24]. Accordingly, the phenomena governing the macro-scale impregnation during commingled fabrics molding can be analysed on the same basis of the processing of powdered thermoplastic polymers, mainly studied in relation with rotational molding of polymers [25, 26.27]. The sintering process is usually associated with two different mechanisms: powder sticking and void removal [28,29].

The aim of this work is the development of theoretical and experimental procedures to monitor and model macro-scale and micro-scale impregnation during compression molding of stackings of thermoplastic films and dry reinforcements and of commingled thermoplastic fabrics. The reduction of composite thickness as a function of time during compression tests at different temperatures and pressures was measured and used to calculate the void fraction. The influence of the macro-scale and micro-scale impregnation in the elimination of voids was evaluated by Scanning Electron Microscopy (SEM) and optical microscopy analysis.

The consolidation during compression molding of alternate stackings of polymer films and glass fiber fabrics was studied by means of a finite element model (FEM) for inter-bundle and intra-bundle matrix flow. FEM analysis, applied to a large number of different random fiber arrangements, was used to calculate the shear rate during the impregnation process, which is frequently disputed in the literature [30]. This aspect is relevant since the non-Newtonian rheological behavior of the thermoplastic matrix was considered [22]. A similar approach, based on randomization of fillers, was recently applied to mass transport and permeability calculation in composites [31] and nanocomposites for gas barrier applications.

The role of sintering and flow of matrix during micro-scale and macro-scale impregnation during molding of commingled fabrics was analyzed comparing non isothermal compression molding results with thermomechanical analysis (TMA). The overall consolidation behavior of the commingled fabric was correlated with the rheological properties of the polymer matrix. A dimensionless analysis was also used to determine the relevance of fiber sintering and matrix flow during consolidation of commingled thermoplastic composites.

MATERIALS AND METHODS

An amorphous PET copolymer (PETg) polymer matrix was reinforced with glass fibers to produce PETg/glass composites. Two different types of materials were used for the production of composite, respectively:

a) A commingled fabric, Comfil ® G, which is a balanced woven fabric made of hybrid yarns containing filaments of E -glass (55% by weight) and PETg. The composite obtained by processing of commingled fabric is labeled as PETg/glass/cy. The PETg matrix was extracted from the composite by grinding, followed by separation in a sodium bromide water solution with a density of 1.9 g/cm^3, which is higher than that of PETg but lower than that of glass fibers. The matrix extracted from the commingled fabric is identified as PETg/cy. The composite material and the matrix were dried before testing.
b) Film stackings of PETg and glass fibers. This composite is labeled as PETg/glass/fs. The matrix used is Embrace Shrink Film supplied by Eastman Chemicals, having an intrinsic viscosity of 0.70 dl/g and a density of 1.30 g/cm^3. This matrix is labeled as PETg/fs, and is different from the matrix PETg/cy, since the intrinsic viscosity, chemical structure and presence of additives in the latter is unknown. The glass fiber reinforcement is a bidirectional unbalanced 90/10 fabric, EE425 provided by SEAL (Italy), with a nominal thickness of 0.26 mm, and a nominal weight of 425 g/m^2. Before molding, the sizing of glass fibers was removed by a thermal treatment at 500°C for 8 h in a forced convection oven. The fabric lost 0.62% of its initial weight. This treatment was necessary due to the uncertainty of its compatibility with the thermoplastic matrix used in this work, as the sizing is usually developed for thermosetting resin.

PETg/fs films (0.15 mm thick), were obtained by compression molding at 50 bar and mold temperature of 30°C, after pre-melting the material at 220°C. Then, the PETg/fs film was stack between two symmetric [0/90] glass fiber fabrics. The composite was molded in the same conditions used for matrix film preparation. All the compression molding experiments on PETg/glass/fs were carried out in a Campana PL 71 parallel plate press. The fiber volume fraction in the composite, as measured by resin burn out tests, was 33 %.

PETg/glass/cy composite was molded using the compression plates of a Lloyd LR5k dynamometer, equipped with a forced convection oven. On-line measurements of the reduction of thickness changes of the commingled fabric during consolidation under a constant applied pressure were made. Four plies of commingled fabric, in a symmetric [0/90] lay -up, were placed between the parallel plates (200 mm diameter), applying a constant compression force during test. The setup of the test is similar to a compression creep test: first the crosshead is moved downwards at a rate of 20 mm/min, up to the set force (which depends on the set pressure and on the sample area); then the reduction of sample thickness is followed for 1 h under a constant force. Tests were performed in isothermal and in dynamic mode at different pressure levels. For the system under investigation, it was shown that temperatures lower than 220°C and pressures lower than 0.5 MPa, are sufficient prevent matrix squeezing during consolidation [24]. In this way the composite thickness reduction can entirely be attributed to a reduction of the composite void fraction, according to eq.1):

$$x_v(t)=1-\frac{\text{Mass}}{A*s(t)*\rho_T} \qquad \text{eq. 1)}$$

where Mass is the composite mass, A is the area of the composite, s(t) is the composite thickness at time t, which is measured as the gap between the tools, and ρ_T is the theoretical composite density, assuming no voids present.

The viscosity of PETg, (PETg/cy extracted from the commingled fabric and PETg/fs used for film stacking), was measured in an ARES cone and plate rheometer from TA Instrument, in a shear rate range between 0.1 and 10 s^{-1} at temperatures ranging from 180°C to 250°C. The ARES rheometer was also equipped with torsion grips for Dynamic Mechanical Analysis (DMA), using 6.28 rad/s frequency, 0.5 % strain, varying the temperature between 25 and 220°C at 0.5 °C/min.

A Nikon Epiphot 200 was used for optical microscopy analysis. Specimens cut from the centre part of molded plates of PETg/glass/fs were polished with SiC papers disks and then with diamond particles dispersion [32].

Thermomecanical analysis (TMA) was performed to study the sintering behavior of PETg/cy matrix using a Perkin Elmer TMA 7 equipped with an expansion probe. Fibers were pulverized with a Retsch Z100 ultracentrifugal mill, equipped with a 0.5 mm sieve. The powders obtained were placed in an aluminum holder, and heated in the TMA apparatus from room temperature to the test temperature at 10°C/min, and then held for 20 minutes at the final temperature. TMA experiments were performed at different pressures. Once the thickness of the powder bed inside the aluminum holder is measured by TMA analysis, the void fraction of the matrix was calculated according to eq.1.

RESULTS AND DISCUSSION

DMA analysis results on neat PETg/cy matrix are reported in Figure 1. PETg shows a distinct glass transition signal, observed as the G" peak, at about 87°C, with an onset at about 70°C, followed by a wide rubbery plateau and a flow region at temperatures higher than 180°C.

The results of rheological measurements, reported in Figure 2 for PETg/fs matrix, show the variation of viscosity as a function of shear rate. PETg/fs shows an initial Newtonian plateau followed by a shear thinning behavior at higher shear rates. In order to model the rheological behavior of the material, the Carreau-Yasuda model [33] was used:

$$\eta=\frac{\eta_0}{[1+(\lambda\dot{\gamma})^2]^{\frac{1-n}{2}}} \qquad \text{eq. 2)}$$

where η_0 is the low shear rate viscosity, n is the power-law index, and λ is the critical value of shear rate which determines the end of the Newtonian plateau and the onset of power law regime.

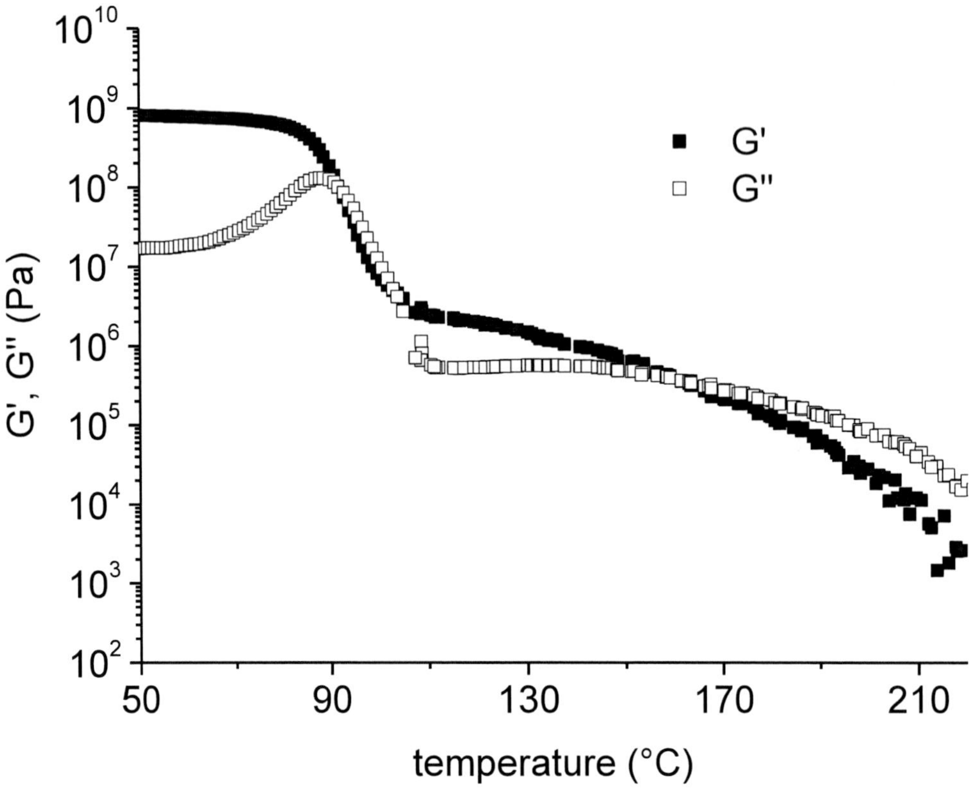

Figure 1. Complex moduli for PETg/cy matrix as a function of temperature.

The Newtonian viscosity was obtained from experimental data as η_o=2160 Pa*s. The non linear curve fitting according to Carreau-Yasuda model, also reported in Figure 2, shows a very good agreement with experimental data for λ=0.834 s and n = 0.662, in the whole range of shear rates between 0.1 and 10 s^{-1}. The corresponding curves for Newtonian behavior and power law behavior are also reported in Figure 2. From comparison between the Carreau-Yasuda model prediction and Newtonian plateau, it can be assumed that the fluid has a Newtonian behavior for $\dot{\gamma} < 0.5\,s^{-1}$, which corresponds to a difference between the two models lower than 2.5%. By using the same arguments, comparing the Carreau-Yasuda and power law models, it can be estimated that the melt behaves like a power law fluid for $\dot{\gamma} > 2.5\,s^{-1}$.

The rheological characteristic of PETg/cy matrix were also measured by cone and plate rheometer in isothermal conditions at different temperatures. In this case, fitting was performed according to power law model. The results for the consistency and power law indices are reported in Table I. The temperature dependence of the consistency index was fitted by an Arrhenius expression, yielding a pre-exponential factor of $2.6*10^{-8}$ Pa*s and an activation energy of 1420 J/mole.

Representative optical micrographs of samples sections cut from the PETg/glass/fs composite are shown in Figure 3. Interbundle spaces are completely filled with matrix, which is indicative of full interbundle impregnation. On the other hand, there are still voids within the fiber bundles, which is indicative of partial intrabundle impregnation. This suggests that impregnation probably occurs along bundle radius into fiber bundles after the interbundle spaces are filled [34].

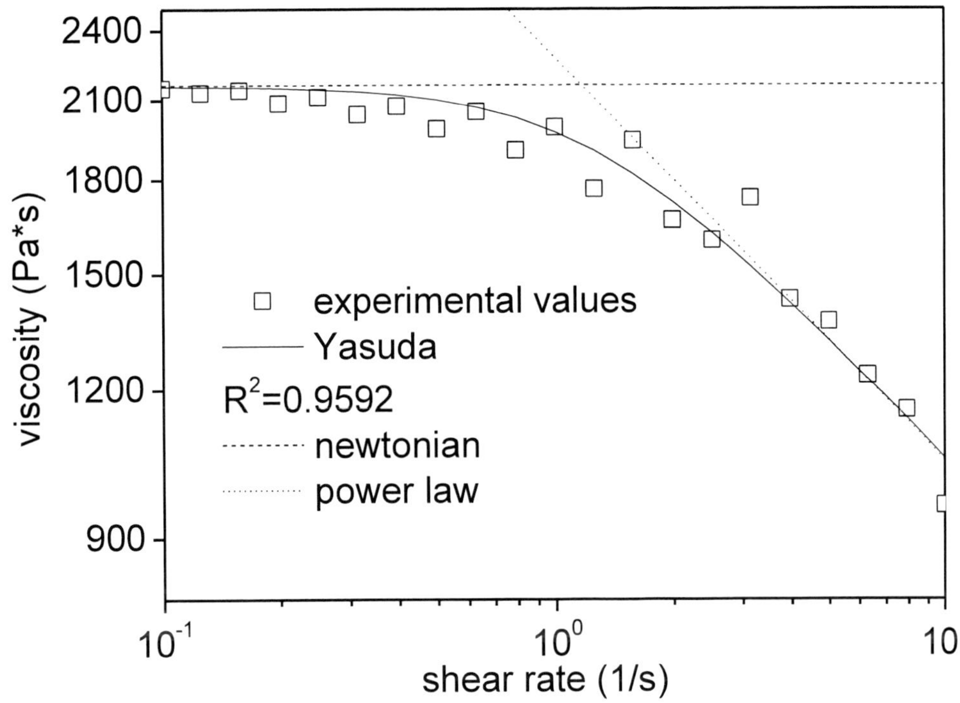

Figure 2. viscosity vs. shear rate at 220°C for PETg/fs.

Table I. Viscosity at constant shear rate and power law index at different temperatures for PETg/cy.

Temperature [°C]	η (shear rate =1s^{-1}) (Pa*s)	n
180	5370	0.62
200	2162	0.66
210	794	0.87
220	489	0.81
250	154	0.85

Several micrographs of polished cross-section were used to determine the intrabundle fiber volume fraction ($V_{f,\ micro}$) [35]. To this purpose, a mass balance was used in order to calculate the weight fraction (W_{bundle}) of a bundle of known length (L_{bundle}). The number of filaments within each bundle was determined according to eq.3):

$$N_{f,intra}=\frac{W_{bundle}}{\rho_{glass}L_{bundle}A_{fiber}} \qquad \text{eq. 3)}$$

Where ρ_{glass} is the density of glass fibers (2.54 g/cm3) and A_{fiber} is the cross section area of each filament (measured diameter= 8 μm). Then the volume fraction of fibers inside each bundle was determined according to eq.4:

$$V_{f,micro}=\frac{N_{f,intra}A_{fiber}}{A_{bundle}} \quad \text{eq. 4)}$$

Where the average area A_{bundle} of each bundle was evaluated from statistical analysis of several pictures like the one shown in Figure 3. From these calculations, an intra-bundle fiber volume fraction of 0.472 was obtained. In a second calculation, each bundle was assumed to consist of an impermeable solid elliptical section cylinder. Thus, the bundle volume fraction ($V_{f,\ macro}$) was obtained from the image analysis as the ratio between the area of the bundles and the total area of the composite. A bundle volume fraction of $V_{f,macro}$ =0.55 was obtained.

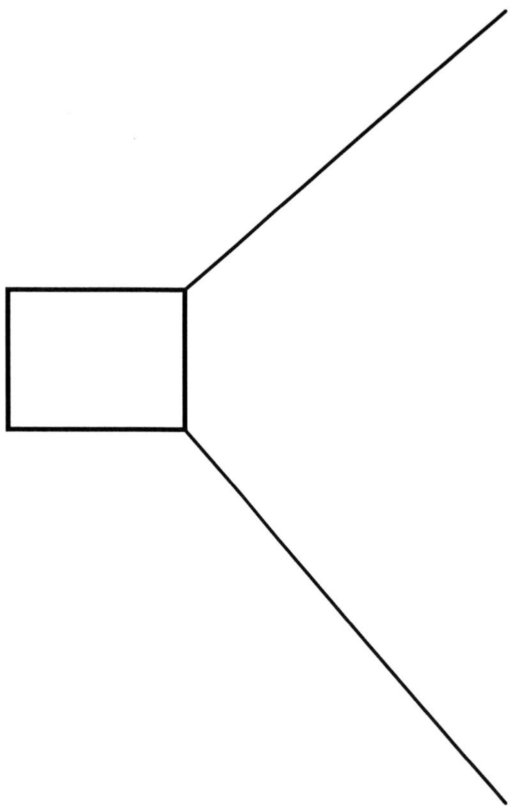

Figure 3 Typical micrographs of polished cross-section of PETg/glass/fs composite.

Optical microscopy pictures (Figure 3) indicated that interbundle spaces are always filled with resin while air pockets (or voids) and unimpregnated fibers are observed inside bundles. This suggests that glass fiber impregnation can take place in two steps. First inter-bundle (macro-scale) flow occurs and air is evacuated along bundles escaping in the laminate plane. Then intra-bundle (micro-scale) impregnation occurs thanks to a radial flow associated with air entrapment into the bundles [8]. The flow of a Newtonian thermosetting reactive mixture of monomers can be expressed by Darcy's law:

$$v=K\frac{\Delta P}{\eta_0 L} \quad \text{eq. 5)}$$

Where v is the average velocity of the RESIN mixture, ΔP is the pressure gradient, η_0 is the Newtonian viscosity, and K is the permeability of the diffusing medium (for example the glass fabric). For a non-Newtonian fluid, whose rheological behavior can be modeled by a power law, the Darcy equation is usually modified according to eq. 6 [21]:

$$v=k'\left(\frac{\Delta P}{mL}\right)^{\frac{1}{n}} \qquad \text{eq. 6)}$$

In which the symbols are the same used in eq.5), except for the parameter K', which depends on the power law index n, and for packed beds of fibers, its value can be estimated using arguments similar to those leading to the Carman-Kozeny equation [36,37].

has been found to agree remarkably well with experiment data in some cases [38, 39, 40,41]. In other cases it has been found that eq. 6) overestimates the flow rate for a given pressure gradient [42,43]. This discrepancy has been attributed to melt elasticity of polymer melts by several authors [11,39].

Neglecting the non-Newtonian behavior of the matrix, the Darcy law (eq.5) can be written as a function of the distance travelled by the molten polymer (s) as:

$$\frac{ds}{dt}=K\frac{\Delta P}{\eta_0 s} \qquad \text{eq. 7)}$$

By integration of eq. 7), a closed form expression for s can be obtained as:

$$s=\sqrt{2\frac{K\Delta P}{\eta_0}}\sqrt{t} \qquad \text{eq. 8)}$$

With reference to the scheme reported in Figure 4, the void fraction can be obtained as a function of s as:

$$x_v=x_{v0}\left(1-\frac{s}{\delta}\right) \qquad \text{eq. 9)}$$

Finally, by combination of eq. 8) and eq. 9), it is possible to express the void fraction as a function of time as:

$$x_v=x_{v0}\left(1-\sqrt{\frac{2K\Delta P}{\delta^2\eta_0}t}\right) \qquad \text{eq. 10)}$$

Therefore, a plot of x_v vs $t^{1/2}$ in isothermal tests should yield a straight line. The results reported in Figure 5 show the reduction of void fraction during isothermal consolidation experiment of PETg/glass/fs sample. The presence of two different linear regimes in Figure 5 indicates that for the PETg/glass/fs system the impregnation is a two stage process. The first process, associated to the higher slope of the curve, occurs at lower times and can be attributed to the macro-impregnation process. The second process, characterized by a lower slope, takes place at longer times, and can be attributed to the micro-scale impregnation process. Each of the two processes is characterized by a characteristic value of the permeability, K, which can be obtained by linear fitting of the data reported in Figure 5 according to eq. 10). The values reported in Table II show that the macro-scale impregnation process is characterized by a permeability 10 times higher than that obtained for the micro-scale impregnation.

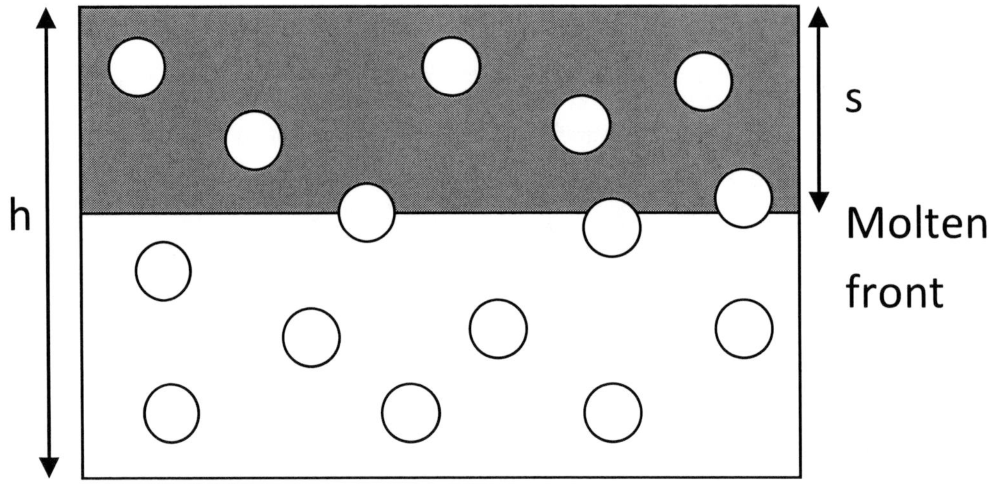

Figure 4. scheme of the impregnation process during film-staking.

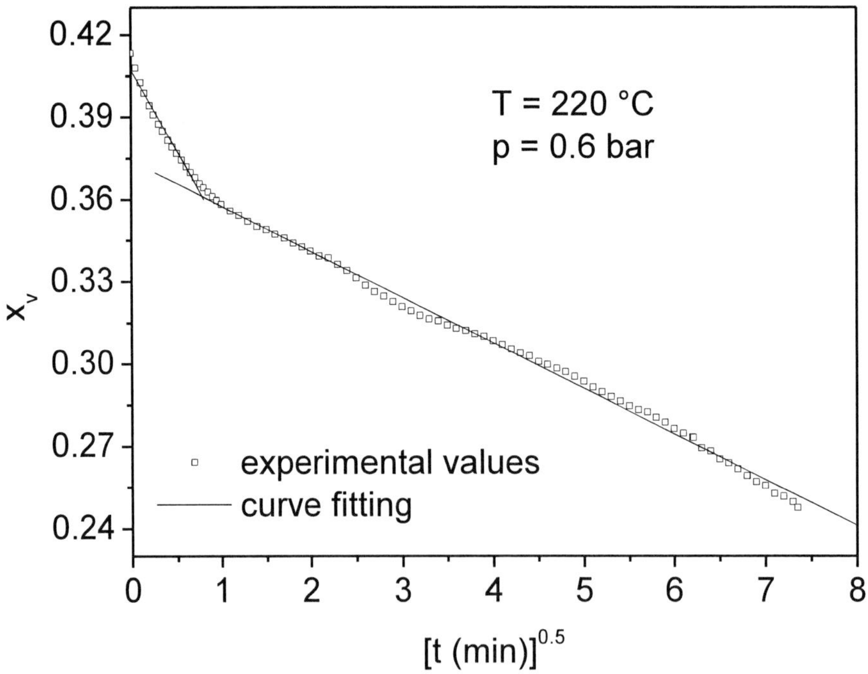

Figure 5. plot of x_v vs $t^{1/2}$ for impregnation of PETg/glass/fs system.

Table II. Experimental Permeability Values For Macro And Micro-Scale Impregnation Of Petg/Glass/Fs System.

P [bar]	$K_{Micro-scale} * 10^{14}$ [m²]	$K_{Macro-scale} * 10^{13}$ [m²]
0.4	0.38	0.78
0.6	0.12	0.12
1	0.33	0.18

According to the observations reported in Figure 3 and Figure 5, the consolidation of the PETg/glass/fs system can be described by the scheme reported in Figure 6.

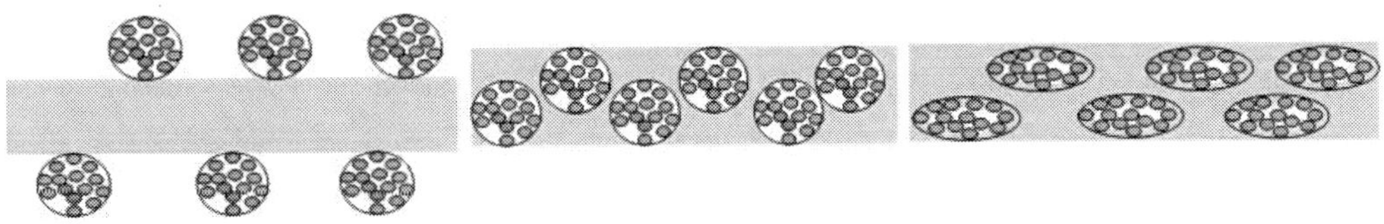

Figure 6. scheme of the consolidation for PETg/glass/fs system.

The evolution of void fraction, X_v, of PETg/glass/cy during consolidation in non-isothermal conditions together with its the derivative, dX_v/dt, is reported in Figure 7.

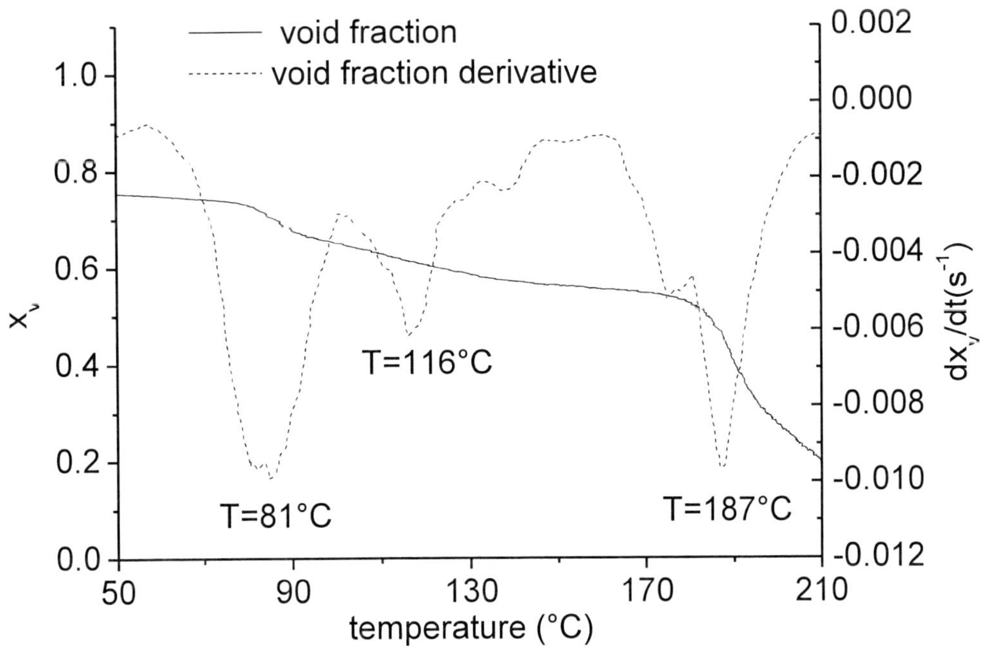

Figure 7. void fraction and its derivative of PETg/glass/cy during dynamic compression test at 1 bar.

As expected, the void fraction decreases with increasing temperature. The derivative of the consolidation curve shows a first peak at about 81 °C, just above the glass transition temperature (Tg) of the matrix (which is about 72°C as observed by differential scanning calorimetry analysis). The second consolidation step is characterized by a peak at 116°C, and the third by a peak at 187°C. The different physical changes occurring during the consolidation of a commingled fabric must include the development of a continuous matrix as the first step, and the impregnation of the glass fibers as the second step. According to the results reported in Figure 7 these processes are not simultaneous but can be observed at different temperature ranges. Therefore, in order to assign each peak in Figure 7 to a specific physical phenomenon, the sintering temperature range of the neat matrix PETg/cy was analyzed by TMA. The results reported in Figure 8, show that sintering of the PETg/cy matrix takes place between 75 °C and 150°C. The derivative of TMA curve, also reported in Figure 8, shows two distinct peaks. The first one, observed at 84°C, can be associated with the one at 81 °C observed in the consolidation experiment of the PETg/glass/cy composite in Figure 7. The second peak, observed at 115°C, corresponds to the second peak of Figure 7 occurring at about the same temperature.

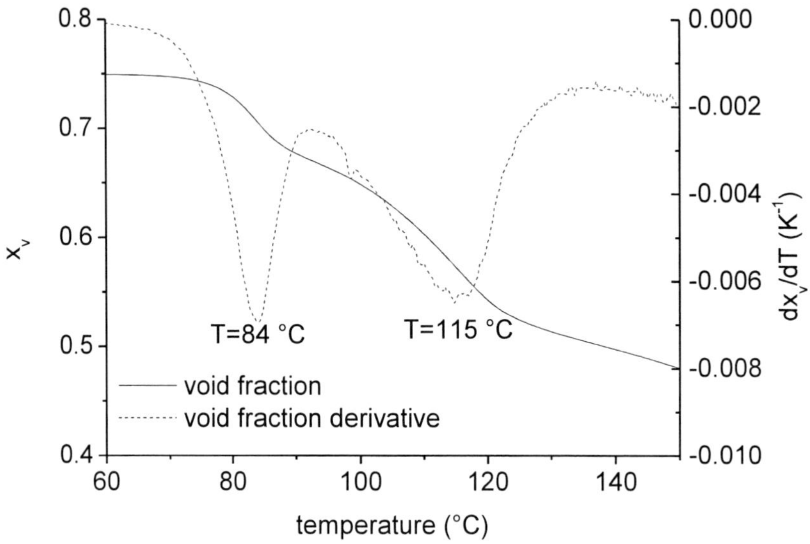

Figure 8. TMA sintering curve and derivative for PETg/cy matrix during a dynamic scan at 0.2 bar.

The SEM image of the PETg/glass/cy composite heated up to 180°C (i.e. a temperature lower than the third consolidation step of Figure 7), reported in Figure 9, shows the presence of a well sintered PETg matrix, but there is no evidence of glass fiber impregnation. The same morphology was observed for the composites heated up to 180°C at different pressures, ranging from 0.5 to 5 bar. Comparing the results shown in Figure 7 through Figure 9, it can be assumed that consolidation of commingled fabric is the results of three distinct physical events:

a) Plastic deformation of PETg fibers just above the T_g (first peak in Figure 7 and Figure 8);
b) PETg fiber sintering and interbundle void removal (macro-impregnation) (second peak in Figure 7 and Figure 8);
c) Intrabundle flow of the PETg associated with micro-scale impregnation of glass fibers and with intrabundle void removal (third peak in Figure 8).

The first two stages, leading to a change of the PETg from fibers to a fully densified matrix, occur according to the mechanisms commonly involved in powder sintering: fibers deformation, sintering and bubble removal [25].

Figure 9. SEM image of PETg/glass/cy after thermal treatment to 180°C.

Isothermal consolidation tests on PETg/glass/cy were also performed at different temperatures (ranging from 180°C to 200°C) for an applied pressure of 1 bar. The temperatures used for isothermal consolidation are in the range of the third consolidation step of Figure 7, and can be therefore entirely attributed to PETg intrabundle flow. Experimental results for the void fraction are reported in Figure 10, show an accelerated kinetic as temperature is increased.

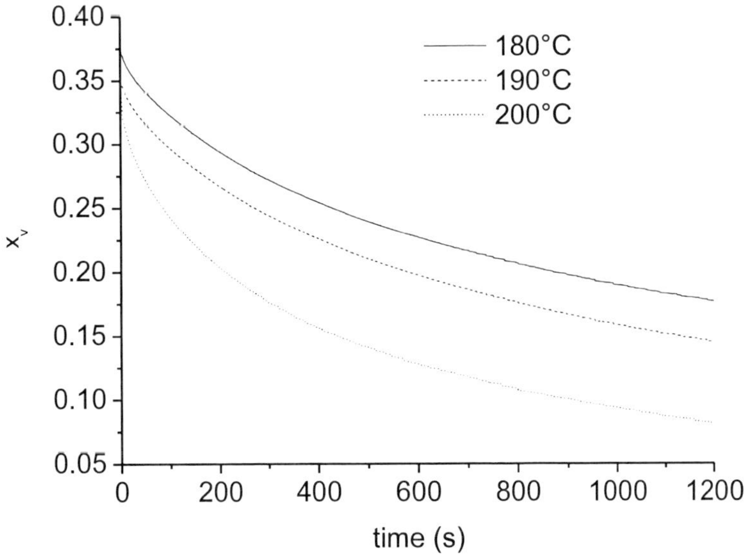

Figure 10. consolidation curve of PETg/glass/cy at a pressure of 1 bar and different temperatures.

A kinetic approach based on an Arrhenius-like dependence on temperature was used to fit the data of Figure 10 [16]:

$$\frac{dx_v}{dt}=A\exp\left(-\frac{E_a}{RT}\right)f(x_v) \qquad \text{eq. 11)}$$

Where A is the pre-exponential factor, E_a is the activation energy for the consolidation process, and $f(x_v)$ is a function of the void fraction. By taking the logarithms on both sides of eq. 11) it is possible to write the Friedman equation [44]:

$$\ln\left(\frac{dx_v}{dt}\right)=\ln(A)+\ln[f(x_v)]-\frac{E_a}{RT} \qquad \text{eq. 12)}$$

Plotting $\ln\left(\frac{dx_v}{dt}\right)$ vs. 1/T at constant values of x_v (i.e. the Friedman plot) it was therefore possible to determine the activation energy of the micro-scale impregnation process, which was calculated to be 1640± 103 J/mole. The activation energy is very close to the value (1420 J/mole) determined for viscosity of the PETg/cy matrix, indicating that the overall micro-scale impregnation is governed by viscous flow.

TMA experiments were also performed on neat PETg/cy at constant pressures (0.2 bar) and at different temperatures. Even in this case, the Friedman plots were used to calculate an activation energy of the sintering process of 1370±22 J/mole, which is in very good agreement with the calculated activation energies for viscosity and consolidation. Therefore, it is possible to conclude that sintering of the matrix (macro-scale impregnation), as well as matrix flow inside bundles (micro-scale impregnation), are both dominated by the viscosity evolution of the matrix with temperature [25,27].

Numerical Analysis of Film Stacking Consolidation

The results reported in Figure 3, Figure 5 and Figure 6, and the relative discussion, indicated that macro and micro-scale impregnation in computer modeling can be modeled separately. The slower mechanism, characterized by a lower value of permeability, is the micro-scale, intra-bundle impregnation. In order to model such process, a 2D geometry was used, modeling glass bundles as a random distribution of long parallel cylindrical filaments, with their axis aligned perpendicularly to the direction of flow [22,32]. The finite element (FE) solutions were obtained with the Comsol Multiphysics software (version 3.5, Comsol AB, Sweden), using the steady state diffusion module. In the calculation, the shear rate dependence of viscosity was taken into account by using the Yasuda model previously discussed.

The model equations were solved with the following boundary conditions:

- no-slip at solid fiber boundaries, i.e. the velocities are zero on the fiber surfaces;
- slip-symmetry at the left and right boundaries, i.e. along the planes of symmetry the normal velocities are zero and the gradients of the velocity in the flow direction is zero;

- ➢ Fixed pressure at the lower boundary (inflow);
- ➢ Zero pressure at the upper boundary (outflow).

After simulation, the average velocity of the fluid in the domain was obtained using a subroutine of Comsol Multiphysics.

According to a previous study [22], a minimum number of fibers N_f=500 and a minimum number of simulations, N_{rd} =20, were considered to provide adequately stable and accurate results for the simulation of intra-bundle flow.

The simulated average shear rate, plotted as a function of the applied pressure in Figure 11, was compared with the rheological behavior reported in Figure 2.

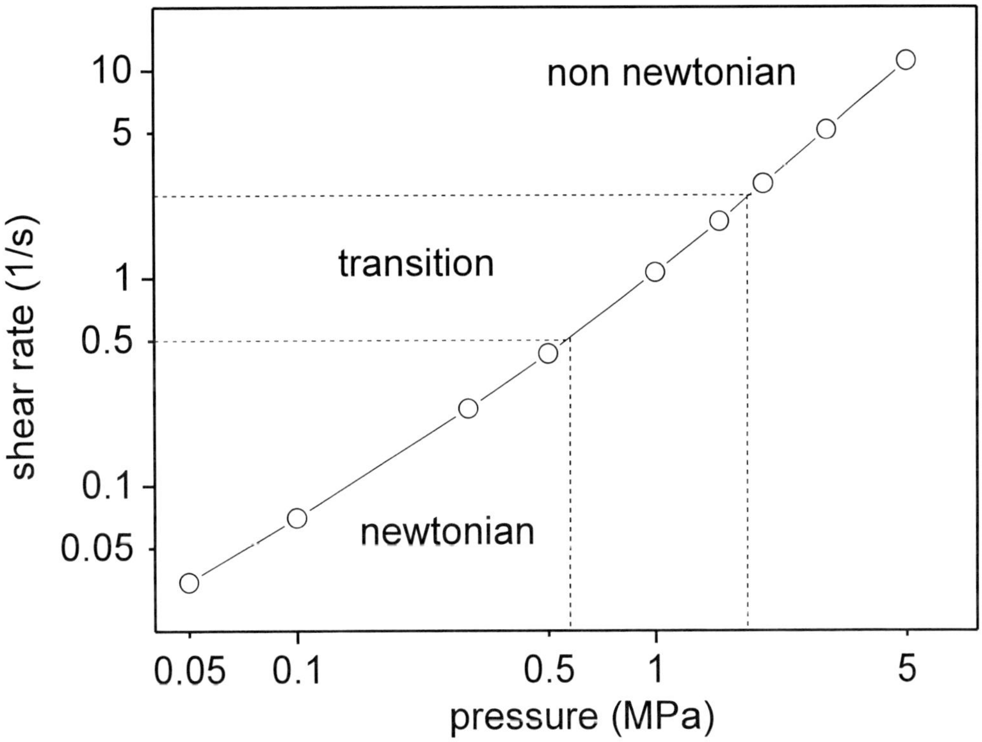

Figure 11. effect of the pressure gradient on the average shear rate.

Referring to the viscosity data reported in Figure 2, the Newtonian behavior of the matrix extends up to $0.5\,s^{-1}$, corresponding in Figure 11 to P< 0.56 MPa. On the other hand, Figure 2 shows that the melt behaves like a power law fluid when the average shear rate is greater than $2.5\,s^{-1}$, corresponding in Figure 11 to P>1.8 MPa. For intermediate values of pressure 0.56<P<1.8 MPa, the fluid is in a transition region. The values of shear rate and pressure corresponding to the limits between the Newtonian, the transition and the power law region are also reported in Figure 11.

The simulated average velocity as a function of the applied pressure is reported in Figure 12.

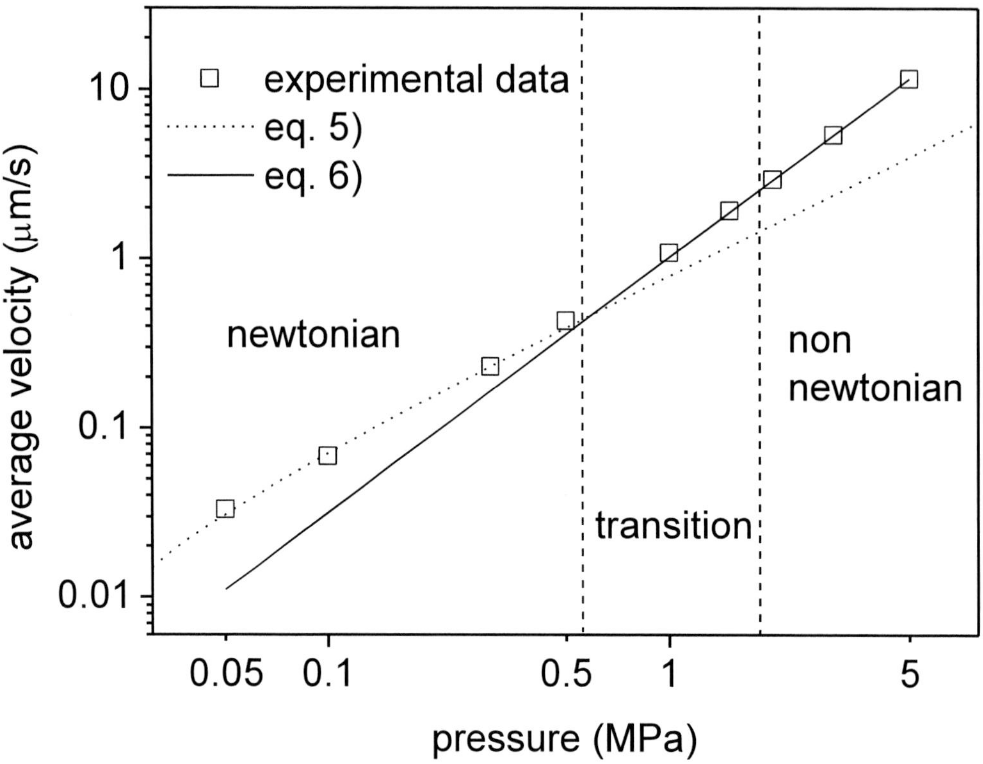

Figure 12. effect of the pressure gradient on the average velocity.

As expected, the velocity increases with applied pressure, showing two different regimes. Below 0.56 MPa the velocity increases linearly with pressure, while at higher pressures this linear behavior is no longer observed. Therefore, in the low pressure regime (P<0.56 MPa), the velocity during impregnation can be accurately described by the Darcy Law (0). The results of linear curve fitting, reported in Figure 12 as a dotted line, show that the Darcy equation provides a very good agreement at low pressures, but fails in predicting the changes in velocity at higher pressures, when the behavior is non-Newtonian. At low pressure levels, linear fitting according to eq. 5) provided a value of K=13E-14 m^2. This value is much higher than that calculated from experimental data and reported in Table II.

Following the approach proposed in ref. [21] for power law fluids, the velocity at high pressure was fitted according to a power law modified Darcy equation (eq. 6) leading to the full line in Figure 12. A good agreement with experimental data is obtained in the power law behavior region (ΔP>1.8MPa). In this case a value of K= 3.1E-16 $[m^{\frac{1}{n}+1} \bullet s^{\frac{1}{n}-1}]$ is obtained, taking n=0.662 from the Yasuda or power law model. This expression, however, leads to permeability values and physical dimensions which are dependent on the parameter n of the rheological model, and therefore are not comparable with those reported in Table II.

Although eq.6) coupled with the Yasuda model was proven to predict accurately the average velocity as a function of applied pressure, a further analysis was performed to verify if non-Newtonian permeability calculated according to eq.6) could be extended to fluids characterized by different n values. Several FEM simulations were run, arbitrarily changing the n values of Yasuda model, holding all the other rheological parameters and the reinforcement geometry constant. The results of these simulations, reported in Figure 13, were fitted according to eq.6) using K as regression parameter and n as independent variable. The poor agreement of the fitting clearly indicates that a single value of K, only dependent on the nature and geometry of the reinforcement, cannot be obtained for a non Newtonian fluid. This is the main evidence of poor significance of permeability when dealing with non Newtonian fluids.

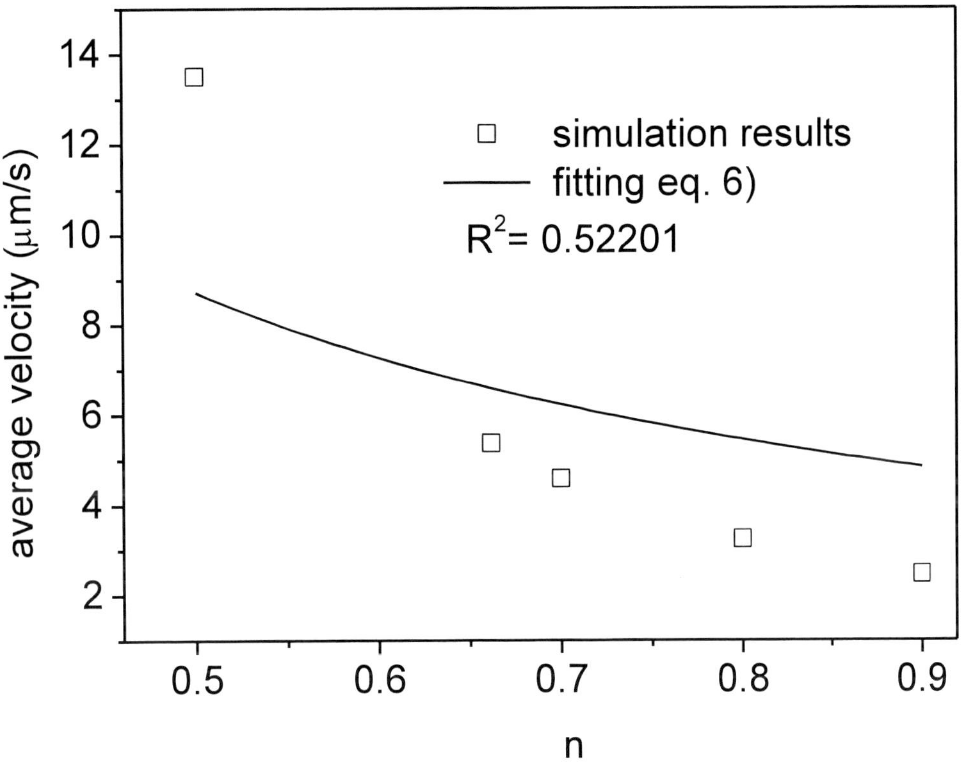

Figure 13. average velocity vs. the Yasuda model parameter n.

Finally, the interbundle impregnation was also modeled using FEM analysis. In this case, a regular array as control volume was adopted, as reported in Figure 14.

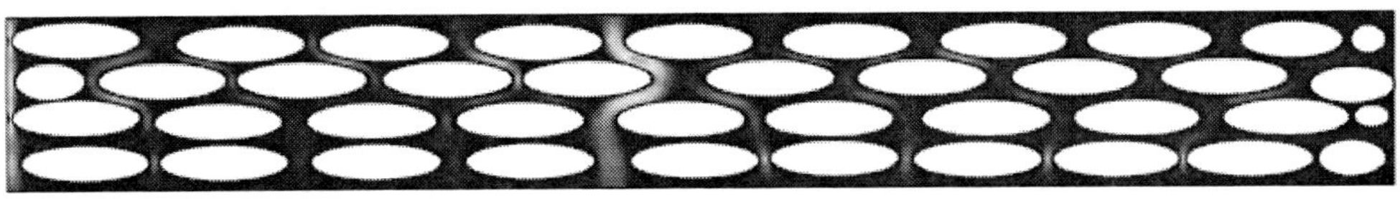

Figure 14. representative volume for inter-bundle impregnation simulations.

In this simulation, assuming that the inter-bundle permeability is much higher than intra-bundle permeability, each bundle was approximated as a solid ellipse, impermeable to the diffusing fluid. The boundary conditions used for the numerical simulation are the same as those used for the intra-bundle simulation. The thickness used for two layers of fabrics was 0.72 mm, while for four layers was 1.5 mm.

The average velocity as a function of applied pressure for two and four fabric layers are reported in Figure 15.

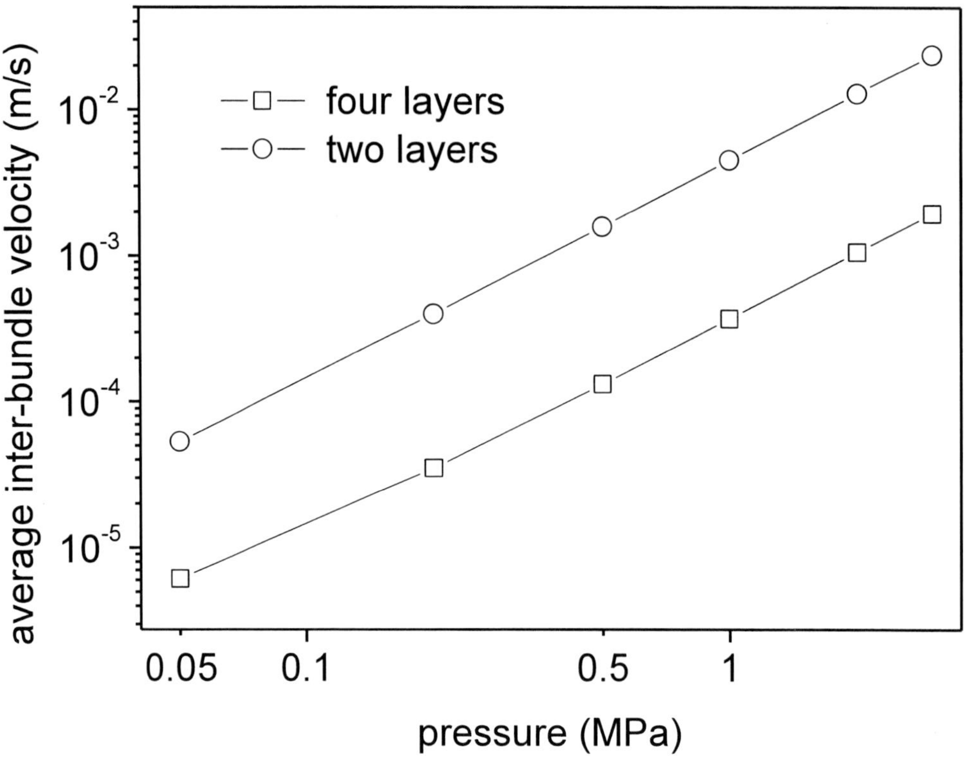

Figure 15. inter-bundle average velocity vs. applied pressure.

In this case, there is no evidence of the existence of a direct proportionality between velocity and ΔP, which suggests that the behavior of the material is significantly non Newtonian for all the applied pressures. FEM simulations in four layers of reinforcement even at 0.2 MPa, led to an average shear rate of 0.85 s^{-1}, which is outside the Newtonian plateau, due to the very high values of velocity compared to the intrabundle simulation.

Numerical Analysis of Commingled Yarn Consolidation

The two processes of sintering and matrix flow during consolidation of commingled composites can be analyzed separately by introducing a proper mathematical modeling.

It was shown that the evolution of void fraction due to matrix sintering can be expressed by a proper kinetic expression, as [24, 25]:

$$\frac{dx_V}{dt} = -\frac{\Gamma}{a_0 \eta} f_2(x_V) \qquad \text{eq. 13)}$$

In which η is the matrix viscosity, Γ is the surface tension of the molten matrix, a_0 is the initial particle radius, and f_2 is a function of the void fraction, which depends by particle shape and sintering geometry. By introducing the dimensionless time $t^*=t/t_{r,sintering}$, eq. 13) can be written as:

$$\frac{dx_v}{dt^*}=-\frac{\Gamma t_{r,sintering}}{a_0\eta}f_2(x_v) \qquad \text{eq. 14)}$$

Where $t_{r,sintering}$ represents a reference time associated with sintering experiment. Therefore, it was possible to introduce a dimensionless number for sintering of the polymer, as:

$$D_1=\frac{\Gamma t_{r,sintering}}{a_0\eta} \qquad \text{eq. 15)}$$

Which represents the ratio between the characteristic time of the experiment $t_{r,sintering}$, and the characteristic sintering time of the material $a_0\eta/\Gamma$.

In the same way, it was possible to write a kinetic expression for the void fraction due to matrix flow, associated to the Darcy law, as [24]:

$$\frac{dx_v}{dt}=-K\frac{\Delta P}{\eta R^2}f_3(x_v) \qquad \text{eq. 16)}$$

Where K is the tow permeability, $\triangle P$ is the applied pressure, R is the flow length across the fabric thickness, and f_3 is a function of the void fraction. By using the dimensionless time $t^*=t/t_{r,flow}$, eq. 16) can be rewritten as reported in eq. 17):

$$\frac{dx_v}{dt^*}=-K\frac{\Delta P t_{r,flow}}{\eta R^2}f_3(x_v) \qquad \text{eq. 17)}$$

Where $t_{r,flow}$ is a characteristic time of the matrix flow experiment and $f_3(x_v)$ a dimensionless function. Therefore, it was possible to introduce a dimensionless number for the flow of molten polymer during yarn impregnation, as:

$$D_2=\frac{K\Delta P t_{r,flow}}{\eta R^2} \qquad \text{eq. 18)}$$

Which represents the ratio between the characteristic time of the experiment, $t_{r,flow}$, and the characteristic time for matrix flow, $\frac{\eta R^2}{K\Delta P}$.

By comparison between eq. 15) and eq. 18), the relevance of sintering and matrix flow can be assessed by introducing the dimensionless ratio D_3, reported in eq. 19):

$$D_3=\frac{D_1}{D_2}=\frac{\Gamma R^2}{a_0 K\Delta P}\frac{t_{r,sintering}}{t_{r,impregnation}} \qquad \text{eq. 19)}$$

In order to calculate the dimensionless numbers D_1, D_2 and D_3, the following values were assumed: Γ=0.045 N/m [45], R= 700 μm (which is the radius of the bundle, measured by image analysis of Figure 9), a_0=10 μm (also measured from image analysis of Figure 9), △P=5 bar. The reference time scale of sintering experiment, $t_{r,sintering}$=360 s, was taken as the ratio between the temperature interval of the second peak of Figure 8, associated with fiber sintering (30°C) and the used heating rate (5°C/min). The reference time for matrix flow, $t_{r,flow}$=600 s was taken as the ratio between the temperature interval of the third peak of Figure 7, associated with matrix flow inside fiber bundle (50°C) and the used heating rate (5°C/min). The permeability K was obtained by means of the Kozeny–Carman equation [46]. By taking a porosity of 0.53 (obtained by optical microscopy), and a value of the Kozeny constant of 150, the permeability of the tow was calculated to be $1.80*10^{-12}$ m^2. Therefore, according to eq. 19), a value of D_3=1470 was obtained, which clearly indicates that for the same viscosity of the resin (and therefore for the same temperature), D_1>>D_2, which in turn indicates the prevalence of sintering compared to matrix flow.

In examining the two processes related to the sintering and flow of the matrix between the fibres, the dimensionless number D_1 was calculated in correspondence of the onset of the second peak of Figure 8, i.e. 88°C. A viscosity of the matrix of $4.15*10^6$ Pa*s at 88°C and 1 s^{-1} was calculated by extrapolation using the Arrhenius fitting data. A value of D_1=0.39 was obtained from eq. 15). This indicates that sintering can take place when the characteristic time for sintering becomes sufficiently low compared to the characteristic time of the experiment. At temperatures lower than 88°C, the high viscosity of the material involves higher values of the characteristic times for sintering, which are much higher than the characteristic time of the experiment, thus preventing sintering to occur. At the same temperature of 88°C, a very low value of D_2 = $2.65*10^{-4}$ was calculated, which indicates that the characteristic time for matrix flow, $\frac{\eta R^2}{K\Delta P}$, is much higher than the characteristic time scale of the experiment, and the matrix flow can be neglected. Therefore, at lower temperatures, sintering is much faster than matrix flow, as confirmed by the consolidation experiments shown above.

The analysis of dimensionless numbers clearly confirms that the second consolidation step observed in Figure 7 and Figure 8 is due to the matrix sintering, leading to a change the PETg/cy from fibers to a fully densified matrix, which occurs according to the mechanisms commonly involved in powder sintering involving in this case matrix fibers deformation and sintering coupled with bubble removal [25]. Referring to the modulus and viscosity evolution of the matrix (Figure 1), these results indicate that sintering of PETg can take place for very high viscosity values, in the rubbery plateau region, where bundle impregnation due to matrix flow can be considered irrelevant. Even if a very high pressure (50 bar) is used during compression molding, the resulting D_3=147 at a temperature of 88°C is too high to make relevant any contribution of pressure forces in comparison with surface tension forces. This again indicates that sintering is more likely to take place before any matrix flow occurs even at higher pressures typical of industrial compression molding processes.

On the other hand, the dimensionless number D_2 calculated in correspondence of the onset of the third consolidation step (183°C calculated from Figure 7) yields a very low value of 0.24, which is roughly the same of D_1 calculated at 88°C. This indicates that in this range of temperatures, the characteristic time for matrix flow, $\frac{\eta R^2}{K\Delta P}$, becomes of the same order of magnitude than the characteristic time of the experiment, thus allowing for micro-scale impregnation to occur as a consequence of bulk matrix flow, which results in the last thickness reduction, or equivalently the last peak, of Figure 7. Therefore it is possible to conclude that, even at very high molding pressures, impregnation of commingled composites occurs first by sintering of the polymer matrix followed by micro scale impregnation of fibers.

Therefore, the consolidation of the PETg/glass/cy system can be represented by the scheme reported in Figure 16. The first step of the process, represents the elastic deformation of fibers observed at temperatures higher than the T_g of the polymer. The second step, due to matrix sintering, involves macro-scale impregnation process and removal of inter-bundle voids. This step takes place without any bulk matrix flow. The last step of the consolidation is the micro-scale impregnation, leading to a reduction of the intra-bundle void fraction [47], which resembles the process already described for the PETg/glass/fs system, being governed by bulk matrix flow.

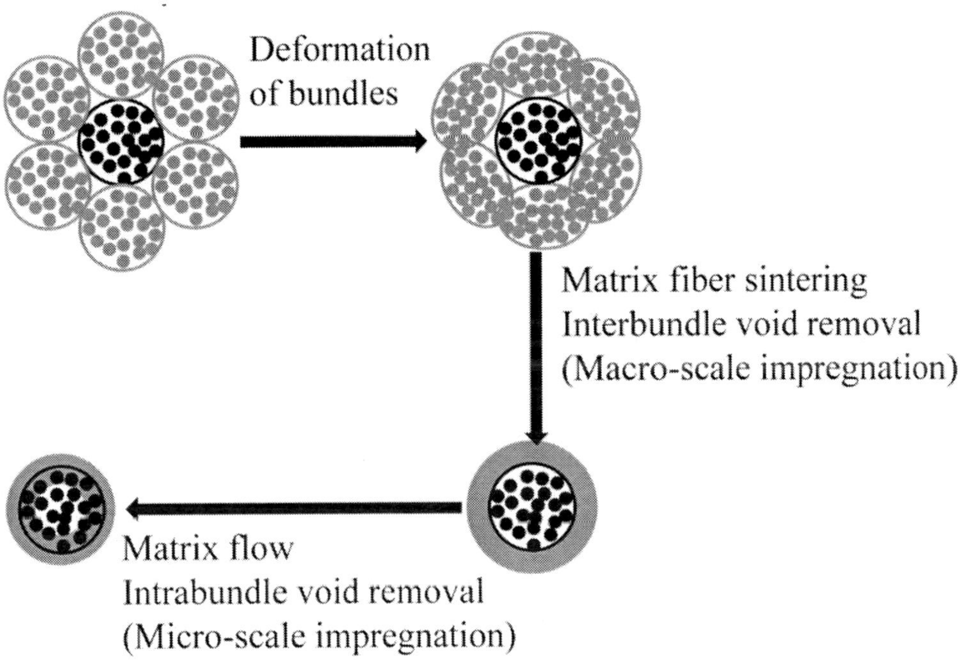

Figure 16. scheme of the consolidation for PETg/glass/cy system.

CONCLUSIONS

In this work, the mechanisms at the basis of the macro and micro-scale impregnation of thermoplastic matrix composite were analyzed through experimental analysis and numerical modeling. Optical and scanning electron microscopy analysis revealed the existence of two different mechanisms for fiber impregnation, the first one associated to a macro-scale, inter-bundle, impregnation and the second one associated to micro-scale, intra bundle, impregnation. The two mechanisms are characteristic of both processing systems considered, i.e. film stacking of matrix film and reinforcement fabric and commingled yarn fabrics. In both cases it was shown that the first step of consolidation involves the formation of a homogeneous polymer phase surrounding each bundle, leaving a high void fraction inside bundles. The second step, which takes place at higher temperatures or longer experimental times, involves the reduction of the intra-bundle voids.

When dry fabric and matrix stacks are used, the two different mechanisms were simulated using either a regular array for inter-bundle flow, either several random arrays for intra-bundle flow. The simulation results showed that at low pressures the Darcy law satisfactorily predicts the micro-scale flow of the matrix through the reinforcement, indicating that permeability only depends on the fiber geometry and architecture. On the other hand, at high pressures and shear rates, the Darcy law and its derivations are not able to predict the matrix flow. At macro-scale, impregnation is associated with higher matrix velocities involving very high shear rates, always beyond of the Newtonian plateau region. In both cases, permeability is not only dependent on the fiber geometry and architecture, but also on the properties of the matrix, and in particular on the power law index.

The two mechanisms of macro and micro-scale impregnation were also studied for the commingled yarn system. In this case, experimental results, coupled with dimensionless analysis, indicated that the macro-scale impregnation cannot be attributed to matrix flow outside the reinforcement bundles, but rather to the formation of a homogeneous polymer pool, governed by sintering of fibers and bubble removal. The second stage of consolidation is actually governed by the matrix flow inside each bundle, which takes place above 180 °C. These conclusions were confirmed by dimensionless analysis, showing that the characteristic time of sintering becomes of the same order of magnitude than the characteristic time of the experiment above T_g, when the matrix is in rubbery plateau region. Dimensionless analysis also showed that in the flow region the characteristic time of matrix flow becomes of the same order of magnitude of the characteristic time of the experiment, finally leading to intrabundle matrix flow and void reduction.

These observations clearly indicate that in most cases, for commingled composites, the consolidation behavior can be only partially explained by the Darcy law, which is usually associated to the matrix flow. In such cases the Darcy law can only be used to model the micro-impregnation, involving flux of the matrix inside reinforcement fiber bundles.

REFERENCES

[1] Pegoretti, A.; Ricco, T. *J. Mat. Sci.* 2001, 36(19), 4637-4641

[2] Passaro, A.; Corvaglia, P.; Manni, O.; Barone, L.; Maffezzoli, A. Polym. *Compos.* 2004, 25, 307-318

[3] Mitschang, P.; Blinzler, M.; Woginger A. *Compos. Sci. Technol.* 2003, 63, 2099-2110

[4] Kim, D.W.; An, Y.S.; Nam, J.D.; Kim, S.W. *Compos. Part A.* 2003, 34, 673-680

[5] Svensson, N.; Shishoo, R.; Gilchrist, M. J. *Thermoplast. Compos. Mater.* 1998, 11, 22-56

[6] Phillips, R.; Akyuz, D.A.; Manson, J.A.E. *Compos. Part A* 1998, 29A, 395-402

[7] Salomi, A.; Greco, A.; Felline, F.; Manni, O.; Maffezzoli, A. *Adv. Polym. Tech.* 2007, 26, 21-32.

[8] Breuer, U.; Neitzel, M.; Ketzer, V.; Reinicke, R. *Polym. Compos.* 1996, 4(17), 643-647.

[9] Breuer, U.; Neitzel, M. *The challenge of stamp forming high-strength thermoplastic composites for transportation.* In 42nd International SAMPE Symposium Proceeding, Anaheim, USA,1997.
[10] Tucker III C.L. *Fundamentals of computer modelling for polymer processing*, Hanser Publishers: Munich, 1989.
[11] William, J.G.; Morris, C.E.M.; Ennis, B.C. *Polym. Eng. Sci.* 1974, 14(6), 413-419
[12] Michaud, V.; Tornqvist, R.; Manson J.A.E. J. *Compos. Mater.* 2001, 35(13), 1174-1200
[13] Long, A.C.; Wilks, C.E.; Rudd, C.D. *Compos. Sci. Technol.* 2001, 61, 1591–1603
[14] Ye, L.; Friedrich, K.; Kastel, *J. Appl. Compos. Mater.* 1995, 1, 415–429.
[15] Klinkmuller, V.; Um, M.K.; Steffens, F.K.; Kim, B.S. *Appl. Compos. Mater.* 1995, 5, 351–371.
[16] Ijaz, M.; Robinson, M.P.; Wright, N.H.; Gibson, A.G. J. *Compos. Mater.* 2007, 41, 243-262
[17] Bernet, N.; Michaud, V.; Bourban, P.E.; Manson, J.A.E. J. *Compos. Mater.* 1999, 33(8), 751-772
[18] Phillips, R.; Akyuz, D.A.; Manson, J.A.E. *Compos. Part A* 1998, 29A, 395-402
[19] Young, W.B. Compos. Sci. Technol. 1995, 54, 299-306
[20] Ye, L.; Friedrich, K.; Kastel, J.; Mai, Y.W. *Compos. Sci. Technol.* 1995, 54, 349-358
[21] Chhabra, R.P.; Comiti, J.; Machac, I. *Chem. Eng. Sci.* 2001, 56, 1-27
[22] Gennaro, R.; Greco, A.; Maffezzoli, A.; *Adv. Polym. Technol.* 2010, 29(2), 122-130
[23] Connor, M.; Toll, S.; Manson, J.A.E.; Gibson, A.G. J. *Thermoplast. Compos. Mater.* 1995, 8, 138-162
[24] Greco, A.; Strafella, A.; La Tegola, C.; Maffezzoli, A.; *submitted to Polym. Compos.*
[25] Bellehumeur, C.T.; Kontopoulou, M.; Vlachopoulos, *J. Rheol. Acta* 1998, 37, 270–278
[26] Bellehumeur, C.T.; Bisaria, M.K.; Vlachopoulos, *J. Polym. Eng. Sci.* 1996, 36(17), 2198-2207
[27] Kontopoulou, M.; Vlachopoulos, *J. Polym. Eng. Sci.* 2001, 41 (2), 155-169
[28] Kontopoulou, M.; Vlachopoulos, *J. Polym. Eng. Sci.* 1999, 39 (7), 1189-1198
[29] Gogos, G. Polym. Eng. Sci. 2004, 44 (2), 388-394
[30] Haffner, S.M.; Friedrich, K.; Hogg, P.J.; Busfield, J.J.C. *Appl. Compos. Mater.* 1998, 5, 237-255
[31] Chen, X.; Papathanasiou, T.D. *Compos. Sci. Technol.* 2007, 67, 1286-1293
[32] Gennaro, R.; Lionetto, F.; Greco, A.; Maffezzoli, A. *Numerical simulation of the microscale impregnation in commingled thermoplastic composite yarns*, ICCM-17 International Conference of Composite Materials, Edinburgh, 2009.
[33] Larson, R.G. *Constitutive Equations for Polymer Melts and Solutions*, Butteworths: New York, 1988.
[34] Woo II L.; Springer, G.S. J. *Compos. Mater.* 1987, 21, 1017-1055.
[35] Trudel-Bucher, D.; Fisa, B.; Denault, J.; Gagnon, P. *Compos. Sci. Tecnol.* 2006, 66, 555-570.
[36] Bird, R.B.; Stewart, W.C.; Lightfoot, E.N. *Transport Phenomena*, John Wiley and Sons: New York, 1960
[37] Sadowski, T.J.; Bird, R.B. *Trans. Soc. Rheol.* 1965, 9, 243-250
[38] Christopher, R.H.; Middleman, S. *Ind. Eng. Chem. Fundam.* 1965, 4, 422-426.
[39] Gaitonde, N.Y.; Middleman, S. *Ind. Eng. Chem. Fundam.* 1967, 6, 145-147
[40] Velten, K.; Lutz, A.; Friedrich, K. *Compos. Sci. Tech.* 1999, 59, 495-504

[41] Sadowski, T. J. *Trans. Soc. Rheol.* 1965, 9, 251-271
[42] Dauben, D.L.; Menzie, D.E. *Flow of Polymer Solutions through Porous Media*, Paper SPE 1688, YPE Symposium on Mechanics of Rheologically Complex Fluids, Houston, Texas, 1966.
[43] Gogarty, W.B. *Soc. Petrol. Eng. J.* 1967, 7(12), 149-160.
[44] Elder, J.P. *Thermochim. Acta* 1996, 272, 41-48
[45] Sauer B.B.; Dee G.T. *Macromol.* 2002, 35, 7024-7030
[46] Bates P.J.; Taylor D.; Cunningham M.F. *Appl. Compos. Mater.* 2001, 8, 163-178
[47] Bernet N, Michaud V, Bourban PE, Manson JAE, *J Compos Mater,* 1999; 33: 751-772

In: Thermoplastic and Thermosetting Polymers and Composites ISBN: 978-1-61209-264-5
Editors: Linda D. Tsai and Matthew R. Hwang

Chapter 7

THERMAL AND CHEMICAL GLASS TRANSITION OF THERMOSETS IN THE PRESENCE OF TWO TYPES OF INORGANIC NANOPARTICLES

J. Baller, M. Thomassey, M. Ziehmer and R. Sanctuary
Laboratory for the Physics of Advanced Materials (LPM),
University of Luxembourg, Luxembourg

ABSTRACT

Composites consisting of epoxies and inorganic nanoparticles are of high technological importance. Despite the fact that these nanocomposites are already widely used, fundamental understanding of the physical and chemical processes before and during epoxy network formation in the presence of nanoparticles is still missing. The present work presents investigations of the thermal and chemical glass transition in epoxies filled with two different types of nanoparticles: hydrophilic alumina and hydrophobic silica. It is shown that macroscopic investigations of static and dynamic thermodynamic properties of the glass transition behaviour allow elucidating network formation in epoxies in the presence of nanoparticles.

INTRODUCTION

Tailor-made composites consisting of epoxies and inorganic particles are nowadays widely used as adhesives to face demands for specific chemical or physical properties of joints [1-12]. Since several years, composites incorporating nanoparticles have gained growing importance, but there is still a lack of understanding concerning the relation between the desired phenomenological properties and the microscopic influence of the nanoparticles. Interaction between inorganic filler particles and polymer matrix in epoxy-based nanocomposites mainly takes place at the giant surface of the nanoparticles. Often these surfaces are coated with a hydrophobic layer e.g. to avoid aggregation of nanoparticles [13, 14]. This coating can dominate the interaction between polymer matrix and dispersed nanoparticles. There are quite a lot of possible interactions imaginable. They range from chemical interactions like catalysis of the epoxy curing by the nanoparticles to physical interactions which may e. g. influence the thermodynamic miscibility of different components constituting the epoxy network. In this chapter we present a study of the effect of two different types of nanoparticles on the physical properties of epoxy resins and on their curing behaviour. The two types of nanoparticles, alumina and silica, mainly differ in their surface properties. Whereas the alumina nanoparticles are used "as is", i.e. without any surface modification, the silica nanoparticles are coated by a hydrophobic layer. The influence of this different surface treatment is mainly examined by thermal analysis.

EXPERIMENTAL

Materials and Sample Preparation

The nanocomposites under study consist of diglycidyl ether of bisphenol A (DGEBA) as matrix material. Figure 1 shows the structural formula of the oligomer DGEBA.

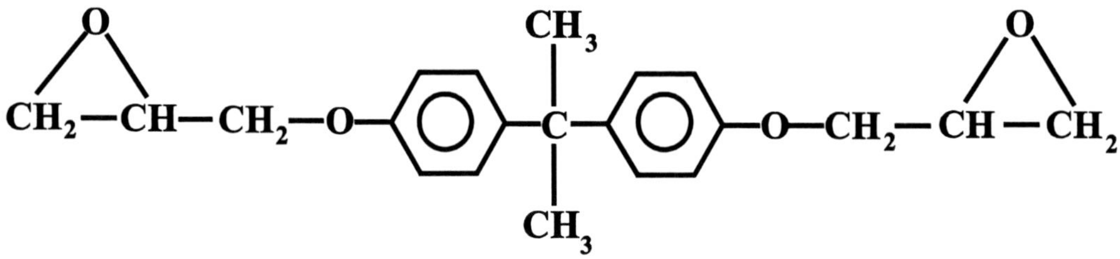

Figure 1. Structural formula of DGEBA

The melting point of DGEBA is at about 315 K [15] but the tendency to crystallize is low and the material easily can be undercooled and kept for weeks without the formation of crystals. The glass transition of the oligomers takes place at a temperature of about 257 K.

All alumina resin nanocomposites have been derived from a master batch of DGEBA (D.E.R. 331 obtained from Dow Plastics, Germany) filled with 28.9 wt% of alumina nanoparticles. The alumina nanoparticles are AEROXIDE Alu C produced by Evonik Industries, Germany.

Alumina nanoparticles (with agglomerates) and DGEBA resin were tempered at 400 K for 8 hours and then mixed together. This guarantees that the epoxy resin is free of crystals and that some of the volatile contaminations of the nanoparticles surfaces are removed. The average diameter of the primary particles is about 13 nm. Due to the production process (noble gas condensation), primary particles sinter together to chemically bonded aggregates (see inset in Figure 2) with linear dimensions between 13 and 200 nm (see Figure 2). The alumina particles are hydrophilic. This is why the aggregates form bigger agglomerates which are bonded by Van der Waals forces. These agglomerates have sizes up to the miocrometer regime. To break the physically bonded agglomerates, mechanical dispersion under vacuum at 369 K using a combined dissolver/bead mill (Torus Mill from VMA-Getzmann, Germany) was used. This dispersion process leads to a relatively homogeneous distribution of the alumina aggregates (see Figure 2). About 30 wt% of alumina particles is the maximum amount which can be incorporated by using this type of dissolver/bead mill.

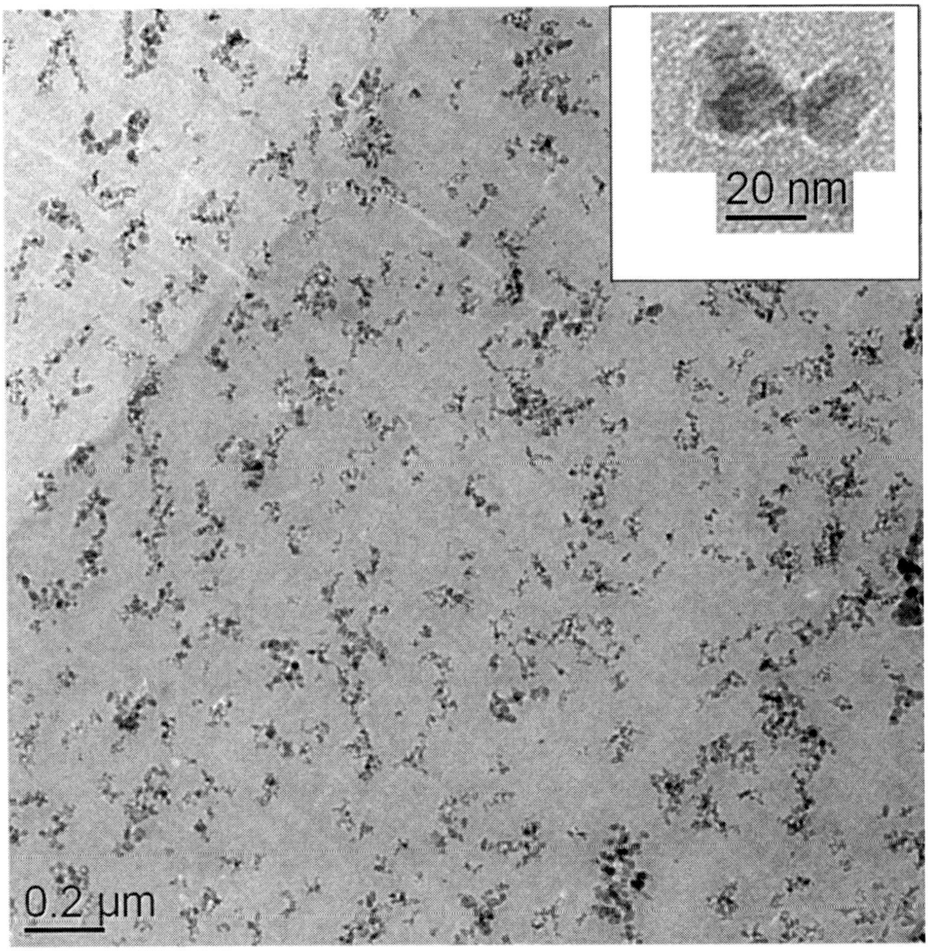

Figure 2. TEM micrograph of a nanocomposite with 7 wt% of alumina nanoparticles. The inset shows two primary particles chemically bonded by a sinter neck; sample cut with microtome, TEM device: JEOL JEM 2011.

The silica resin nanocomposites have also been derived from a master batch: a distilled type of DGEBA with 40 wt% of silica nanoparticles (Nanopox A410, nanoresins AG, Geesthacht, Germany). The silica nanoparticles have a diameter of about 20 nm. They are coated by a hydrophobic silane layer. The particles have a narrow size distribution and are homogeneously distributed inside the oligomer matrix [5, 6, 8, 16, 17]. Due to the production by a sol-gel process [18, 19] and due to the silane layer, there is no aggregation of particles [5, 6, 8, 16, 17]; see Figure 3. According to the manufacturer, Nanopox A410 contains the same nanoparticles as the nanocomposites (e.g. Nanopox F400) described in references [5, 6, 8, 16, 17].

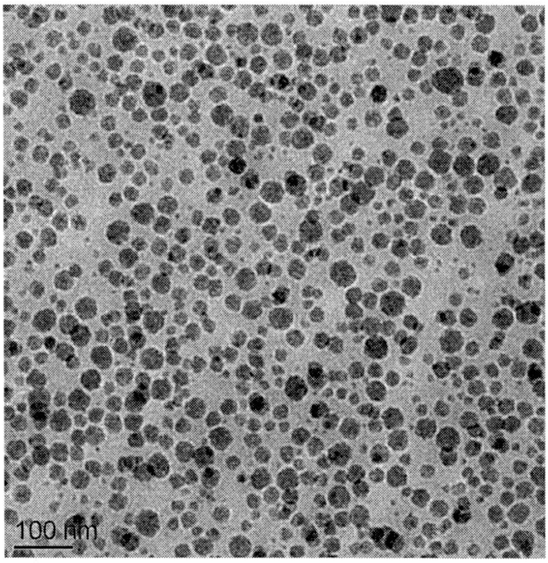

Figure 3. TEM micrograph of a nanocomposite with 40 wt% of silica nanoparticles, TEM device: JEOL JEM 2011.

Both alumina and silica masterbatches were diluted with the appropriate type of DGEBA (D.E.R. 331 from Dow Plastics and DGEBA distilled from nanoresins AG). Dilution yielded samples with different weight concentrations x of nanoparticles (np):

$$x = \frac{m_{\mathrm{np}}}{m_{\mathrm{DER331}} + m_{\mathrm{np}}}$$

The nanoparticles themselves as well as the production process (e.g. mechanical dispersion) could lead to a reduction of the number of functional epoxy groups in the DGEBA matrix. Therefore we measured the Epoxy Equivalent Weight (EEW) of the nanocomposites by two different kinds of wet analysis [20]: the hydrogen bromide method [20, 21] was used for the silica nanocomposites and the pyridinium chloride method [21, 22] for the alumina nanocomposites. Figure 4 shows the measured values EEW_{meas} together with calculated values EEW_{calc}:

$$EEW_{calc}(x) = \frac{EEW_{meas}(x=0)}{(1-x)}$$

The measured EEW of the alumina nanocomposites is slightly higher than the calculated values. This means that the number of epoxy groups is lower due to the presence of the alumina nanoparticles and/or due to the preparation/dispersion process. For the correction of heat flows and reaction heats, the measured EEW values are used to calculate values specific to mol epoxy groups [23].

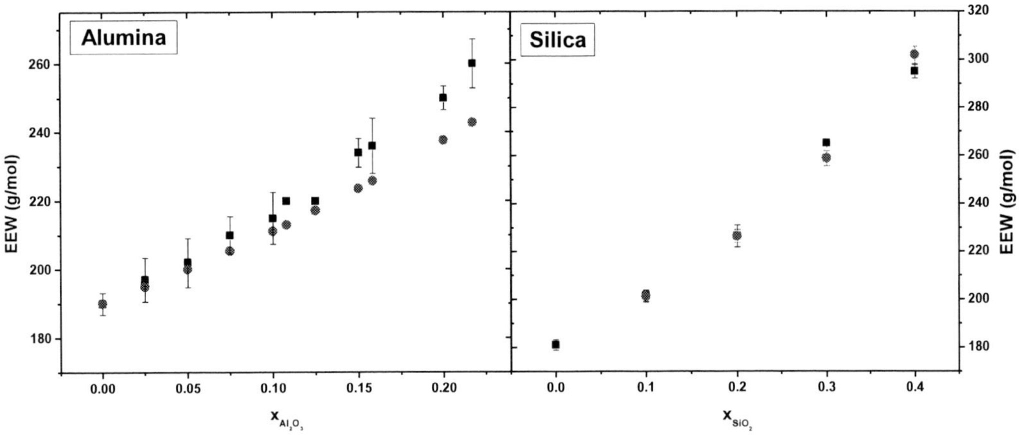

Figure 4. Epoxy Equivalent Weight of alumina and silica nanocomposites as a function of wt% of nanoparticles. Squares: measured values; circles: theoretical values.

Curing Process

The resin nanocomposites were cured with diethylene triamine (DETA, purity > 97%) obtained from Sigma/Aldrich, France. A ratio $s = m_{DETA} / m_{DEGEBA} = 0.142 \pm 0.001$ was applied [23, 24]. As temporal reference, the point in time $t_{cure}=0$ represents the moment when the hardener (DETA) was injected into the samples at a temperature of 298.2 K; all components (samples, laboratory tools) used for sample preparation have also been put to 298.2 K by using a temperature-controlled chamber. Injection of hardener was followed by stirring the samples by hand for five minutes.

Rheologic Measurements

The viscosity coefficients of the resins have been determined by means of a HAAKE MARS II rheometer following a standard procedure described in [25, 26]. We used plate–plate geometry with a diameter of 35 mm and a gap of 1 mm. The temperature of the measuring cell was stabilized at T = 298.0 ± 0.1K . The viscosity coefficients $\eta_0 = \lim_{\dot{\gamma} \to 0} \eta(\dot{\gamma})$ were obtained by extrapolating the first Newtonian range to $\dot{\gamma}=0$.

Calorimetric Measurements

Thermal investigations were performed using Temperature Modulated Differential Scanning Calorimetry (TMDSC) on a DSC823e and a modified DSC821e (Mettler Toledo, Switzerland) [27, 28]. The instruments were temperature and heat flow calibrated using water, indium, naphthalene and benzoic acid. Sample masses were between 15 and 25 mg.

The calorimeter furnace is controlled by the following temperature program

$$T(t) = T_0 + \beta t + T_\mathrm{a} \sin \omega t$$

During the isothermal curing experiments, $T_0 = 298\ \mathrm{K}$, $\beta = 0$, the temperature amplitude was $T_\mathrm{a} = 0.5\ \mathrm{K}$ and the angular frequency $\omega = \frac{2\pi}{t_\mathrm{p}} = 0.052\ \mathrm{rad/s}$, $t_\mathrm{p} = 120\ \mathrm{s}$ representing the modulation period.

In the frame of linear response theory the heat flow Φ into the sample can be written as [29]

$$\Phi = \Phi^u + \Phi^a \cos(\omega t - \varphi)$$

This signal can be deconvolved in order to get the underlying heat flow Φ^u, the amplitude Φ^a of the modulated heat flow and the phase angle φ between the measured heat flow and the temperature rate. Φ^u is related to the conventional DSC curve. In case of an isothermal cure experiment Φ^u is proportional to the calorimetric reaction rate. The modulus of the specific complex heat capacity at constant pressure $\left|c_\mathrm{p}^*\right|$ is determined from the respective amplitudes of the temperature T_a and the heat flow Φ^a

$$\left|c_\mathrm{p}^*\right| = \frac{K \cdot \Phi^a}{m \cdot T_\mathrm{a} \cdot \omega}$$

where *m* is the sample mass. The calibration factor *K* is obtained using an aluminum standard with well known specific heat capacity. Calorimetric measurements of the curing process have been started about 6 to 8 minutes after injecting (t_{cure}=0) the hardener into the resin nanocomposites. Measured data have been linearly extrapolated to t_{cure}=0. The calorimetric measurements have been carried out with a sampling rate of one data point per second. In most of the following figures, symbols are only used for the sake of clarity; data points lie much closer and are represented by solid line unless stated otherwise.

RESULTS AND DISCUSSION

Resin Nanocomposites

The first investigations elucidate the change of dynamic properties of epoxy oligomers by the presence of alumina or silica nanoparticles. Interactions between nanoparticles and epoxy oligomers should change the dynamics of the oligomers in the neighborhood of the nanoparticles. Transport properties such as the viscosity should also be affected by the presence of fillers. Table 1 shows the viscosity coefficients of the different nanocomposite systems at 298 K. The systems with 0% Al_2O_3 and 0 wt% SiO_2 correspond to pure D.E.R. 331 and pure distilled DGEBA, respectively (see experimentals).

Table 1. Viscosity coefficients of the resin nanocomposites at 298 ± 0.1 K.

	0 wt% Al_2O_3	10 wt% Al_2O_3	0 wt% SiO_2	10 wt% SiO_2
η (Pa s)	14.76	47.26	9.96	21.8

As expected, the viscosity of the resins with nanoparticles is higher than the viscosity of the pure resins. The viscosity of the alumina system is twice as high as the viscosity of the silica system. The differences in the viscosities are a first sign for differences in matrix/filler interactions between the alumina and the silica systems. Epoxy oligomers are canonical glass formers. Due to the fact that their tendency to crystallize is very low, the glass transition behaviour can easily be examined, e.g. by calorimetric measurements [30, 31]. A big part of the dynamics of the epoxy oligomers is represented by the intrinsic structural relaxation process, the so-called α process.

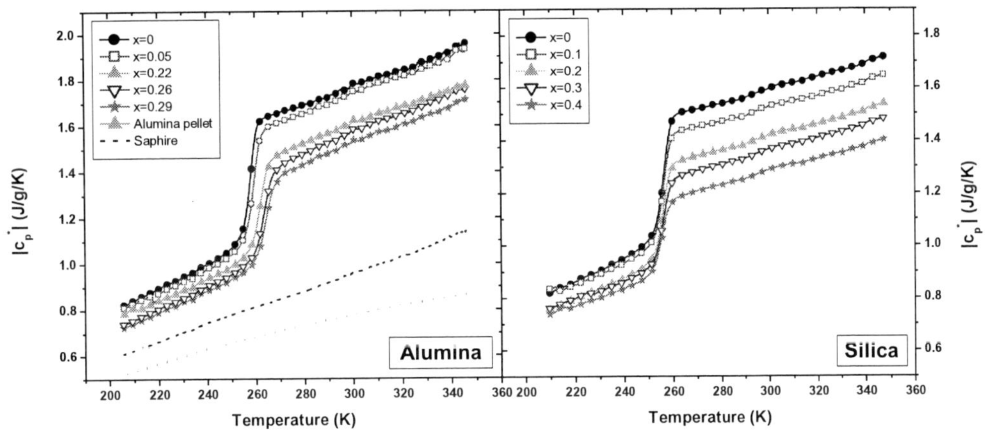

Figure 5. Modulus of the complex specific heat capacities of the resin nanocomposites with different weight concentrations x of fillers. Dashed line: measured c_p of pressed alumina nanopowder (pellet). Dotted line: c_p of sapphire taken from literature [32].

Since the α process is responsible for the glass transition, its investigation yields insight into the dynamics of the glass-forming molecules (see e.g. [33-35]). Figure 5 shows the complex specific heat capacities of resins nanocomposites with alumina and silica fillers measured by TMDSC. At first sight, the main difference between alumina and silica fillers is the shift in the glass transition temperature which takes place in the alumina nanocomposites with increasing filler content. In principle there are three different contributions to the specific heat capacity $\left|c_p^*\right|$: i) epoxy matrix (c_p^{resin}) ii) nanoparticles (c_p^{np}) iii) interphase between epoxy oligomers and nanoparticles. The contribution of i) is well determined for the alumina as well as for the silica systems by the curves with x=0 in Figure 5. Since the alumina nanoparticles are available as a powder, we were able to produce a pellet with a laboratory press. The dashed line in Figure 5 shows the specific heat capacity of the alumina nanoparticles; as a reference, the specific heat capacity of sapphire is indicated by the dotted line. If we assume that the contribution of iii) to the nanocomposite's specific heat capacity is small, we can calculate the real part of the specific heat capacity of the resin part of the nanocomposite by applying a simple mixing rule:

$$c_p^{'resin} = \frac{\left|c_p^*\right|\cos(\varphi) - x\,c_p^{np}}{(1-x)} = \frac{c_p^{'} - x\,c_p^{np}}{(1-x)}$$

For the silica nanocomposites, the situation is more complex. Due to the fabrication process, the silica nanoparticles are only available dispersed inside the epoxy matrix. Therefore it is not possible to determine the specific heat capacity of the silica nanoparticles experimentally. We used the complex heat capacity data of the silica nanocomposites along with the assumption that the contribution of iii) can be neglected to calculate an effective heat capacity $c_{p_{eff}}^{SiO_2}$ of the silica nanoparticles (for details see [23]). The results are shown in Figure 6. For both systems it can be stated that the specific heat capacities outside the glass transition region coincide for every concentration x of nanoparticles. This is on one hand a confirmation for the assumption that the contribution of iii) to the specific heat capacity of the nanocomposites is negligible. On the other hand this procedure allows determining the unknown specific heat capacity of the silica nanoparticles (with coating) inside the nanocomposites (see Figure 6). Inside the glass transition region, the situation is different. Whereas the silica nanocomposites don't show any change in the glass transition behaviour with increasing filler content, the glass transition temperature of the alumina nanocomposites is systematically shifted to higher temperatures with higher filler content.

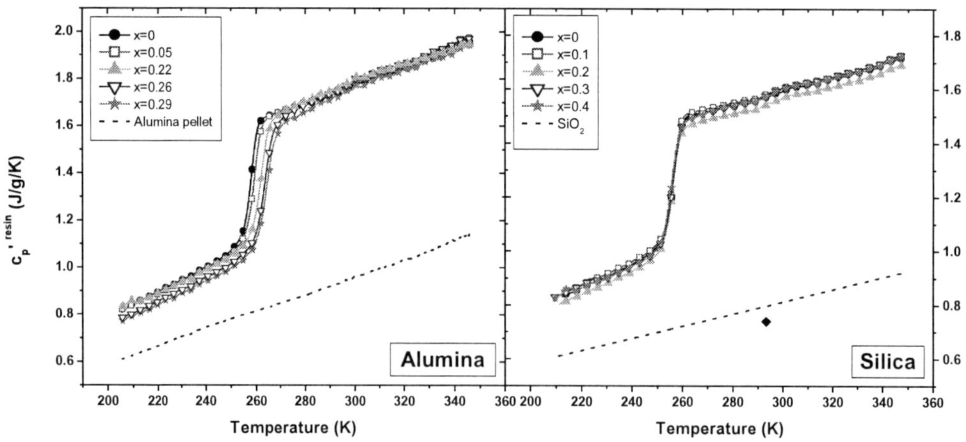

Figure 6. Real part of the complex specific heat capacities of the nanocomposite resins with different weight concentrations x of fillers. Values corrected to the epoxy part (see text). Dashed lines: measured c_p of pressed alumina nanopowder (pellet) and calculated effective heat capacity $c_{p_{eff}}^{SiO_2}$ of silica nanoparticles, respectively. Rhombus: c_p of silica glass at 293K taken from literature [36].

This result is also supported by the behaviour of the imaginary part of the complex specific heat capacities shown in Figure 7.

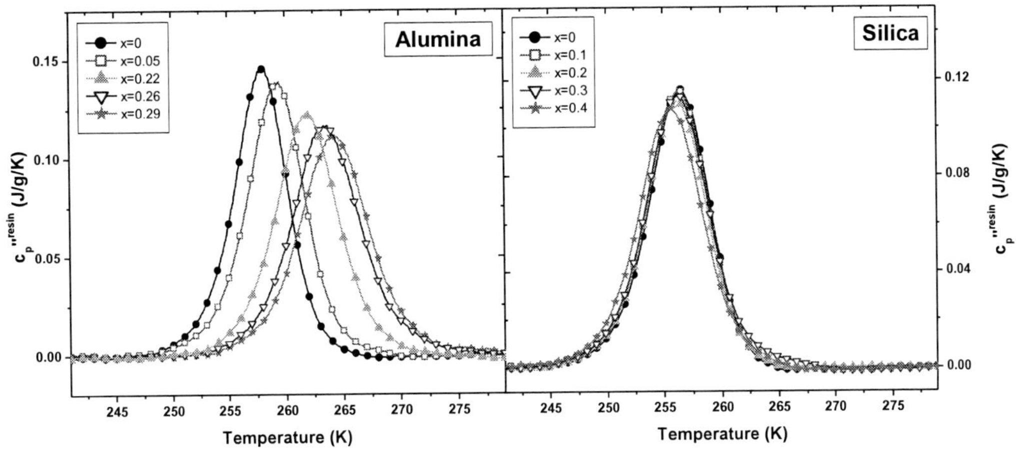

Figure 7. Imaginary part of the complex specific heat capacity of the nanocomposite resins with different weight concentrations x of fillers. Values corrected to the epoxy part (see text).

Since the imaginary part of the specific heat capacity is only caused by the relaxing epoxy oligomers around the thermal glass transition, the contribution of ii) to the imaginary part is zero and the imaginary part of the specific heat capacity can easily be corrected to the resin part of the nanocomposites (shown in Figure 7):

$$c_p^{''\,\mathrm{resin}} = \frac{\left|c_p^*\right|\sin(\varphi)}{(1-x)} = \frac{c_p''}{(1-x)}$$

For the silica systems it is obvious that the relaxation process of the epoxy oligomers which is visible by the peak in c_p'' is not significantly affected by the nanoparticles. This goes in line with the observation of the behaviour of c_p' (Figures 5 and 6) and leads to the conclusion that the interactions between silica nanoparticles and oligomer matrix molecules are very weak.

In case of the alumina particles, position and shape of the peak in c_p'' change with increasing nanoparticles content: The more filler particles the smaller the peak height and the broader the peak width. The integral area under the peak in c_p'' remains almost constant. This signifies that the number of relaxators is *not* reduced by the nanoparticles. Therefore we assume that the interactions between alumina nanoparticles and oligomer molecules are small (physical not chemical). For more details please see references [23, 37].

Curing

Curing reactions in the epoxy systems take place between DGEBA with two functional epoxy groups and DETA with three functional amine groups. These diepoxide-triamine systems are known to yield homogeneous systems without phase separation [38]. In the examined temperature regime two types of reactions can be found: epoxy groups reacting with primary or secondary amine groups [39, 40]. During these reactions, -OH groups are produced which strongly catalyze the reactions [40, 41]. Since the catalysed version of these reactions have significantly higher reaction rates than the un-catalysed form, they are responsible for the autocatalytic appearance of the epoxy/amine curing process [41]. Beside the hydroxyl groups, tertiary amines are also known to catalyse epoxy/amine reactions.

In a first series of experiments, the curing process is analyzed with classical DSC scans from 210 to 450 K to determine the total reaction enthalpy. Table 2 shows the results.

Table 2. Reaction enthalpies per mol of epoxy groups for different mass concentrations x of fillers.

Alumina		Silica	
x	$\Delta H^{epoxy\ group}$ (kJ/mol)	x	$\Delta H^{epoxy\ group}$ (kJ/mol)
0	123±7	0	124±5
0.07	122±7	0.1	123±6
0.12	123±7	0.2	122±5
0.16	119±8	0.3	124±6
		0.4	119±6

The total reaction enthalpies per mol of epoxy groups do not change by the presence of the fillers. Please note that the small variation of the total reaction enthalpies in reference [23] is due to the fact that data in [23] are presented with respect to gram epoxy. The raw data presented here are the same as in [23]. All values for the reaction enthalpies are close to values of 110-118 kJ/mol found for a variety of epoxy/amine systems (see [40] and citations therein). This result is a first argument for the assumption that the incorporation of alumina and silica nanoparticles into the diepoxy/triamine systems doesn't give rise to the appearance of additional reactions with significant endothermic or exothermic effects.

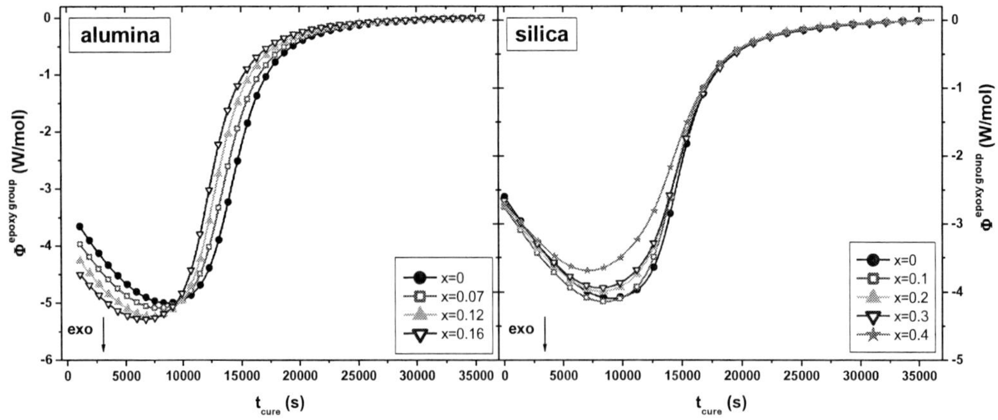

Figure 8. Heat flow during isothermal curing at 298.15 K for alumina and silica nanocomposites with different weight concentrations x of fillers; values specific to mol epoxy groups.

To examine the curing process in more detail, it is often studied under isothermal conditions [23, 24, 42]. In the following, isothermal curing of alumina and silica nanocomposites is examined by TMDSC experiments. In order to be able to analyze the results for different epoxy and composite systems, most quantities are given in units specific to the number of mol of epoxy (oxirane) groups. Figure 8 shows the heat flow during isothermal curing at 298.15 K.

In the first stages of isothermal curing, chemical reactions dominate the temporal evolution of the curing process [40, 41, 43-45]. The increasing heat flow at the beginning of the reactions is caused by the autocatalytic effect of the hydroxyl groups which are by-products of the epoxy/amine reactions. In the later stages of isothermal curing, diffusion processes dominate the temporal evolution of network formation because the reactants mobility is significantly reduced by growing network fragments [23, 24, 42]. This leads to a decreasing heat flow and to the transition to the glassy state before the maximum conversion could be reached.

The differences between the alumina and silica systems are obvious:

In the first stages of curing, the silica particles seem to have no effect on the curing process whereas the alumina particles accelerate the curing from the beginning. After about 5000 seconds, the heat flow curves differ substantially: whereas the heat flow curves of the alumina system remain similar for all concentrations, the silica nanoparticles systematically slow down the curing process. Kinetic analysis of the curing processes [23, 24] has shown that the character of the chemical reactions does not change by the incorporation of silica or alumina nanoparticles. They can be modelled by semi - empirical Kamal equations taking into account autocatalytic n^{th} order kinetics [46, 47]. In the case of the silica nanocomposites the change of the heat flow curves by the nanoparticles can be interpreted by the assumption that the reduction of the reactants mobility by the nanoparticles leads to an earlier transition from the reaction controlled to the diffusion controlled regime [23, 42, 48].

To understand the accelerating effect of the alumina nanoparticles on the isothermal curing process, we examined another filler which is known to have an accelerating effect: water [49]. As described above, hydroxyl groups catalyze the epoxy/amine curing reaction [40]. Figure 9 compares the heat flow curves during isothermal curing of alumina nanocomposites and composites with different amounts of water as fillers.

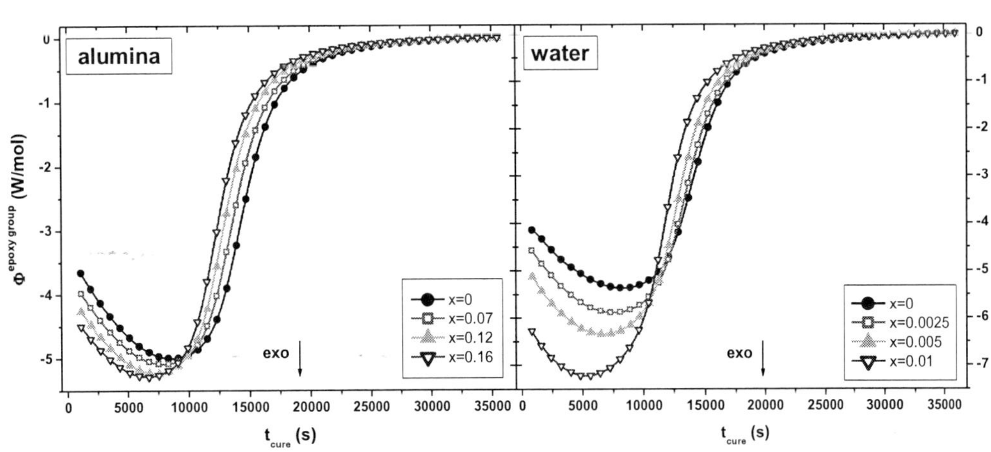

Figure 9. Heat flow during isothermal curing at 298.15 K for alumina and water composites with different weight concentrations x of fillers; values specific to mol epoxy groups.

In both systems, the fillers have a catalytic effect which leads to a higher initial reaction rate, a higher and earlier reaction maximum and an earlier occurrence of the inflection point. Taking the maximum of the reaction rate as a reference one can scale the curves vertically by a factor. This is justified since it is clear that the strength of the catalytic effect of such different fillers as alumina nanoparticles and water is different leading to different increase of the maximum reaction rate with filler concentration. The result of this vertical scaling is shown in Figure 10.

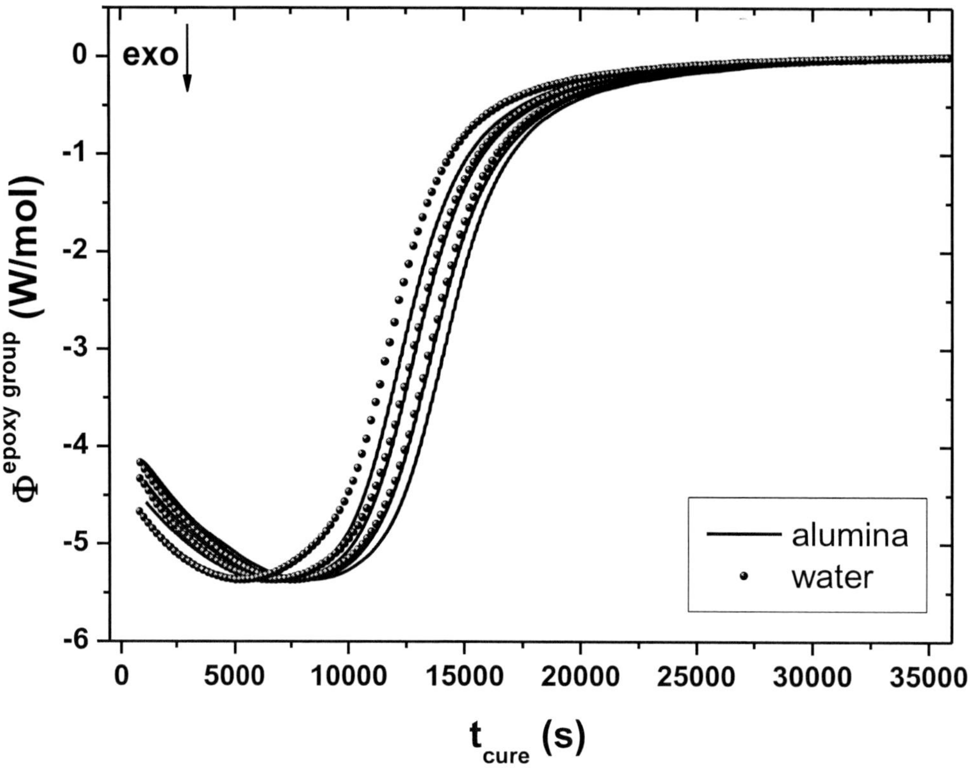

Figure 10. Heat flow curves of Fig. 9 scaled vertically (see text). $t_{IP}(0)$: inflection point of the heat flow curve for x=0.

It has now become evident that the heat flow curves of the alumina and water systems describe the same processes, the influence of both fillers on curing is very similar. If there are other reactions involved between epoxy and alumina or water fillers, they are not reflected in the heat flow curves, i.e. they are not endothermic or exothermic. We therefore conclude that the hydroxyl groups on the hydrophilic alumina particles surfaces have the same effect as the hydroxyl groups in the water. This argumentation is also supported by the fact, that the same analysis can also be done for the specific heat capacities of alumina and water composites during isothermal curing [49]. This also leads to very similar curves independent of the type of filler.

CONCLUSION

The calorimetric investigation of the thermal glass transition of epoxy resins filled with alumina and silica nanoparticles gave no hints for a distinct interphase with significantly changed thermal properties. There is only one glass transition detectable which in the case of the silica nanoparticles is identical for filled and unfilled systems. For the alumina nanocomposites, the thermal investigation depicted a change of the dynamic glass transition temperature and of the shape of the relaxation process. This is a sign for interactions between resin oligomers and alumina nanoparticles. Since the number of relaxators remains unchanged by the nanoparticles, the interactions are assumed to be of physical (Van der Waals) nature. These results go in line with the hydrophilic character of the alumina particles and the hydrophobic character of the silica particles. In addition, the thermal investigations of the resin nanocomposites allowed calculating the specific heat capacity of the silica nanoparticles, a quantity which is not accessible to direct experimental measurements. Investigations of the isothermal curing process of the two nanocomposites also show the difference of the surface treatment between alumina and silica systems. The silica nanoparticles don't seem to affect the chemical reactions but speed up the transition from the reaction controlled to the diffusion controlled regime by mobility restrictions. The alumina nanoparticles have an accelerating effect on the curing process. Comparison with epoxies filled with small amounts of water led to the conclusion that the influence of the alumina nanoparticles on the curing process is mainly caused by the catalytic effect of hydroxyl groups on the nanoparticles surfaces. Determination of the total reaction enthalpy per mol of functional groups by non-isothermal measurements showed that the maximum chemical conversion in the epoxy nanocomposites is independent of the filler content.

ACKNOWLEDGMENT

This work was kindly supported by the National Research Fund, Luxembourg.

REFERENCES

[1] Dudkin BN, Zainullin GG, Krivoshapkin PV, Krivoshapkina EF, and Ryazanov MA. *Glass Physics and Chemistry* 2008;34(2):187-191.

[2] Gorbatkina YA, Ivanova-Mumzhieva VG, and Ul'yanova TM. *Polymer Science Series* C 2007;49:131-134.

[3] Han JT and Cho K. *Journal of Materials Science* 2006;41(13):4239-4245.

[4] Ji QL, Zhang MQ, Rong MZ, Wetzel B, and Friedrich K. *Journal of Materials Science* 2004;39(21):6487-6493.

[5] Johnsen BB, Kinloch AJ, Mohammed RD, Taylor AC, and Sprenger S. *Polymer* 2007;48(2):530-541.

[6] Kinloch AJ, Lee JH, Taylor AC, Sprenger S, Eger C, and Egan D. *Journal of Adhesion* 2003;79(8-9):867-873.

[7] Kinloch AJ, Masania K, Taylor AC, Sprenger S, and Egan D. *Journal of Materials Science* 2008;43(3):1151-1154.
[8] Kinloch AJ, Mohammed RD, Taylor AC, Eger C, Sprenger S, and Egan D. *Journal of Materials Science* 2005;40(18):5083-5086.
[9] Wetzel B, Haupert F, and Zhang MQ. *Composites Science and Technology* 2003;63(14):2055-2067.
[10] Zhai LL, Ling G, and Wang YW. *International Journal of Adhesion and Adhesives* 2008;28(1-2):23-28.
[11] Zhai LL, Ling GP, Li J, and Wang YW. *Materials Letters* 2006;60(25-26):3031-3033.
[12] Zhang H, Zhang Z, Friedrich K, and Eger C. *Acta Materialia* 2006;54(7):1833-1842.
[13] Guo ZH, Pereira T, Choi O, Wang Y, and Hahn HT. *Journal of Materials Chemistry* 2006;16(27):2800-2808.
[14] Li HY, Zhang ZS, Ma XF, Hu M, Wang XY, and Fan PF. *Surface & Coatings Technology* 2007;201(9-11):5269-5272.
[15] Flippen-Anderson JL and Gilardi R. *Acta Crystallographica Section B* 1981;37:1433-1435.
[16] Blackman BRK, Kinloch AJ, Lee JS, Taylor AC, Agarwal R, Schueneman G, and Sprenger S. *Journal of Materials Science* 2007;42(16):7049-7051.
[17] Kinloch AJ and Taylor AC. *Journal of Materials Science* 2006;41(11):3271-3297.
[18] *HanseChemie.* 2002.
[19] Sprenger S, Eger C, Kinloch AJ, Taylor AC, Lee JH, and Egan D. *Adhäsion Kleben Dichten* 2003;3:24e28.
[20] Lee H and Neville K. *Handbook of epoxy resins.* New York: McGraw-Hill, 1967.
[21] Jungnickel JL, Peters ED, Polgar A, and Weiss FT. *Organic Analysis.lnterscience.* New York. 1953: Val. I. pp. 127 et seq. 1953;1:127-154.
[22] Dobinson B, Hofmann W, and Stark BP. *The determination of epoxide groups.* Oxford: Pergamon Press, 1969.
[23] Baller J, Becker N, Ziehmer M, Thomassey M, Zielinski B, Müller U, and *Sanctuary R. Polymer* 2009;50(14):3211-3219.
[24] Sanctuary R, Baller J, Zielinski B, Becker N, Kruger JK, Philipp M, Muller U, and Ziehmer M. J*ournal of Physics: Condensed Matte*r 2009;21(3):035118.
[25] Macosko CH. *Rheology: principles, measurements and applications.* New York: Wiley-VCH, 1994.
[26] Mezger TG. *The rheology handbook for users of rotational and oscillatory rheometer.* Hannover: Vencentz, 2006.
[27] Krüger JK, Bohn KP, Le Coutre A, and Mesquida P. *Measurement Science and Technology* 1998;9(11):1866-1872.
[28] Mesquida P, Krüger SP, and Krüger J-K. USER COM. *Informationen für Anwender von METTLER TOLEDO Thermoanalysen-Systemen* 1998;6:12-16.
[29] Schawe JEK. *Thermochimica Acta* 1995;260:1-16.
[30] Schawe JEK. *Thermochimica Acta* 1995;261:183-194.
[31] Schawe JEK. *Journal of Thermal Analysis* 1996;47(2):475-484.
[32] *NBS Certificate for Standard Reference Material* 720. 1982.
[33] Angell CA, Ngai KL, McKenna GB, McMillan PF, and Martin SW. *J. Appl. Phys.* 2000;88(6):3113.
[34] Donth E. *The Glass Transition*: Springer Berlin Heidelberg, 2001.

[35] Elliott SR. *Physics of amorphous materials 2nd ed.*: Longman, 1990.
[36] Blachnik R. *D'Ans-Lax Taschenbuch für Chemiker und Physiker*. Berlin: Springer, 1998.
[37] Sanctuary R, Baller J, Krüger JK, Schaefer D, Bactavatchalou R, Wetzel B, Possart W, and Alnot P. *Thermochimica Acta* 2006;445(2):111-115.
[38] Dušek K. *Advances in Polymer Science* 1986;78.
[39] Zvetkov VL. *Thermochimica Acta* 2005;435(1):71-84.
[40] Rozenberg BA. *Advances in Polymer Science* 1986;75.
[41] Mijovic J and Wijaya J. *Macromolecules* 1994;27(26):7589-7600.
[42] Van Assche G, Van Hemelrijck A, Rahier H, and Van Mele B. *Thermochimica Acta* 1995;268:121-142.
[43] Smith IT. *Polymer* 1961;2:95.
[44] Paz-Abuin S, Lopez-Quintela A, Pellin MP, Varela M, and Prendes P. *Journal of Polymer Science Part a-Polymer Chemistry* 1998;36(6):1001-1016.
[45] Min BG, Stachurski ZH, and Hodgkin JH. *Polymer* 1993;34(23):4908-4912.
[46] Kamal MR. *Polymer Engineering and Science* 1974;14(3):231-239.
[47] Kamal MR and Sourour S. *Polymer Engineering and Science* 1973;13(1):59-64.
[48] Schawe JEK. *Thermochimica Acta* 2002;388(1-2):299-312.
[49] Baller J, Thomassey M, Ziehmer M, and Sanctuary R. *Thermochimica Acta* in press; DOI: 10.1016/j.tca.2011.01.029.

INDEX

C

H

I

J

K

L

M

R

S

T

U

V

W

X

Y

Z